Boundary Value Problems in the Spaces of Distributions

Mathematics and Its Applications

Volume 498

Boundary Value Problems in the Spaces of Distributions

by

Yakov Roitberg

Department of Mathematical Analysis,
Chernigov State Pedagogical University,
Chernigov, Ukraine

KLUWER ACADEMIC PUBLISHERS

DORDRECHT / BOSTON / LONDON

A C.I.P. Catalogue record for this book is available from the Library of Congress.

ISBN 978-90-481-5343-5

Published by Kluwer Academic Publishers,
P.O. Box 17, 3300 AA Dordrecht, The Netherlands.

Sold and distributed in North, Central and South America
by Kluwer Academic Publishers,
101 Philip Drive, Norwell, MA 02061, U.S.A.

In all other countries, sold and distributed
by Kluwer Academic Publishers,
P.O. Box 322, 3300 AH Dordrecht, The Netherlands.

Printed on acid-free paper

TABLE OF CONTENTS

PREFACE

The present monograph is devoted to the theory of the solvability in generalized functions of general boundary value problems of mathematical physics. It is the continuation of the author's book [R1], where elliptic boundary value problems have been studied in complete scales of spaces of generalized functions.

From the early sixties, in the works of Lions and Magenes [LiM] and Yu. Berezanskii, S. Krein and Ya. Roitberg [BKR] the theorems on complete collection of isomorphisms have been established. These theorems, roughly speaking, mean that the operator generated by an elliptic boundary value problem establishes an isomorphism between spaces of functions which 'have s derivatives' and spaces functions which 'have $s-r$ derivatives' (here s is an arbitrary real number, r is the order of the elliptic problem). The close results were also obtained by Schechter [Sche]. These results and some of their applications are contained in the book of Lions and Magenes [LiM2] (see also the survey of Magenes [Mag]) and Yu. Berezanskii [Ber].

Further progress in the theory under consideration was connected, first, with the completion of the class of elliptic problems for which the theorems on complete collection of isomorphisms hold, and, hence, with the development of new methods of proving of these theorems, and, second, with the increase of a number of applications of the isomorphism theorems.

In the author's monograph [R1] the last years' investigations on the isomorphism theorems and some of their applications have been presented. There the theorems on isomorphisms have been established, in particular, for general elliptic boundary value problems for systems of mixed order, or, the same, Douglis–Nirenberg systems, and for parameter-elliptic problems. These theorems gave a possibility to construct in [EZh] the parabolic theory in complete scale of spaces of generalized functions.

The goal of the present monograph is to summarize the results on the solvability of elliptic, parabolic and hyperbolic problems of mathematical physics in complete scales of spaces of generalized functions.

The reading of the book requires of the reader familiarity with the elements of functional analysis, the theory of partial differential equations, and the theory of generalized functions. In this book the notions introduced in [R1], and statements proved there are used essentially. For the covenience of the reader the corresponding ones are defined and explained clearly.

Working in the field to which this book is devoted for more that 35 years I was under the beneficial influence of Yu. M. Berezanskii. For all this period I was largely stimulated by the deep permanent interest and careful attention of S. G. Krein.

I discussed some problems presented in the book with M. S. Agranovich, R. Denk, S. D. Eidelman, G. Geimonat, M. L. Gorbachuk, V. I. Gorbachuk, V. A. Kondratiev, A. N. Kozhevnikov, G.-L. Lions, E. Magenes, M. M. Malamud, V. A. Marchenko, V. G. Mazja, R. Mennicken, S. Nazarov, L. Nirenberg, L. P. Nyzhnik, O. A. Oleinik, B. P. Panejah, I. Ya. Roitberg, M. Schechter, B.-W. Schulze, Z. G. Sheftel, M. A. Shubin, P. E. Sobolevskii, M. Z. Solomjak, V. A. Solonnikov, M. I. Vishik, L. R. Volevich, W. L. Wendland, and N. V. Zhitarashu.

It should be noted here that the investigations of Sections 1.2 and 1.3 were stimulated by works of Agranovich and Kozhevnikov on the spectral theory, and the investigations of Section 1.7 were stimulated by Volevich.

I am grateful to all these mathematicians.

I am also greatful to my daughter Dr. Inna Roitberg who translated and typeset the book.

December, 1998 *Ya. A. Roitberg*

CHAPTER 0

INTRODUCTION

Let us describe shortly in this chapter the main questions presented in the monograph.

0.1. Green's Formulas and Theorems on Isomorphisms for General Elliptic Boundary Value Problems for Systems of Douglis–Nirenberg Structure

In Chapter 1 we deduce the Green's formula for general elliptic boundary value problems for systems of Douglis–Nirenberg structure. This formula enables us to prove various theorems on complete collection of isomorphisms for such problems. In this chapter elliptic problems with a parameter for general systems of equations are investigated, and the Cauchy problem for general parameter-elliptic systems is studied.

0.1.1. Green's formula for general elliptic problem for systems of Douglis–Nirenberg structure

In the bounded domain $G \subset \mathbf{R}^n$ with the boundary $\partial G \in C^\infty$ we consider the elliptic boundary value problem

$$l(x,D)u = f(x) \qquad (x \in G), \tag{0.1.1}$$

$$b(x,D)u(x) = \phi(x) \qquad (x \in \partial G). \tag{0.1.2}$$

Here

$$l(x,D) = \Big(l_{rj}(x,D)\Big)_{r,j=1,\dots,N}, \quad \operatorname{ord} l_{rj} \le s_r + t_j,$$

$t_1,\dots,t_N, s_1,\dots,s_N$ are given integers, $|S| + |T| = \sum_{j=1}^{N}(s_j + t_j) = 2m$;

$$b(x,D) = (b_{hj}(x,D))_{\substack{h=1,\dots,m\\ j=1,\dots,N}}, \quad \operatorname{ord} b_{hj} \le \sigma_h + t_j,$$

$\sigma_1, \dots \sigma_n$ are given integers.

Denote

$$æ = \max\{0, \sigma_1 + 1, \dots, \sigma_m + 1\}, \tag{0.1.3}$$

$$\tau = (\tau_1, \ldots, \tau_N),\ |\tau| = \tau_1 + \cdots + \tau_N,\ \tau_j = t_j + æ\ (j = 1, \ldots, N);$$

the number τ_j is the maximal order of differentiation of the function u_j in problem (0.1.1)–(0.1.2).

We identify every element $u_j \in C^\infty(\overline{G})$ with the vector

$$u_j = \begin{pmatrix} u_{j0} \\ \vdots \\ u_{j,\tau_j} \end{pmatrix} = \begin{pmatrix} u_{j0} \\ U_j \end{pmatrix},$$

where $u_{j0} = u_j|_{\overline{G}}$, $u_{jk} = D_\nu^{k-1} u_j|_{\partial G}$, and U_j is the vector of the Cauchy data of the element u_j. Here $D_\nu = i\partial/\partial\nu$, and ν is a normal to ∂G.

Let us now represent the element $u = \begin{pmatrix} u_1 \\ \vdots \\ u_n \end{pmatrix} \in \left(C^\infty(\overline{G})\right)^N$ in the form $u = \begin{pmatrix} u_0 \\ U \end{pmatrix}$, where $u_0 = \begin{pmatrix} u_{10} \\ \vdots \\ u_{N0} \end{pmatrix}$, and $U = \begin{pmatrix} U_1 \\ \vdots \\ U_N \end{pmatrix}$ is the vector of the Cauchy data of the element u.

Similarly, let $f = \begin{pmatrix} f_0 \\ F \end{pmatrix}$, where $f_0 = \begin{pmatrix} f_{10} \\ \vdots \\ f_{N0} \end{pmatrix}$, $F = \begin{pmatrix} F_1 \\ \vdots \\ F_N \end{pmatrix}$, $f_{j0} = f_j|_{\overline{G}}$, $f_{jk} = D_\nu^{k-1} f_j|_{\partial G}$, and $F_j = \begin{pmatrix} f_{j1} \\ \vdots \\ f_{j,æ-s_j} \end{pmatrix}$ is the Cauchy data vector of the element f_j.

Then the element $u \in \left(C^\infty(\overline{G})\right)^N$ is a solution of system (0.1.1) if and only if the following equalities hold:

$$\begin{aligned} &l(x, D)u_0 = f_0(x) && (x \in G), \\ &D_\nu^{k-1} l_r u|_{\partial G} = f_{rk} \quad (r = 1, \ldots, N,\ k = 1, \ldots, -s_r + æ). \end{aligned} \tag{0.1.4}$$

Here $l_r = (l_{r1}, \ldots, l_{rN})$ is a line with the index r of the matrix $l(x, D)$. It is easy to see that the left hand side of (0.1.4) is quite defined by the Cauchy data vector U. Let us write relations (0.1.4) in the form

$$e(x', D')U = F, \tag{0.1.5}$$

where $e(x', D')$ is a $(\sum_{j=1}^{N}(æ - s_j) \times \sum_{j=1}^{N} t_j)$–matrix.

Further, it is clear that $b(x,D)u|_{\partial G}$ is completely defined by the Cauchy data vector U. We write boundary conditions (0.1.2) in the form

$$b(x',D')U = \phi(x') \quad (x' \in \partial G). \tag{0.1.6}$$

Let us consider the matrix

$$c(x,D) = (c_{hj}(x,D))_{\substack{h=1,\dots,m\\ j=1,\dots,N}}$$

of order $\operatorname{ord} c_{hj} \le \sigma_h^c + t_j$. It is clear that the element $c(x,D)u|_{\partial G}$ is quite defined by the Cauchy data vector U, therefore one can rewrite the equation $c(x,D)u|_{\partial G} = \psi$ in the form

$$c(x',D')U(x')|_{\partial G} = \psi(x') \quad (x' \in \partial G). \tag{0.1.7}$$

It turns out that if problem (0.1.1)–(0.1.2) is elliptic, then there exists a matrix $c(x,D)$ with $\sigma_h^c < 0$ $(h = 1,\dots,m)$ such that problem (0.1.5)–(0.1.7) is a Douglis–Nirenberg elliptic problem on ∂G. Let us write this problem in the form

$$E(x',D')U = \Phi, \tag{0.1.8}$$

where

$$E(x',D') = \begin{pmatrix} e(x',D') \\ b(x',D') \\ c(x',D') \end{pmatrix}, \qquad \Phi = \begin{pmatrix} F \\ \phi \\ \psi \end{pmatrix}.$$

The elements of the matrix E are, generally speaking, pseudo-differential operators on ∂G.

Since system (0.1.8) is elliptic on ∂G, the operator

$$E : U \mapsto E(x',D')U$$

is Noetherian. In addition, the kernel $\mathfrak{N} = \mathfrak{N}(E)$ and the cokernel $\mathfrak{N}^* = \mathfrak{N}(E^*)$ are finite-dimentional and consist of infinitely smooth elements.

For simplicity, let us assume in Introduction that

$$\mathfrak{N} = \mathfrak{N}^* = \{0\}. \tag{0.1.9}$$

Then the operator E has the continuous inverse operator E^{-1}, and it follows from (0.1.8) that

$$U = E^{-1}\Phi, \tag{0.1.10}$$

where E^{-1} is a matrix pseudo-differential operator on ∂G. This enables us to deduce the Green's formula for elliptic problem (0.1.1)–(0.1.2).

By integrating by parts we get that

$$(lu,v) - (u,l^+v) = \langle U, Mv\rangle \quad \left(u,v \in \left(C^\infty\left(\overline{G}\right)\right)^N\right) \tag{0.1.11}$$

On the other hand,

$$\langle U, Mv\rangle = \langle E^{-1}\Phi, Mv\rangle = \langle \Phi, (E^{-1})^* Mv\rangle = \left\langle \begin{pmatrix} e(x', D') \\ b(x', D') \\ c(x', D') \end{pmatrix} U, (E^{-1})^* Mv \right\rangle,$$

and the Green's formula

$$\begin{aligned}(lu, v) \ &+ \ \sum_{h=1}^{m} \langle b_h u, c'_h v\rangle + \sum_{r=1}^{N} \sum_{k=1}^{-s_r+\text{æ}} \langle D_\nu^{k-1} l_r u, e'_{kr} v\rangle \\ &= \ (u, l^+ v) + \sum_{h=1}^{m} \langle c_h u, b'_h v\rangle \quad \left(u, v \in \left(C^\infty\left(\overline{G}\right)\right)^N\right) \end{aligned} \quad (0.1.12)$$

follows immediately.

If condition (0.1.9) does not hold, then the Green's formula contains the additional projection operators onto the finite-dimentional kernel $\mathfrak{N}$ (see formula (1.2.17)).

The problem

$$l^+ v(x) = g(x) \quad (x \in G), \qquad b'v\big|_{\partial G} = \psi \qquad (0.1.13)$$

is called the formally adjoint to problem (0.1.1)–(0.1.2) with respect to Green's formula (0.1.12).

It turns out that problem (0.1.13) is elliptic if and only if problem (0.1.1)–(0.1.2) is elliptic.

Let $\mathfrak{N}^+ \subset \left(C^\infty\left(\overline{G}\right)\right)^N$ be a finite-dimentional kernel of problem (0.1.13). The Green's formula implies that problem (0.1.1)–(0.1.2) is solvable in the corresponding space if and only if the relations

$$(f_0, v) + \sum_{r=1}^{N} \sum_{k=1}^{-s_r+\text{æ}} \langle f_{rk}, e'_{kr} v\rangle + \sum_{h=1}^{m} \langle \phi_h, c'_h v\rangle \quad (\forall v \in \mathfrak{N}^+) \qquad (0.1.14)$$

hold.

0.1.2. Functional spaces. Theorem on complete collection of isomorphisms

Assume that $p, p' \in (1, \infty)$, and $1/p + 1/p' = 1$.

We benote by $H^{s,p}(G)$ $s \geq 0$ the space of Bessel potentials, and by $H^{-s,p}(G)$ the space dual to $H^{s,p'}(G)$ with respect to the extension of the scalar product in $L_2(G)$.

We denote by $B^{s,p}(\partial G)$ $s \in \mathbf{R}$ the Besov space. The spaces $B^{s,p}(\partial G)$ and $B^{-s,p'}(\partial G)$ are dual to each other with respect to the extension of the scalar product in $L_2(\partial G)$.

For each $s \in \mathbf{R}$ we denote by $\widetilde{H}^{t_j+s,p,(\tau_j)}(G)$ the closure of the set of smooth vectors $\begin{pmatrix} u_{j0} \\ U_j \end{pmatrix}$ (see the beginning of Subsection 0.1.1) in the space

$$H^{t_j+s,p}(G) \times \prod_{k=1}^{\tau_j} B^{t_j+s-k+1-1/p,p}(\partial G) = \mathcal{H}^{t_j+s}(G, \partial G),$$

and by $\widetilde{H}^{T+s,p,(\tau)}$ the closure of smooth vectors $\begin{pmatrix} u_0 \\ U \end{pmatrix}$ in $\prod_{j=1}^{N} \mathcal{H}^{t_j+s}(G, \partial G)$. It is clear that

$$\widetilde{H}^{T+s,p,(\tau)} = \prod_{j=1}^{N} \widetilde{H}^{t_j+s,p,(\tau_j)}(G).$$

Similarly, we denote by $\widetilde{H}^{s-s_j,p,(æ-s_j)}(G)$ the closure of the set of smooth vectors $\begin{pmatrix} f_{j0} \\ \vdots \\ f_{j,æ-s_j} \end{pmatrix}$ in the space

$$H^{s-s_j,p}(G) \times \prod_{j=1}^{æ-s_j} B^{s-s_j-k+1-1/p,p}(\partial G).$$

Assume that

$$\widetilde{H}^{s-S,p,(æ-S)} = \prod_{j=1}^{N} \widetilde{H}^{s-s_j,p,(æ-s_j)}(G).$$

For each $s \in \mathbf{R}$, and $p \in (1, \infty)$ the closure $A = A_{s,p}$ of the mapping

$$u \mapsto \left(lu|_{\overline{G}}, \left\{ (l_j u|_{\partial G}, \ldots, D_\nu^{æ-s_j-1} l_j u|_{\partial G}) \;:\; j = 1, \ldots, N \right\}, bu|_{\partial G} \right)$$

$$\left(u \in (C^\infty(\overline{G}))^N, \quad l_j u = \sum_{k=1}^{N} l_{jk}(x, D) u_k \right)$$

acts continuously in the pair of spaces

$$\widetilde{H}^{T+s,p,(\tau)} \to K_{s,p} := \widetilde{H}^{s-S,p,(æ-s)} \times \prod_{h=1}^{m} B^{s-\sigma_h-1/p,p}(\partial G). \qquad (0.1.15)$$

It turns out that ([R1, Ch. X]) the operator $A_{s,p}$ is Noetherian. In addition, the kernel $\mathfrak{N}$ and the cokernel $\mathfrak{N}^*$ are finite-dimentional, do not depend on s and p, and consist of infinitely smooth elements. If the problem has no defect (this means that $\mathfrak{N} = 0$ and $\mathfrak{N}^* = 0$), then the operator $A_{s,p}$ realizes an isomorphism between spaces (0.1.15). In the general case the isomorphism between the corresponding subspaces of spaces (0.1.15) with the finite-dimensional defects independent of s and p is realized by the restriction of the operator $A_{s,p}$.

This statement is a theorem on complete collection of isomorphisms for elliptic problem (0.1.1)–(0.1.2). This theorem enables us to investigate problem (0.1.1)–(0.1.2) in the case where the right hand sides have arbitrary power singularities on manifolds of various dimensions, to construct and to investigate the regularity properties of the Green's matrix of the problem under consideration, to study strongly degenerating elliptic problems, et cetera (see [R1]).

Let us show two simple methods which enable us to obtain another theorems on isomorphisms.

Let B_1 and B_2 be Banach spaces and let T be a linear operator mapping isomorphically the space B_1 onto the space B_2. Let E_1 be a subspace of B_1, and let $E_2 = TE_1$. Then it is clear that the operator T in a natural way defines a linear operator T_1 that maps isomorphically the quotient space B_1/E_1 onto the quotient space B_2/E_2.

Further, let Q_2 be a Banach space, and let $Q_2 \subset B_2$ (the imbedding is algebraic and topological). Then $Q_1 = T^{-1}Q_2$ is a linear (generally speaking, nonclosed) subset of the space B_1. However, the space Q_1 becomes a Banach space (denoted by Q_{1T}) with respect to the graph norm

$$\|x\|_{Q_{1T}} = \|x\|_{B_1} + \|Tx\|_{Q_2} \quad (x \in Q_1).$$

The restriction of the operator T onto Q_1 establishes the isomorphism $Q_{1T} \to Q_2$.

These procedures those are called **'Pasting' (or factorization) method** and **graph method** together with the Green's formula give us a possibility to obtain for problem (0.1.1)–(0.1.2) the analogs of all isomorphism theorems known for the simplest case of one equation with normal boundary conditions.

In particular, we get the inequalities useful in the spectral theory of the problems under consideration. These investigation were stimulated by works of M. Agranovich and A. Kozhevnikov.

0.1.3. Parameter-elliptic problems

All the results described above are also obtained for elliptic problems with a parameter for general systems of equations. Now, instead of problem (0.1.1)–(0.1.2) we consider the problem

$$l(x, D, q_1)u(x) = f(x) \quad (x \in G), \qquad b(x, D, q_1)u\big|_{\partial G} = \phi(x'). \qquad (0.1.16)$$

Here $q_1 = qe^{i\theta}$, $q \in \mathbf{R}$, $\theta \in [\theta_1, \theta_2]$ (the case where $\theta = \theta_1 = \theta_2$ is not excluded), numbers $t_1, \ldots, t_N, s_1, \ldots, s_N, \sigma_1, \ldots, \sigma_m$ are given integers such that

$$s_1 + \cdots + s_N + t_1 + \cdots + t_N = 2m,$$

and

$$t_1 \geq \cdots \geq t_N \geq 0 = s_1 \geq \ldots \geq s_N, \qquad \sigma_1 \geq \ldots \geq \sigma_m.$$

Moreover,

$$l = \left(l_{rj}(x, D, q_1)\right)_{r,j=1,\ldots,N}, \quad \operatorname{ord} l_{rj} \leq s_r + t_j,$$

$$b = \left(b_{rj}(x, D, q_1)\right)_{\substack{r=1,\ldots,m \\ j=1,\ldots,N}}, \quad \operatorname{ord} b_{rj} \leq \sigma_r + t_j.$$

Problem (0.1.16) is assumed to be elliptic with a parameter in the sense of Agranovich–Vishik [AgV].

For this problem we established all the results described in Subsections 0.1.1 and 0.1.2, but here, instead of the Noetherian property of corresponding problems, it is proved their unique solvability for sufficiently large values of the parameter q.

0.1.4. Cauchy problem for general parameter-elliptic systems of the Douglis–Nirenberg structure

In the last section of Chapter 1 we investigate the Cauchy problem for general parameter-elliptic systems, and construct Calderón projections.

In the works of L.Volevich and A. Shirikyan [VSh1], [VSh2] Calderón projections connected with Dirichlet problem for one $(2m)$-order parameter elliptic equation were studied; the results were used to prove the solvability of corresponding quasi-linear problems.

In the present monograph the similar results are obtained for general parameter-elliptic boundary value problems for systems of mixed order (or, the same, Douglis-Nirenberg systems). Here the results on Green's formula and on theorems on complete collection of isomorphisms for parameter elliptic problems for general systems of mixed order are used essentially.

In $G = \mathbf{R}^n_\pm = \{x = (x', x_n) : x_n \gtrless 0\}$ we consider parameter-elliptic boundary value problem (0.1.16) for Douglis-Nirenberg (T, S)-system.

Let

$$\tau_j = t_j + æ \quad (j = 1, \ldots, N), \qquad æ = \max\{0, \sigma_1 + 1, \ldots, \sigma_m + 1\}.$$

Then there exists (see [R1, Ch. X]) a number $q_0 > 0$ such that for $q \geq q_0$ the closure $A^\pm = A^\pm_s$ of the mapping

$$u \mapsto \left(lu, bu|_{\partial G}\right) \quad \left(u \in \left(C_0^\infty\left(\overline{\mathbf{R}}^n_\pm\right)\right)^N\right)$$

realizes an isomorphism between the corresponding spaces.

If the element

$$u = (u_1, \dots, u_N), \quad u_j \in C_0^{\tau_j}(\overline{\mathbf{R}^n_\pm})$$

is a solution of problem (0.1.16) with $f = 0$, then it is nesessary that

$$D_\nu^{j-1} l_r(x, D, q_1)u\big|_{x_n=0} = D_\nu^{j-1} f_r(x)\big|_{x_n=0} = 0 \quad (1 \le j \le æ - s_r).$$

Let us rewrite this condition in the form

$$e(x, D, q_1)U = 0,$$

where U is the vector of the Cauchy data of the solution of (0.1.16).

In $\mathbf{R}^n_\pm$ we consider the Cauchy problem

$$l(x, D, q_1)u^\pm = 0 \qquad (\text{in } \mathbf{R}^n_\pm), \tag{0.1.17}$$

$$D_n^{j-1} u_k^\pm\big|_{x_n=0} = u_{kj} \in B^{t_k+s-j+1-1/p,p}(\mathbf{R}^{n-1}, q) \tag{0.1.18}$$

$$(k = 1, \dots, N, \quad j = 1, \dots, \tau_j),$$

where $B^{s,p}(\mathbf{R}^{n-1}, q)$ is a Besov space whose norm depends on the parameter $q \in \mathbf{R}$. Let

$$U = \{u_{kj}\} \in \prod_{j=1}^{N} \prod_{k=1}^{\tau_j} B^{t_k+s-j+1-1/p,p}(\mathbf{R}^{n-1}, q) =: B^{T+s,p}$$

be a Cauchy data vector, and let $\mathfrak{U}_{s,p} = \{U \in B^{T+s,p} : eU = 0\}$ be a subspace of $B^{T+s,p}$. Denote by $\mathfrak{U}^\pm_{s,p}$ the subspace of the space $\mathfrak{U}_{s,p}$ of the Cauchy data of problem (0.1.17)–(0.1.18). It turns out that

$$\mathfrak{U}_{s,p} = \mathfrak{U}^+_{s,p} \dot{+} \mathfrak{U}^-_{s,p}, \quad \mathfrak{U}^+_{s,p} \bigcap \mathfrak{U}^-_{s,p} = \{0\}.$$

The norms of the projection operators

$$P^\pm : \mathfrak{U}_{s,p} \to \mathfrak{U}^+_{s,p}$$

are bounded by the constant independent of q ($q \ge q_0 > 0$).

All the results of Chapter 1 have been obtained by author jointly with I. Roitberg.

0.2. Elliptic Boundary Value Problems for General Systems of Equations with Additional Unknown Functions Defined at the Boundary of the Domain

In the second chapter we study elliptic boundary value problems for general systems of equations for the case where the boundary conditions contain additional unknown functions defined on the boundary of the domain.

In problems of elasticity theory and hydrodynamics, for example in the works of Aslanyan, Vassiliev and Lidskii [AVL], Garlet [Gar], Nazarov and Pileckas [NaP], there arise boundary value problems for general elliptic systems whose boundary conditions contain both the functions $u_1, \ldots, u_N$ from the system, and the additional unknown functions $u'_{N+1}, \ldots, u'_{N+k}$ defined at the boundary. The number of the boundary conditions increases respectively.

Chapter 2 is devoted to the investigation of the solvability of these problems in complete scales of Banach spaces. These results belong to I. Roitberg. To explain them we follow to the works [RI1]–[RI5].

Let $G \subset \mathbf{R}^n$ be a bounded domain with the boundary $\partial G \in C^\infty$. We consider the following boundary value problem:

$$l(x, D)u = f(x) \quad (x \in G); \tag{0.2.1}$$

$$b(x, D)u\big|_{\partial G} + b'(x', D')u' = \phi(x') \quad (x' \in \partial G). \tag{0.2.2}$$

The system

$$lu := (l_{rj}(x, D))_{r,j=1,\ldots,N}$$

and the boundary differential expressions

$$b(x, D) := \big(b_{hj}(x, D)\big)_{\substack{h=1,\ldots,m+k'' \\ j=1,\ldots,N}}, \quad b'(x', D') := \big(b'_{hj}(x', D')\big)_{\substack{h=1,\ldots,m+k'' \\ j=N+1,\ldots,N+k'}}$$

are described by the following relations:
$\operatorname{ord} l_{rj} \le s_r + t_j$ for $s_r + t_j \ge 0$, $l_{rj} = 0$ for $s_r + t_j < 0$, $r, j = 1, \ldots, N$;
$\operatorname{ord} b_{hj} \le \sigma_h + t_j$ for $\sigma_h + t_j \ge 0$, $b_{hj} = 0$ for $\sigma_h + t_j < 0$, $j = 1, \ldots, N$, $h = 1, \ldots, m + k''$;
$\operatorname{ord} b'_{hj} \le \sigma_h + t_j$ for $\sigma_h + t_j \ge 0$, $b'_{hj} = 0$ for $\sigma_h + t_j < 0$, $j = N+1, \ldots, N+k'$, $h = 1, \ldots, m + k''$.

Here $t_1, \ldots, t_{N+k'}, s_1, \ldots, s_N, \sigma_1, \ldots, \sigma_{m+k''}$ are given integers such that

$$s_1 + \cdots + s_N + t_1 + \cdots + t_N = 2m$$

and

$$t_1 \ge \cdots \ge t_N \ge 0 = s_1 \ge \ldots \ge s_N, \ \sigma_1 \ge \ldots \ge \sigma_{m+k''}.$$

In addition, $u(x) = (u_1(x), \ldots, u_N(x))$, $x \in \overline{G}$, and, for $x' \in \partial G$,

$$u'(x') = \big(u'_{N+1}(x'), \ldots, u'_{N+k'}(x')\big)', \ u'_j(x') = \big(u'_{j1}(x'), \ldots, u_{j,\sigma_1+t_j+1}(x')\big)'.$$

Here and in what follows we assume that the coefficients of all differential expressions and the boundary ∂G are infinitely smooth.

The definition of the notion of ellipticity of problem (0.2.1)–(0.2.2) is given, corresponding functional spaces are introduced, and theorem on complete collection of isomorphisms is proved for the problem under consideration, some applications of this theorem are presented. Then, for this problem we deduce the Green's formula similar to (0.1.12), and study the formally adjoint problem to (0.2.1)–(0.2.2) with respect to the Green's formula, the ellipticity of this problem is proved.

In addition, in this chapter we consider parameter-elliptic problems with additional unknown functions on the boundary. For such problems we prove the analogs of all the statements mentioned above. Note that here instead of Noetherity the unique solvability is taken place for sufficiently large values of parameter.

0.3. The Sobolev Problem

Chapter 3 is devoted to study of the Sobolev problem in complete scale of Banach spaces of generalized functions. The boundary conditions of this problem are given by linear differential (or pseudo-differential) expressions on the manifolds of various dimensions.

The Sobolev problem has been studied in the classes of sufficiently smooth functions by Sobolev, Sternin (see [Sob], [St1], [St2] and bibliography there).

In the present monograph the restrictions of these papers are thrown out, and the Sobolev problem is studied in complete scales of Banach spaces of generalized functions.

Let $G \subset \mathbf{R}^n$ be a bounded domain with the boundary $\partial G = \Gamma_0 \cup \Gamma_1 \cup \dots \Gamma_{\overline{k}} \in C^\infty$, and let Γ_0 be an $(n-1)$-dimensional compact set that is the exterior boundary of the domain G. Denote by Γ_j $(j = 1, \dots, \overline{k})$ the i_j-dimensional manifold without boundary lying inside of Γ_0, $0 \leq i_j \leq n-1$. Let $i'_j = n - i_j$ denotes the codimensionality of Γ_j. Assume that $\Gamma_j \in C^\infty$ $(j = 0, \dots, \overline{k})$, and $\Gamma_j \cap \Gamma_k = \emptyset$ for $j \neq k$.

We consider the Sobolev problem

$$L(x, D)u(x) = f(x) \qquad (x \in G;\ \operatorname{ord} L = 2m) \tag{0.3.1}$$

$$B_{j0}(x, D)u\big|_{\Gamma_0} = \varphi_{j0} \qquad (j = 1, \dots, m;\ \operatorname{ord} B_{j0} = q_{j0}), \tag{0.3.2}$$

$$B_{jk}(x, D)u\big|_{\Gamma_k} = \varphi_{jk} \quad (k = 1, \dots, \overline{k};\ j = 1, \dots, m_k;\ \operatorname{ord} B_{jk} = q_{jk}). \tag{0.3.3}$$

In the natural way we introduce the notion of the generalized solution of this problem, define functional spaces, and prove the theorem on complete collection of isomorphisms. For simplisity, we formulate all the results for the case of one equation (0.3.1), but remark that these results are true for the system of Douglis–Nirenberg structure. The possible applications of the results obtained are noted.

The solvability of the Sobolev problem in complete scales of Banach spaces is obtained also for elliptic problems with a parameter (Section 3.2) and parabolic problems. We consider also a number of applications of this theory.

The main results of the chapter are obtained by the author jointly with my post-graduate student V. Los [LR1]–[LR3], the results concerning with parameter-elliptic and parabolic problems are found by V. Los [Los]. In the simple case where the boundary conditions are normal the isomorphism theorem for the Sobolev problem is proved by the author joinly with A. Sklyarets.

0.4. Hyperbolic Problems for General Systems of Equations in Complete Scales of Spaces of Sobolev Type

Chapter 4 and Chapter 5 deal with the Cauchy problem, and boundary value and mixed problems for general hyperbolic systems.

Boundary value and mixed problems for hyperbolic equations have been studied in classes of sufficiently smooth functions by Sakamoto [Sak], Kreiss [Kre], Agranovich [Agr1], Chazarain and Piriou [ChP], Volevich and Gindikin [VoG] (see also the survey of Volevich and Ivrii [VIv]). In the present monograph boundary value and mixed problems for general hyperbolic systems in the Leray–Volevich sense are studied in a complete scale of spaces of Sobolev type depending on parameters $s, \tau \in \mathbf{R}$; s characterizes the smoothness of a solution in all variables, while τ describes additional smoothness in the tangential variables.

Let $(t, x) = (t, x_1, \dots, t_n) \in \mathbf{R}^{n+1}$, and let $(\sigma, \xi) = (\sigma, \xi_1, \dots, \xi_n)$ be the dual variables. Assume that

$$l = l(D_t, D_x) = (l_{kj}(d_t, D_x))_{k,j=1,\dots,N}, \tag{0.4.1}$$

where l is a matrix differential expression and the l_{kj} are homogeneous differential expressions of orders $s_k + t_j$ with constant coefficients, with $l_{kj} = 0$ if $s_k + t_j < 0$. Here $D_t = i\partial/\partial t$, $D_x = D_1 \dots D_n$, $D_j = i\partial/\partial x_j$; $s_1, \dots, s_N$ and $t_1, \dots, t_N$ are integers, and $t_1 \geq \dots \geq t_N \geq 0 = s_1 \geq \dots \geq s_N$. Let $s_1 + \dots + s_N + t_1 + \dots + t_N = r$, and let

$$L(\sigma, \xi) = \det(l(\sigma, \xi)) = \sum_{j+|\alpha|\leq r} a_{j\alpha}\sigma^j\xi^\alpha. \tag{0.4.2}$$

Expression (0.4.1) is said to be *strictly hyperbolic in the Leray–Volevich sense* if polynomial (0.4.2) is strictly hyperbolic: the coefficients $a_{r,0,\dots,0}$ of σ^r in (0.4.2) are nonzero, and for each $\xi \in \mathbf{R}^n \setminus \{0\}$ the roots of the equation $L(\sigma, \xi) = 0$ relative to σ are real and distinct.

From the strict hyperbolicity of $l(D_t, D_x)$ it follows that for each $\gamma > 0$ the equation

$$L(\sigma + i\gamma, \xi', \xi_n) = 0 \tag{0.4.3}$$

has no real roots relative to ξ_n. Let

$$\zeta_1(\sigma+i\gamma,\xi'),\ldots,\zeta_r(\sigma+i\gamma,\xi') \quad \Big((\sigma+i\gamma,\xi')\neq(0,0),\ \gamma\geq 0\Big) \tag{0.4.4}$$

be the ξ_n-roots of equation (0.4.3). To be specific, we suppose that for $\gamma>0$ the first m roots have negative imaginary parts, while the remainder have positive imaginary parts. We set

$$L(\sigma+i\gamma,\xi',\xi_n)=L_-(\sigma+i\gamma,\xi',\xi_n)L_+(\sigma+i\gamma,\xi',\xi_n),$$
$$L_-(\sigma+i\gamma,\xi',\xi_n)=\prod_{1\leq j\leq m}(\xi_n-\zeta_j(\sigma+i\gamma,\xi')).$$

In this note we investigate the solvability in $\mathbf{R}^{n-1}$ of the problem

$$l(D_t,D_x)u=f \tag{0.4.5}$$

$$\Big(u=(u_1,\ldots,u_N),\quad f=(f_1,\ldots,f_N)\Big),$$

and also of the problem

$$\Big(l(D_t,D_x)+l'(t,x,D_t,D_x)\Big)u=f \tag{0.4.6}$$

obtained by perturbation of system (0.4.5) by lower-order terms with infinitely smooth coefficients all of whose derivatives are bounded.

In the half-space $G=\Big\{(t,x)=(t,x',x_n)\in\mathbf{R}^{n+1}: x_n>0\Big\}$ we study the boundary value problem

$$l(D_t,D_x)u=f\ \ (\text{in } G),\qquad b(D_t,D_x)u=\phi\ \ (\text{on } \partial G); \tag{0.4.7}$$

here $b(x,D)=(b_{hj}(x,D))_{\substack{h=1,\ldots,m\\ j=1,\ldots,N}}$ is a matrix of linear homogeneous differential expressions with constant complex coefficients of orders σ_h+t_j respectively ($b_{hj}=0$ if $\sigma_h+t_j<0$), and $\sigma_1,\ldots,\sigma_m$ are given integers. We assume everywhere that problem (0.4.7) is hyperbolic. This means that system (0.4.1) is strictly hyperbolic, $a_{r,0,\ldots,0}\neq 0$, the number of boundary conditions is equal to the number m of roots of (0.4.4) with negative imaginary parts, and the Lopatinskii condition is satisfied: for each $(\sigma+i\gamma,\xi')\neq(0,0)$, $\gamma\geq 0$, the rows of the matrix

$$L(\sigma+i\gamma,\xi',\xi_n)b(\sigma+i\gamma,\xi',\xi_n)l^{-1}(\sigma+i\gamma,\xi',\xi_n),$$

whose elements are considered as polynomials in ξ_n, are linearly independent modulo $L_-(\xi_n)=L_-(\sigma+i\gamma,\xi',\xi_n)$.

We also investigate the problem

$$\Big(l(D_t, D_x) + l'(t, x, D_t, D_x)\Big)u = f \quad (\text{ in } G), \tag{0.4.8}$$

$$\Big(b(D_t, D_x) + b'(t, x, D_t, D_x)\Big)u = \phi \quad (\text{ on } \partial G) \tag{0.4.9}$$

obtained by perturbation of problem (0.4.7) by lower terms with infinitely smooth coefficienats all of whose derivatives are bounded.

We also note that it will be shown that if the right hand sides of problems (0.4.7) and (0.4.8)–(0.4.9) vanish for $t \leq 0$, then the solutions of these problems are also equal to zero for $t \leq 0$. Therefore, theorems on the solvability of the corresponding mixed problems in $G_+ = \Big\{(t, x) \in G : t > 0\Big\}$ with homogeneous (zero) initial data at $t = 0$ follows from the theorems on solvability of problems (0.4.7) and (0.4.8)–(0.4.9).

All these problems are studied in complete scales of Hylbert spaces with the norms depending on the real parameter γ.

Let us describe the results obtained for the case of the Cauchy problem.

Let s, τ, $\gamma \in \mathbf{R}$. We denote by $H^{s,\tau}(\mathbf{R}^{n+1}, \gamma)$ the space of distributions f with the norm

$$\|f, \mathbf{R}^{n+1}, \gamma\|_{s,\tau} = \Big(\int \Big(1 + \gamma^2 + \sigma^2 + |\xi|^2\Big)^s \times \Big(1 + \gamma^2 + |\xi|^2\Big)^\tau |\widetilde{f}(\sigma, \xi)|^2 \, d\sigma \, d\xi\Big)^{1/2}, \tag{0.4.10}$$

where $\widetilde{f}(\sigma, \xi)$ is the Fourier transform of the element f, and the integration goes over the entire space. It is clear that for each fixed $\gamma \in \mathbf{R}$ the norm $\|f, \mathbf{R}^{n+1}, \gamma\|_{s,\tau}$ is equivalent to the norm $\|f, \mathbf{R}^{n+1}, 0\|_{s,\tau} = \|f, \mathbf{R}^{n+1}\|_{s,\tau}$, and the set $H^{s,\tau}(\mathbf{R}^{n+1}, \gamma)$ thus does not depend on γ. However, in this note it is convenient to consider only norms equivalent to (0.4.10) for which the constants in the corresponding two-sided estimates can be chosen not to depend on γ.

Let $\Omega = \mathbf{R}^{n+1}_+ = \Big\{(t, x) \in \mathbf{R}_{n+1} : t > 0\Big\}$.

We denote by $H^{s,\tau}(\mathbf{\Omega}, \gamma)$, s, τ, $\gamma \in \mathbf{R}$, $s \geq 0$, the set of restrictions of functions in $H^{s,\tau}(\mathbf{R}^{n+1}, \gamma)$ to $\mathbf{\Omega}$ with the norm of quotient space:

$$\|w, \mathbf{\Omega}, \gamma\|_{s,\tau} = \inf \|u, \mathbf{R}^{n+1}, \gamma\|_{s,\tau} \quad (s, \tau, \gamma \in \mathbf{R}, \ s \geq 0),$$

where the infimum is taken over all functions $u \in H^{s,\tau}(\mathbf{R}^{n+1}, \gamma)$ equal to w in $\mathbf{\Omega}$. We denote by $H^{-s,-\tau}(\mathbf{\Omega}, \gamma)$, s, τ, $\gamma \in \mathbf{R}$, $s \geq 0$, the space dual to

$H^{s,\tau}(\Omega,\gamma)$ with respect to the extension $(\cdot,\cdot) = (\cdot,\cdot)_\Omega$ of the scalar product in $L_2(\Omega)$;

$$\|u,\Omega,\gamma\|_{-s,-\tau} = \sup_{v\in H^{s,\tau}(\Omega,\gamma)} |(u,v)|\,\|v,\Omega,\gamma\|_{s,\tau}^{-1}$$

is a norm in $H^{-s,-\tau}(\Omega,\gamma)$, $s \geq 0$.

For $s,\gamma \in \mathbf{R}$, let $H^s(\partial\Omega,\gamma)$ denote the space of distributions g on $\partial\Omega$ with the norm

$$\|g,\partial\Omega,\gamma\|_s = \left(\int_{\partial\Omega} |\widehat{g}(\xi)|^2 \left(1+|\xi|^2+\gamma^2\right)^s d\xi\right)^{1/2} < \infty,$$

where $\widehat{g}(\xi) = (F_{x\to\xi}g)(\xi)$ is the Fourier transform of g.

We fix a natural number r and suppose that s, τ, $\gamma \in \mathbf{R}$, $s \neq k+1/2$, $k = 0,\dots,r-1$. We denote by $\widetilde{H}^{s,\tau,(r)}(\Omega,\gamma)$ the completion of $C_0^\infty(\overline{\Omega})$ in the norm

$$|||u,\Omega,\gamma|||_{s,\tau,(r)} = \left(\|u,\Omega,\gamma\|_{s,\tau}^2 + \sum_{j=1}^{r} \|D_t^{j-1}u,\partial\Omega,\gamma\|_{s-j+\tau-1/2}^2\right)^{1/2}, \tag{0.4.11}$$

where $C_0^\infty(\overline{\Omega})$ denotes the set of restrictions to $\overline{\Omega}$ of functions in $C_0^\infty(\mathbf{R}^{n-1})$. A similar space was introduced by the author in [R10] and studied in [R11] (see also [Ber, Ch. 3, §6.8], and [RS1]). For $s = k+1/2$ ($k = 0,\dots,r-1$) the space $\widetilde{H}^{s,\tau,(r)}(\Omega,\gamma)$ and the norm $|||u,\Omega,\gamma|||_{s,\tau,(r)}$ are defined by interpolation. It follows from (0.4.11) that the closure S of the mapping

$$u \mapsto \left(u|_{\overline{\Omega}}, u|_{\partial\Omega},\dots,D_t^{r-1}u|_{\partial\Omega}\right)$$

($u \in C_0^\infty(\overline{\Omega})$) establishes an isometry between $\widetilde{H}^{s,\tau,(r)}(\Omega,\gamma)$ and a subspace of the direct product

$$H^{s,\tau}(\Omega,\gamma) \times \prod_{j=1}^{r} H^{s+\tau-j+1/2}(\partial\Omega,\gamma).$$

We henceforth agree to identify an element $u \in \widetilde{H}^{s,\tau,(r)}(\Omega,\gamma)$ with the element $Su = (u_0,\dots,u_r)$. We write

$$u = (u_0,\dots,u_r) \in \widetilde{H}^{s,\tau,(r)}(\Omega,\gamma)$$

for each $u \in \widetilde{H}^{s,\tau,(r)}(\Omega,\gamma)$. If $r = 0$, then we set

$$\widetilde{H}^{s,\tau,(0)}(\Omega,\gamma) := H^{s,\tau}(\Omega,\gamma), \quad |||u,\Omega,\gamma|||_{s,\tau,(0)} := \|u,\Omega,\gamma\|_{s,\tau}.$$

We further introduce the spaces

$$\mathcal{H}^{s,\tau}(\mathbf{R}^{n+1},\gamma), \qquad \mathcal{H}^{s,\tau}(\boldsymbol{\Omega},\gamma), \qquad \widetilde{\mathcal{H}}^{s,\tau,(r)}(\boldsymbol{\Omega},\gamma),$$

and denote the respective norms in these spaces by

$$|u,\mathbf{R}^{n+1},\gamma|_{s,\tau}, \qquad |u,\boldsymbol{\Omega},\gamma|_{s,\tau}, \qquad |u,\boldsymbol{\Omega},\gamma|_{s,\tau,(r)}$$

in the following way:

$$\mathcal{H}^{s,\tau}(\mathbf{R}^{n+1},\gamma) := \Big\{u : e^{-\gamma t}u \in H^{s,\tau}(\mathbf{R}^{n+1},\gamma)\Big\},$$

$$|u,\mathbf{R}^{n+1},\gamma|_{s,\tau} := \|e^{-\gamma t}u,\mathbf{R}^{n+1},\gamma\|_{s,\tau}\,;$$

if here $\mathbf{R}^{n+1}$ is replaced by $\boldsymbol{\Omega}$, we obtain the definition of $\mathcal{H}^{s,\tau}(\boldsymbol{\Omega},\gamma)$ and the norm in it; in an entirely similar way,

$$\widetilde{\mathcal{H}}^{s,\tau,(r)}(\boldsymbol{\Omega},\gamma) := \Big\{u : e^{-\gamma t}u \in \widetilde{H}^{s,\tau,(r)}(\boldsymbol{\Omega},\gamma)\Big\},$$

$$|u,\boldsymbol{\Omega},\gamma|_{s,\tau,(r)} := |||e^{-\gamma t}u,\boldsymbol{\Omega},\gamma|||_{s,\tau,(r)}\,.$$

It has been proved that

i) For any $s,\tau \in \mathbf{R}$ the closures l and $l+l'$ of the respective mappings $u \mapsto lu$ and $u \mapsto (l+l')u$, $\left(u \in (C_0^\infty(\mathbf{R}^{n+1},\gamma))^N\right)$ acts continuously in the pair of spaces

$$\mathcal{H}^{T+s,\tau}(\mathbf{R}^{n+1},\gamma) \to \mathcal{H}^{s-S,\tau}(\mathbf{R}^{n+1},\gamma), \tag{0.4.12}$$

$$\mathcal{H}^{T+s,\tau}(\mathbf{R}^{n+1},\gamma) := \prod_{j=1}^{N} \mathcal{H}^{t_j+s,\tau}(\mathbf{R}^{n+1},\gamma),$$

$$\mathcal{H}^{s-S,\tau}(\mathbf{R}^{n+1},\gamma) := \prod_{k=1}^{N} \mathcal{H}^{s-s_k,\tau}(\mathbf{R}^{n+1},\gamma),$$

and

$$|lu,\mathbf{R}^{n+1},\gamma|_{s-S,\tau} \le c\,|u,\mathbf{R}^{n+1},\gamma|_{T+s,\tau}\,, \tag{0.4.13}$$

$$|(l+l')u,\mathbf{R}^{n+1},\gamma|_{s-S,\tau} \le c\,|u,\mathbf{R}^{n+1},\gamma|_{T+s,\tau}\,, \tag{0.4.14}$$

where $|f,\mathbf{R}^{n+1},\gamma|_{s-S,\tau}$ and $|u,\mathbf{R}^{n+1},\gamma|_{T+s,\tau}$ are the respective norms in the spaces of images and pre-images of mapping (0.4.12); the constant $c>0$ does not depend on u and γ. The question arises of the invertibility of the operators l and $l+l'$.

ii) For each $f \in \mathcal{H}^{s-S,\tau}(\mathbf{R}^{n+1},\gamma)$ $(s,\tau,\gamma \in \mathbf{R},\ |\gamma| \ge \gamma_0 > 0)$ there exists one and only one element $u \in \mathcal{H}^{T+s,\tau-1}(\mathbf{R}^{n+1},\gamma)$ such that $lu=f$. There exists a constant $c>0$ independent of f, u, and γ $(|\gamma| \ge \gamma_0 > 0)$ such that

$$|u,\mathbf{R}^{n+1},\gamma|_{T+s,\tau-1} \le \frac{c}{|\gamma|}\,|f,\mathbf{R}^{n+1},\gamma|_{s-S,\tau}. \tag{0.4.15}$$

If $\operatorname{supp} f \subset \overline{\Omega} = \{(t,x) \in \mathbf{R}^{n+1} : t \geq 0\}$ then also $\operatorname{supp} u \subset \overline{\mathbf{\Omega}}$.

Comparison of (0.4.14) and (0.4.15) shows that the transition $f \mapsto u$ 'loses one unit of smoothness in the tangential direction'; moreover, the norm of this operator can be estimated in terms of $c\,|\gamma|^{-1}$ and is small for large $|\gamma|$. From statement ii) we therefore obtain the statement presented below.

iii) We associate the Cauchy problem

$$lu = f \quad (\text{ in } \mathbf{\Omega}), \qquad D_t^{k-1} u_j\big|_{t=0} = u_{jk} \tag{0.4.16}$$

$$(\forall j : t_j \geq 1,\, k = 1, \ldots, t_j)$$

with the operator $A = A_{s,\tau}$ $(s, \tau \in \mathbf{R})$ which is the closure of the mapping

$$u \mapsto \left(lu, \{D^{k-1} u_j \ (j : t_j \geq 1,\, k = 1, \ldots, t_j)\}\right) \quad \left(u \in \left(C_0^\infty\left(\overline{\Omega}\right)\right)^N\right)$$

acting continuously from the entire of

$$\widetilde{\mathcal{H}}^{T+s,\tau,(T)}(\mathbf{\Omega},\gamma) := \prod_{1 \leq j \leq N} \widetilde{\mathcal{H}}^{t_j+s,\tau,(t_j)}(\mathbf{\Omega},\gamma)$$

into the space

$$\widetilde{K}^{s,\tau} := \prod_{1 \leq j \leq N} \widetilde{\mathcal{H}}^{s-s_j,\tau,(-s_j)}(\mathbf{\Omega},\gamma) \times \prod_{j : t_j \geq 1} \prod_{1 \leq k \leq t_j} H^{t_j+s+\tau-k+1/2}(\mathbf{\Omega},\gamma)$$

$$= \widetilde{\mathcal{H}}^{s-S,\tau,(-S)}(\mathbf{\Omega},\gamma) \times B^{s,\tau}(\mathbf{\Omega},\gamma).$$

In (0.4.16) we set

$$f = (f_1, \ldots, f_N) \in \widetilde{\mathcal{H}}^{s-S,\tau,(-S)}(\mathbf{\Omega},\gamma),$$

$$f_k = (f_{k0}, \ldots f_{k,-s_k}) \in \widetilde{\mathcal{H}}^{s-s_k,\tau,(-s_k)}(\mathbf{\Omega},\gamma),$$

$$U = \left\{u_{jk} \ (j : t_j \geq 1,\ k = 1, \ldots, t_j)\right\} \in B^{s,\tau}(\mathbf{\Omega},\gamma) \quad (s, \tau \in \mathbf{R}).$$

The following statements are true.

Let $s, \tau, \gamma \in \mathbf{R}$, $|\gamma| \geq \gamma_0 > 0$, $F = (f, U) \in K^{s,\tau}$ and the certain compatibility conditions are satisfied. Then the Cauchy problem (0.4.16) has one and only one solution $u \in \widetilde{H}^{T+s,\tau-1,(T)}(\mathbf{\Omega},\gamma)$. There exists a constant $c > 0$ independent of F, u, and γ $(|\gamma| \geq \gamma_0 > 0)$ such that

$$\|u\|_{\widetilde{H}^{T+s,\tau-1,(T)}(\mathbf{\Omega},\gamma)} \leq |\gamma|^{-1} \|F\|_{K^{s,\tau}}. \tag{0.4.17}$$

Let $s, \tau \in \mathbf{R}$. There exists a number $\gamma_0 > 0$ such that for $|\gamma| \geq \gamma_0$ for each $F = (f, U) \in K^{s,\tau}$ satisfying the compatibility conditions the Cauchy problem for the equation

$$\Big(l(D_t, D_x) + l'(t, x, D_t, D_x)\Big)u = f \tag{0.4.18}$$

has one and only one solution $u \in \widetilde{\mathcal{H}}^{T+s,\tau-1,(T)}(\mathbf{\Omega}, \gamma)$. Moreover, estimate (0.4.17) holds.

In Chapter 5 the similar statements have been proved for boundary value and mixed problems.

The results both of Chapter 4 and Chapter 5 belong to the author. In these chapters l is assumed to be a strictly hyperbolic expression. The case where the strict hyperbolicity is broken was considered in the work of the author's post-graduate students Lapa and Movsha [LaMo], and Doropienko [Dor].

0.5. Green's Formula and Density of Solutions for General Parabolic Boundary value Problems in Functional Spaces on Manifolds

In Chapter 6 the Green's formula and theorems on isomorphisms have been found for parabolic boundary value problems. For simplicity, we explain all the results and proofs for the case of one equation and for L_2-spaces of generalized functions. It has been noted that the results of Chapter 1 about parameter-elliptic problems for systems of equations enable us to obtain the corresponding results for parabolic problems for general systems in L_p-spaces of generalized functions. The results of this chapter were obtained by the author joinly with I. Roitberg.

GREEN'S FORMULAS AND THEOREMS ON COMPLETE COLLECTION OF ISOMORPHISMS FOR GENERAL ELLIPTIC BOUNDARY VALUE PROBLEMS FOR SYSTEMS OF DOUGLIS–NIRENBERG STRUCTURE

1.1. General Elliptic Boundary Value Problems for Systems of Equations

1.1.1. In the bounded domain $G \subset \mathbf{R}^n$ with the boundary $\partial G \in C^\infty$ we consider the elliptic boundary value problem

$$l(x, D)u(x) = f(x) \qquad (x \in G), \tag{1.1.1}$$

$$b(x, D)u(x) = \varphi(x) \qquad (x \in \partial G). \tag{1.1.2}$$

Here

$$l(x, D) = \left(l_{rj}(x, D)\right)_{r,j=1,\dots,N}, \quad \operatorname{ord} l_{rj} \leq s_r + t_j,$$

numbers $t_1, \dots, t_N, s_1, \dots, s_N$ are given integer,

$$|S| + |T| = s_1 + \cdots + s_N + t_1 + \cdots + t_N = 2m;$$

$$l_{rj}(x, D) = \begin{cases} \sum_{|\alpha| \leq s_r + t_j} a_\alpha(x) D^\alpha & \text{for} \quad s_r + t_j \geq 0 \\ 0 & \text{for} \quad s_r + t_j < 0 \end{cases}$$

($\alpha = (\alpha_1, \dots, \alpha_n)$, $D^\alpha = D_1^{\alpha_1} \cdots D_n^{\alpha_n}$, $D_j = i\partial/\partial x_j$; $|\alpha| = \alpha_1 + \cdots + \alpha_n$). Since $s_k + t_j = (s_k - r) + (t_j + r)$, without restriction of generality, one can suppose that $\max\{s_j\} = 0$. Then the maximal order of differentiation of the function u_j in the system (1.1.1) is not larger than t_j. In what follows we will assume that

$$t_1 \geq \cdots \geq t_N \geq 0 = s_1 \geq \cdots \geq s_N$$

(one can obtain this by changing the numbers both of the functions and of the equations).

Furthermore,

$$b(x,D) = (b_{hj}(x,D))_{\substack{h=1,\dots,m \\ j=1,\dots,N}}, \quad \operatorname{ord} b_{hj} \le \sigma_h + t_j,$$

$$b_{hj}(x,D) = \begin{cases} \sum_{|\alpha|\le\sigma_h+t_j} b_\alpha(x)D^\alpha, & \text{for } \sigma_h + t_j \ge 0, \\ 0, & \text{for } \sigma_h + t_j < 0, \end{cases}$$

$\sigma_1, \dots \sigma_n$ are given integer.

It is assumed, for simplisity, that the coefficients of all differential expressions are infinitely smooth.

The ellipticity of problem (1.1.1)-(1.1.2) means that the system (1.1.1) is properly elliptic in $\overline{G}$, and the boundary conditions (1.1.2) satisfy Lopatinskii condition on ∂G ([ADN], [Vol], [Sol], [R1, Ch. X]). The orders of the differential expressions are arbitrary.

Denote

$$æ = \max\{0, \sigma_1 + 1, \dots, \sigma_m + 1\}, \tag{1.1.3}$$

$$\tau = (\tau_1, \dots, \tau_N),\ |\tau| = \tau_1, \dots, \tau_N,\ \tau_j = t_j + æ\ (j = 1, \dots, N).$$

1.1.2. Let us introduce functional spaces (see [R1, Ch.I, II]). Let $p, p' \in]1, \infty[$, $1/p + 1/p' = 1$, and let $\Omega \subset \mathbf{R}^n$ be a domain with the boundary $\partial\Omega$. In the work we consider only the cases
i) where $\Omega = G$ is a bounded domain with the boundary $\partial\Omega = \partial G$, and
ii) where

$$\Omega = \mathbf{R}^n_- = \{x = (x', x_n) \in \mathbf{R}^n : x_n < 0\},$$

or

$$\Omega = \mathbf{R}^n_+ = \{x = (x', x_n) \in \mathbf{R}^n : x_n > 0\},$$

with the boundary

$$\partial\Omega = \{x \in \mathbf{R}^n : x_n = 0\} = \mathbf{R}^{n-1}.$$

Denote by $H^{s,p}(\Omega)$ $s \ge 0$ the space of Bessel potentials (Liouville classes); $H^{-s,p}(\Omega)$ denotes the space dual to $H^{s,p'}(\Omega)$ with respect to the extension $(\cdot,\cdot)$ of the scalar product in $L_2(\Omega)$. The norm in $H^{s,p}(\Omega)$ is denoted as $\|\cdot\|_{s,p} = \|\cdot,\Omega\|_{s,p}$ $(s \in \mathbf{R})$.

Denote by $B^{s,p}(\partial\Omega)$ $(s \in \mathbf{R})$ the Besov space with the norm $\langle\langle\cdot\rangle\rangle_{s,p} = \langle\langle\cdot,\partial\Omega\rangle\rangle_{s,p}$. The spaces $B^{s,p}(\partial\Omega)$ and $B^{-s,p'}(\partial\Omega)$ are dual to each other with respect to the extension $\langle\cdot,\cdot\rangle$ of the scalar product in $L_2(\partial\Omega)$.

In what follows we denote by $(\cdot,\cdot)$ and by $\langle\cdot,\cdot\rangle$ the scalar products (and their extensions) in $L_2(\Omega)$ and $L_2(\partial\Omega)$ respectively, and the scalar products (and their extensions) in the direct products $(L_2(\Omega))^N$ and $(L_2(\partial\Omega))^{|\tau|}$ as well.

Denote by $C_0^\infty(\overline{\Omega})$ the set of restrictions of functions from $C_0^\infty(\mathbf{R}^n)$ to $\overline{\Omega}$. It is clear that $C_0^\infty(\overline{G}) = C^\infty(\overline{G})$ for a bounded domain G.

Let r be a fixed natural number, $1 < p < \infty$, $s \in \mathbf{R}$, and $s \neq k + 1/p$ ($k = 0, \dots, r-1$). By $\widetilde{H}^{s,p,(r)} = \widetilde{H}^{s,p,(r)}(\Omega)$ we denote the completion of $C_0^\infty(\overline{\Omega})$ in the norm

$$|||u|||_{s,p,(r)} = |||u,\Omega|||_{s,p,(r)} = \Big(\|u\|_{s,p}^p + \sum_{j=1}^{p} \langle\langle D_\nu^{j-1} u\rangle\rangle_{s-j+1-1/p,p}\Big)^{1/p}, \quad (1.1.4)$$

where $D_\nu = i\partial/\partial\nu$ and ν is the vector normal to $\partial\Omega$.

We identify every element $u \in C_0^\infty(\overline{\Omega})$ with the vector

$$\begin{gathered} u = (u_0, u_1, \dots, u_r), \\ u_0 = u|_{\overline{\Omega}}, \; u_j = D_\nu^{j-1} u|_{\partial\Omega} \quad (j = 1, \dots, r). \end{gathered} \quad (1.1.5)$$

Then the space $\widetilde{H}^{s,p,(r)}(\Omega)$ is isometrically equivalent to the closure $\mathcal{H}_0^{s,p,(r)}(\Omega)$ of the space of vectors (1.1.5) $\{u = (u_0, u_1, \dots, u_r) \in C_0^\infty(\overline{\Omega})\}$ in the direct product

$$\mathcal{H}^{s,p,(r)}(\Omega) = H^{s,p}(\Omega) \times \prod_{j=1}^{r} B^{s-j+1-1/p,p}(\partial\Omega).$$

Therefore (see [R1, Ch. II])

$$\begin{aligned} \widetilde{H}^{s,p,(r)}(\Omega) = \mathcal{H}_0^{s,p,(r)}(\Omega) = \Big\{&(u_0, u_1, \dots, u_r) \in \mathcal{H}^{s,p,(r)}(\Omega) : \\ &u_0 \in H^{s,p}(\Omega), \; u_j = D_\nu^{j-1} u_0\big|_{\partial\Omega} \; (\forall j : s - j + 1 - 1/p > 0); \\ &\text{if } s - j + 1 - 1/p < 0 \text{ then } u_j \in B^{s-j+1-1/p,p}(\partial\Omega) \\ &\text{does not depend on } u_0\Big\}. \end{aligned}$$

In particular, if $s < 1/p$, then

$$\widetilde{H}^{s,p,(r)}(\Omega) = \mathcal{H}^{s,p,(r)}(\Omega) = H^{s,p}(\Omega) \times \prod_{j=1}^{r} B^{s-j+1-1/p,p}(\partial\Omega).$$

Hence, the space $\widetilde{H}^{s,p,(r)}(\Omega)$ consists of the vectors $u = (u_0, u_1, \dots, u_r)$ where

$$\begin{gathered} u_0 \in H^{s,p}(\Omega), \qquad u_j \in B^{s-j+1-1/p,p}(\partial\Omega) \quad (j = 1, \dots, r), \\ u_j = D_\nu^{j-1} u_0|_{\partial\Omega} \quad (\forall j : s - j + 1 - 1/p > 0). \end{gathered}$$

For any sequence $u_k \in C_0^\infty(\overline{\Omega})$ convergent to $u = (u_0, u_1, \dots, u_r)$ in $\widetilde{H}^{s,p,(r)}(\Omega)$ the sequence $D_\nu^{j-1} u_k|_{\partial\Omega}$ convergent to u_j in $B^{s-j+1-1/p,p}(\partial\Omega)$. In this (strong) sense the equality

$$u_j = D_\nu^{j-1} u|_{\partial\Omega} \qquad (j = 1, \dots, r;\ u \in \widetilde{H}^{s,p,(r)}(\Omega)).$$

is true.

If $s > r - 1/p$, then norm (1.1.4) is equivalent to the norm $\|\cdot, \Omega\|_{s,p}$ and $\widetilde{H}^{s,p,(r)}(\Omega) = H^{s,p}(\Omega)$. In this case all the components of the vector $(u_0, u_1, \dots, u_r) \in \widetilde{H}^{s,p,(r)}(\Omega)$ are defined by the first component, i.e.,

$$u_j = D_\nu^{j-1} u_0\big|_{\partial\Omega} \qquad (j = 1, \dots, r).$$

For $s = k + 1/p$ $(k = 0, \dots, r-1)$ we define the space $\widetilde{H}^{s,p,(r)}$ and norm (1.1.4) by the method of complex interpolation.

Finally, if $r = 0$, then we set

$$\widetilde{H}^{s,p,(0)}(\Omega) := H^{s,p}(\Omega), \quad |||u|||_{s,p,(0)} := \|u\|_{s,p}.$$

Recall here ([R1, Ch. II]) that if

$$M(x, D) = \sum_{|\alpha| \le q} a_\alpha(x) D^\alpha \quad (a_\alpha(x) \in C_0^\infty(\overline{\Omega})) \tag{1.1.6}$$

is a differential expression of order $q \le r$, then the closure M of the mapping

$$u \mapsto M(x, D)u \qquad (u \in C_0^\infty(\overline{\Omega}))$$

acts continuously in the pair of spaces

$$\widetilde{H}^{s,p,(r)}(\Omega) \to H^{s-q,p}(\Omega);$$

and if $q \le r - 1$ then the closure $M_{\partial\Omega}$ of the mapping

$$u \mapsto M(x, D)u|_{\partial\Omega} \qquad (u \in C_0^\infty(\overline{\Omega}))$$

acts continuously in the pair of spaces

$$\widetilde{H}^{s,p,(r)}(\Omega) \to B^{s-q-1/p,p}(\partial\Omega).$$

It implies that the closure $\mathbf{M}$ of the mapping

$$u \mapsto (Mu|_{\overline{\Omega}}, Mu|_{\partial\Omega}, \dots, D_\nu^{r-q-1} Mu|_{\partial\Omega}) \qquad (u \in C_0^\infty(\overline{\Omega}))$$

acts continuously in the pair of spaces

$$\widetilde{H}^{s,p,(r)}(\Omega) \to \widetilde{H}^{s-q,p,(r-q)}(\Omega).$$

Rewrite expression (1.1.6) in the form of

$$M(x.D) = \sum_{j=0}^{q} M_j(x, D')D_\nu^j,$$

where $M_j(x, D')$ is a $(q-1)$th-order tangential operator. By integration by parts we obtain that

$$(M(x, D)u, v) = (u, M^+(x, D)v) - i\sum_{j=1}^{q}\sum_{k=1}^{j}\langle D_\nu^{k-1}u, D_\nu^{j-k}M_j^+ v\rangle$$

$$(u, v \in C_0^\infty(\overline{\Omega})).$$

Here expressions M^+ and M_j^+ are formally adjoint to the expressions M and M_j. By passing to the limit It implies that if $u = (u_0, \dots, u_r) \in \widetilde{H}^{s,p,(r)}(\Omega)$, then $Mu = f \in H^{s-q,p}(\Omega)$ if and only if

$$(f, v) = (u_0, M^+ v) - i\sum_{j=1}^{q}\sum_{k=1}^{j}\langle u_k, D_\nu^{j-k}M_j^+ v\rangle \quad (\forall v \in C_0^\infty(\overline{\Omega})). \qquad (1.1.7)$$

Formula (1.1.7) gives us a rule of calculation of the element $Mu = f \in H^{s-q,p}(\Omega)$ according to the element $u = (u_0, \dots, u_r) \in \widetilde{H}^{s,p,(r)}(\Omega)$.

Rewrite (1.1.7) in the form of

$$\left(M(x, D)u\right)_+ = M(x, D)u_{0+} - i\sum_{j=1}^{q}\sum_{k=1}^{j} M_j D_\nu^{j-k}u_k \times \delta(\partial\Omega).$$

Here $\left(M(x, D)u_0\right)_+$ and u_{0+} are the extensions by zero of the functions $M(x, D)u$ and u_0 onto $\mathbf{R}^n$, and $\delta(\partial\Omega)$ is the Dirac measure concetrated on $\partial\Omega$:

$$\left(M_j D_\nu^{j-k}(u_k \times \delta(\partial\Omega)), v\right) := \langle u_k, D_\nu^{j-k}M_j^+ v\rangle \quad (v \in C_0^\infty(\mathbf{R}^n)).$$

In the set of the expressions of the form

$$M(x, D) = \sum_{j=1}^{q} M_j(x, D')D_\nu^j$$

we introduce the operator J such that

$$JM(x, D) = \begin{cases} 0, & \text{for } q = 0 \\ \sum_{j=1}^{q} M_j(x, D')D_\nu^{j-1}, & \text{for } q \geq 1. \end{cases}$$

Now one can rewrite (1.1.7) in the form of

$$f_+ = (Mu)_+ = Mu_{0+} - i\sum_{k=1}^{q} J^k M(x,D)(u_k \times \delta(\Omega)). \qquad (1.1.7)'$$

Here f_+, $(Mu)_+$ and u_{0+} are the extensions of the functions f, Mu and u_0 onto $\mathbf{R}^n$. The equality (1.1.7)' considered in whole $\mathbf{R}^n$ gives us a formula for calculation of the element f_+ according to the vector $(u_0, \dots, u_r)$.

In a similar way it is easy to verify that

$$Mu|_{\partial\Omega} = \varphi \in B^{s-q-1/p,p}(\partial\Omega) \qquad (q \le r-1)$$

if

$$\sum_{j=0}^{q} M_j(x,D')u_{j+1} = \varphi. \qquad (1.1.8)$$

1.1.3. It follows from all mentioned in 1.1.2 that for any $s \in \mathbf{R}$ and any $p \in]1,\infty[$ the closure $A = A_{s,p}$ of the mapping

$$u \mapsto \left(lu|_{\overline{G}}, \{(l_j u|_{\partial G}, \dots, D_\nu^{æ-s_j-1} l_u|_{\partial G}) : j = 1, \dots, N\}, bu|_{\partial G}\right)$$

$$\left(u \in (C^\infty(\overline{G}))^N;\ l_j u = \sum_{k=1}^{N} l_{jk}(x,D)u_k\right)$$

acts continuously in the pair of spaces

$$\widetilde{H}^{T+s,p,(\tau)} \to K_{s,p}, \qquad (1.1.9)$$

$$\widetilde{H}^{T+s,p,(\tau)} := \prod_{j=1}^{N} \widetilde{H}^{t_j+s,p,(\tau_j)}$$

$$K_{s,p} := \prod_{j=1}^{N} \widetilde{H}^{s-s_j,p,(æ-s_j)} \times \prod_{h=1}^{m} B^{s-\sigma_h-1/p,p}(\partial G).$$

Definition 1.1.1. *Element*

$$u = (u_1, \dots, u_N) \in \widetilde{H}^{T+s,p,(\tau)}$$

$$\left(u_j = (u_{j0}, \dots, u_{j,\tau_j}) \in \widetilde{H}^{t_j+s,p,(\tau_j)} \quad (j = 1, \dots, N)\right)$$

such that

$$Au = F = (f, \varphi_1, \dots, \varphi_m),\ f = (f_1, \dots, f_N),\ f_j = (f_{j0}, \dots, f_{j,æ-s_j}) \qquad (1.1.10)$$

is called a generalized solution of problem (1.1.1)-(1.1.2).

It turns out that [R6] following theorem on complete collection of isomorphisms is true.

Theorem 1.1.1. *Let (1.1.1) be an elliptic problem. Then*
i) for any $s \in \mathbf{R}$ and any $p \in (1, +\infty)$ the operator $A = A_{s,p}$ is Noetherian. This means that the kernel $\mathfrak{N}$ and the cokernel $\mathfrak{N}^*$ are finite dimensional and do not depend on s and p,

$$\mathfrak{N} = \{u \in (C^\infty(\overline{G}))^N \ : \ Au = 0\},$$

$$\mathfrak{N}^* \subset \Big\{V = (V_1, \ldots, V_N, \psi_1, \ldots, \psi_m) \ : \ V_j = (V_{j0}, \ldots, V_{j,æ-s_j}) \in$$

$$C^\infty(\overline{G}) \times (C^\infty(\partial G))^{æ-s_j}; \ \psi_j \in C^\infty(\partial G) \ (j = 1, \ldots, m)\Big\}.$$

Problem (1.1.10) is solvable in $\widetilde{H}^{T+s,p,(\tau)}$ if and only if the relation

$$[F, V] := \sum_{j=1}^{N} (f_{j0}, V_{j0}) + \sum_{r=1}^{N} \sum_{k=1}^{æ-s_r} \langle f_{rk}, V_{rk} \rangle + \sum_{h=1}^{m} \langle \varphi_h, \psi_h \rangle = 0 \qquad (1.1.11)$$

$$(\forall V \in \mathfrak{N}^*)$$

holds;
ii) the restriction $\widehat{A}_{s,p}$ of the operator $A_{s,p}$ realizes an isomorphism

$$P\widetilde{H}^{T+s,p,(\tau)}(G) \to Q^+ K_{s,p} \quad (s \in \mathbf{R}, \ 1 < p < \infty).$$

Here

$$P\widetilde{H}^{T+s,p,(\tau)}(G) = \{u \in \widetilde{H}^{T+s,p,(\tau)} \ : \ (u_0, v) = 0 \ (\forall v \in \mathfrak{N})\}$$

is a subspace of the space $\widetilde{H}^{T+s,p,(\tau)}$, and

$$Q^+ K_{s,p} = \{F \in K^{s,p} \ : \ [F, V] = 0 \quad (\forall V \in \mathfrak{N}^*)\}$$

is a subspace of $K_{s,p}$.

It is clear that for $s_2 < s_1$ the operators $A_{s_2,p}$ and $\widehat{A}_{s_2,p}$ are extensions in continuity of the operators $A_{s_1,p}$ and $\widehat{A}_{s_1,p}$ respectively. Therefore all operators under consideration are restrictions of the operators $A_{s_2,p} = A$ and $\widehat{A}_{s_2,p} = \widehat{A}$ with sufficiently negative s_2. Therefore one can write A and $\widehat{A}$ in place of $A_{s_1,p}$ and $\widehat{A}_{s_1,p}$ respectively.

1.2. Green's Formula

1.2.1.

Definition 1.2.1. *Matrix $B(x, D)$ of the system of $|\tau| = \tau_1 + \cdots + \tau_N$ boundary conditions*

$$B_h u(x) = \sum_{j=1}^{N} B_{hj}(x, D) u_j(x) \quad (h = 1, \ldots, |\tau|) \tag{1.2.1}$$

is called a τ-complete matrix ([R1], [RR1]) if:
a)

$$B_{hj}(x, D) = \sum_{s=1}^{\tau_j} \Lambda_{h,js}(x, D') D_\nu^{s-1} \quad (x \in \partial G,\ j = 1, \ldots, N,\ h = 1, \ldots, |\tau|),$$

where $\Lambda_{h,js}(x, D')$ are tangential, generally speaking, pseudo-differential operators. Moreover there exist integer $\sigma_h < 0$ $(h = 1, \ldots, |\tau|)$ such that $\operatorname{ord} B_{hj} \le \sigma_h + \tau_j$ for $\sigma_h + \tau_j > 0$, and $B_{hj} \equiv 0$ for $\sigma_h + \tau_j \le 0$;
b) For every real vector $\gamma \ne 0$ tangential to ∂G at the point x the determinant of the square $|\tau| \times |\tau|$-matrix

$$\left(\Lambda^0_{h,js}(x, \gamma)\right)_{\substack{h=1,\ldots,|\tau| \\ s=1,\ldots,\tau_j,\ j=1,\ldots,N}} \tag{1.2.2}$$

does not equal to zero at every point $x \in \partial G$. Here $\Lambda^0_{h,js}(x, D')$ is the principal part of the expression $\Lambda_{h,js}(x, D')$ which includs only $(\tau_j + \sigma_h - s + 1)$th-order differentiations. In addition $\Lambda^0_{h,js} = 0$ for $\tau_j + \sigma_h - s + 1 < 0$.

This definition implies that if system (1.2.1) is a τ-complete matrix, then the system of the equations

$$\sum_{j=1}^{N} \sum_{s=1}^{\tau_j} \Lambda_{h,js}(x, D') \eta_{js} = \varphi_h \quad (h = 1, \ldots, |\tau|;\ x \in \partial G) \tag{1.2.3}$$

or, in short,

$$\Lambda \eta = \varphi$$

is an elliptic in the Douglis–Nirenberg sense on the manifold ∂G.

In fact, if we associate the number $\tau_{js} = \tau_j - s + 1$ with the function η_{js}, and the number σ_h with the equation with index h, then $\operatorname{ord} \Lambda_{h,js} \le \tau_{js} + \sigma_h$ for $\tau_{js} + \sigma_h \ge 0$, and $\Lambda_{h,js} \equiv 0$ for $\tau_{js} + \sigma_h < 0$. Thus, the ellipticity of system (1.2.3) on ∂G follows from the fact that the determinant of matrix (1.2.2) does not equal to zero.

It is clear that the formally adjoint system

$$\sum_{h=1}^{|\tau|} \Lambda^{+}_{h,js}(x, D')\zeta_h = \psi_{js} \quad (s = 1, \ldots, \tau_j;\ j = 1, \ldots, N;\ x \in \partial G) \tag{1.2.4}$$

or, in short,

$$\Lambda^{+}\zeta = \psi$$

(where $\Lambda^{+}_{h,js}$ is the expression formally adjoint to the expression $\Lambda_{h,js}$) is a Douglis-Nirenberg elliptic system. In addition, Green's formula

$$\langle \Lambda\eta, \zeta \rangle = \langle \eta, \Lambda^{+}\zeta \rangle \quad (\eta, \zeta \in (C^{\infty}(\partial G))^{|\tau|}) \tag{1.2.5}$$

is true.

For any $q \in \mathbf{R}$ the closure $\Lambda = \Lambda_q$ of the mapping

$$\eta \mapsto \Lambda\eta \qquad (\eta \in (C^{\infty}(\partial G))^{|\tau|})$$

acts continuously in the pair of spaces

$$U^{q,\tau,p} := \prod_{j=1}^{N} \prod_{s=1}^{\tau_j} B^{q+\tau_j-s,p}(\partial G) \to V^{q,\sigma,p} := \prod_{h=1}^{|\tau|} B^{q-\sigma_h-1,p}(\partial G). \tag{1.2.6}$$

The kernel $\mathfrak{N}_{\Lambda}$ of this operator is finite-dimensional and does not depend on q, i.e.,

$$\mathfrak{N}_{\Lambda} \subset (C^{\infty}(\partial G))^{|\tau|}.$$

The range $\mathcal{R}(\Lambda_q)$ of values of operator Λ_q is closed in the space $V^{q,\sigma,p}$ and has a finite codimension independent of q ([R2], [R1, Ch. X]).

The similar statement is true also for the operator Λ^{+}_q defined by problem (1.2.4) with the kernel $\mathfrak{N}_{\Lambda^{+}}$. In addition, the cokernel (the kernel) of the operator Λ is the kernel (the cokernel) of the operator Λ^{+}.

It easily follows from finite dimensionality of $\mathfrak{N}_{\Lambda}$ that every element

$$\eta \in \bigcup_{q=-\infty}^{+\infty} U^{q,\tau,p}$$

can be uniquely represented in the form

$$\eta = \eta' + \eta'', \quad \eta'' \in \mathfrak{N}_{\Lambda}, \qquad \langle \eta', \mathfrak{N}_{\Lambda} \rangle = 0. \tag{1.2.7}$$

It is clear ([RR1], [R1, Ch. V, Ch. X]) that the restriction $\widehat{\Lambda}_q$ of the operator Λ_q onto $P_{\Lambda}U^{q,\tau,p}$ realizes an isomorphism

$$\widehat{\Lambda}_q : P_{\Lambda}U^{q,\tau,p} \to P_{\Lambda^{+}}V^{q,\sigma,p}. \tag{1.2.8}$$

The operator

$$\Pi_q = \widehat{\Lambda}_q^{-1} P_{\Lambda+}$$

acts continuously from the entire $V^{q,\sigma,p}$ into $U^{q,\tau,p}$. Since for $q_1 \leq q_2$ the operators Λ_{q_1}, $\widehat{\Lambda}_{q_1}$ and Π_{q_1} are the extension by continuity of the operators Λ_{q_2}, $\widehat{\Lambda}_{q_2}$ and Π_{q_2} respectively, in what follows we will write Λ, $\widehat{\Lambda}$ and Π in place of Λ_q, $\widehat{\Lambda}_q$ and Π_q.

Note that the following statement is true.

Lemma 1.2.1. *For any $\varphi \in V^{q,\sigma,p}$ there exists an element $u \in \widetilde{H}^{T+q,p,(\tau)}$ such that $Bu|_{\partial G} = \varphi$ if and only if $\mathfrak{N}_{\Lambda+} = 0$. In addition there exists a linear continuous operator*

$$T_q : \varphi \mapsto u \quad (V^{q,\sigma,p} \to \widetilde{H}^{T+q,p,(\tau)}).$$

For $q_2 < q_1$ the operator T_{q_2} is an extension by continuity of the operator T_{q_1}. If $\varphi \in (C^\infty(\partial G))^{|\tau|}$, then $u \in (C^\infty(\overline{G}))^N$.

Proof. Let $\mathfrak{N}_{\Lambda+} = 0$. Then for any $\varphi \in V^{q,\sigma,p}$ the element $\eta' = \widehat{\Lambda}^{-1}\varphi$ belongs to $U^{q,\sigma,p}$. Therefore, there exists the extension operator (see, for example, [R1])

$$Q_q : \eta' \mapsto u \quad (V^{q,\sigma,p} \to \widetilde{H}^{T+q,p,(\tau)})$$

$$(\eta'_{js} = D_\nu^{s-1} u_j|_{\partial G} \quad (s = 1, \ldots, \tau_j,\ j = 1, \ldots, N)).$$

For $q_2 < q_1$ the operator Q_{q_2} is an extension of the operator Q_{q_1}. Since $Bu|_\Gamma = \Lambda\eta' = \varphi$ and $T_q = Q_q\widehat{\Lambda}^{-1}$, the necessity is proved.

Sufficiency. Under the condition of the Lemma, for any element $\varphi \in V^{q,\sigma,p}$ there exists an element $u \in \widetilde{H}^{T+q,p,(\tau)}$ such that $Bu|_{\partial G} = \varphi$. Since $Bu|_{\partial G} = \Lambda\eta$, equation (1.2.3) is solvable for any φ. Therefore $\mathfrak{N}_{\Lambda+} = 0$. This completes the proof of the Lemma. □

For any element $\varphi \in (C^\infty(\partial G))^{|\tau|}$ there exist the element $u \in (C^\infty(\overline{G}))^N$ such that $Bu|_{\partial G} = \varphi$ if and only if the same condition $\mathfrak{N}_{\Lambda+} = 0$ holds.

1.2.2. Let us now obtain the Green's formula. Let

$$l_{rj}(x, D) = \sum_{k=0}^{s_r+t_j} l_k^{rj}(x, D') D_\nu^k \quad (r, j : s_r + t_j \geq 0) \qquad (1.2.9)$$

in some neighborhood of the boundary ∂G in $\overline{G}$. Here $l_k^{rj}(x, D')$ are tangential operators whose orders do not exceed $s_r + t_j - k$. If we use the integration by parts and interchange the order of the summation, we obtain that

$$(lu,v)-(u,l^+v)=\sum_{j:t_j\geq 1}\sum_{s=1}^{t_j}\langle D_\nu^{s-1}u_j, M_j^s v\rangle,$$

$$M_j^s v=-i\sum_{r:s_r+t_j\geq s}\sum_{k=s}^{s_r+t_j} D_\nu^{k-j}(l_k^{rj}(x,D'))^+ v_r. \tag{1.2.10}$$

We associate the vector

$$\eta=(\eta_1,\dots,\eta_N),\ \ \eta_j=(\eta_{j1},\dots,\eta_{j,\tau_j}),\ \ \eta_{js}=D_\nu^{s-1}u_j|_{\partial G}$$

$$(s=1,\dots,\tau_j,\quad j=1,\dots,N)$$

with the element $u=(u_1,\dots,u_N)$, and the vector

$$Mv=(\zeta_1,\dots,\zeta_N),\quad \zeta_j=(\zeta_{j1},\dots,\zeta_{j,\tau_j}),$$

$$\zeta_{jk}=\begin{cases} M_j^k v|_{\partial G} & \text{for } k=1,\dots,t_j,\\ 0 & \text{for } t_j<k\leq\tau_j.\end{cases}$$

with the element $v=(v_1,\dots,v_N)$.

Then $Bu|_{\partial G}=\Lambda\eta$, and one can rewrite formula (1.2.10) in the form

$$(lu,v)-(u,l^+v)=\langle\eta,Mv\rangle\quad (u,v\in(C^\infty(\overline{G}))^{|\tau|}). \tag{1.2.11}$$

By (1.2.7), using the fact that $\Pi\Lambda\eta=\eta'$, rewrite (1.2.11) in the form

$$\begin{aligned}(lu,v)-(u,l^+v) &= \langle\Pi\Lambda\eta,Mv\rangle+\langle\eta'',Mv\rangle\\ &= \langle\Lambda\eta,\Pi^+Mv\rangle+\langle\eta'',Mv\rangle\\ &= \langle\Lambda\eta,P_{\Lambda^+}\Pi^+Mv\rangle+\langle\eta'',Mv\rangle.\end{aligned}$$

Thus, the following theorem is proved.

Theorem 1.2.1. *There holds the Green's formula*

$$(lu,v)-(u,l^+v)=\langle Bu,B'v\rangle+\langle\eta'',Mv\rangle$$
$$(u,v\in(C^\infty(\overline{G}));\ \eta''=\eta-P\eta\in\mathfrak{N}_\Lambda) \tag{1.2.12}$$

where

$$B'v=P_{\Lambda^+}\Pi^+Mv|_{\partial G} \tag{1.2.13}$$

is, generally speaking, pseudo-differential matrix expression.

In particular, if $\mathfrak{N}_\Lambda = 0$, then the last term in the right-hand side of (1.2.12) is missing. It will be, for example, if either the matrix B is a parameter-elliptic matrix, or B is a Dirichlet matrix ([RSh2]). In the last case, the fact that the matrix B is differential implies that the matrix B' is also differential.

1.2.3. Let us rewrite the Green's formula in more convenient form. To do this, we complement the matrix $b(x, D)$ (see (1.1.2)) to the τ-complete system $B(x, D)$ by new rows.
Denote:

$$c(x, D) = \left(c_{hj}(x, D)\right)_{\substack{h=1,\dots,m,\\ j=1,\dots,N}} \tag{1.2.14}$$

where $\operatorname{ord} c_{hj} \le \sigma_h^c + t_j$ for $\sigma_h^c + t_j \ge 0$, and $c_{hj} \equiv 0$ for $\sigma_h^c + t_j < 0$;

$$c_h(x, D) = \left(c_{h1}(x, D), \dots, c_{hN}(x, D)\right), \quad h = 1, \dots, m,$$

is a row with index h of the matrix $c(x, D)$.
Let

$$e(x, D) = \left(D_\nu^{k-1} l_r(x, D)\right)_{\substack{r=1,\dots,N\\ k=1,\dots,-s_r+\varpi}}. \tag{1.2.15}$$

Here

$$l_r(x, D) = \left(l_{r1}(x, D), \dots, l_{rN}(x, D)\right)$$

is a row with index r of the matrix $l(x, D)$.
The matrix $c(x, D)$ consists of m rows, and the matrix $e(x, D)$ consists of $|\tau| - 2m$ rows. It turns out that the considerations from [R1, Ch.X], imply that the following statement is true.

Lemma 1.2.2. *Let (1.1.1)-(1.1.2) be an elliptic problem. Then there exists the matrix* $c(x, D)$ *with* $\sigma_h^c < 0$, $h = 1, \dots, m$ *such that the* $|\tau| \times N$*-matrix*

$$B(x, D) = \begin{pmatrix} b(x, D) \\ e(x, D) \\ c(x, D) \end{pmatrix} \tag{1.2.16}$$

is τ*-complete.*

The proof of this lemma will be given below in Section 1.4
Lemma 1.2.2 and Theorem 1.2.1 directly imply the validity of the following theorem.

Theorem 1.2.2. *Let problem the (1.1.1)-(1.1.2) be elliptic, and let the matrix $b(x, D)$ be complemented to the τ-complete matrix $B(x, D)$ (1.2.16) by matrixes (1.2.14) and (1.2.15). Then there holds the Green's formula*

$$(lu, v) + \sum_{h=1}^{m} \langle b_h u, c'_h v\rangle + \sum_{r=1}^{N} \sum_{k=1}^{-s_r+æ} \langle D_\nu^{k-1} l_r u, e'_{kr} v\rangle$$

$$= (u, l^+ v) + \sum_{h=1}^{m} \langle c_h u, b'_h v\rangle + \langle \eta'', Mv\rangle, \qquad (1.2.17)$$

$$u, v \in (C^\infty(\overline{G}))^N, \quad \eta'' = \eta - P\eta \in \mathfrak{N}_\Lambda.$$

To obtain formula (1.2.17), it is necessary to represent matrix (1.2.13) in the form of the matrix $\begin{pmatrix} -c'(x,D) \\ -e'(x,D) \\ b'(x,D) \end{pmatrix}'$, transposed to $\begin{pmatrix} -c'(x,D) \\ -e'(x,D) \\ b'(x,D) \end{pmatrix}$.

1.2.4. In this subsection let us additionally assume that $\mathfrak{N}_{\Lambda^+} = 0$. In other words, assume that, for any $\varphi \in (C^\infty(\partial G))^{|\tau|}$, there exists an element $u \in (C^\infty(\overline{G}))^N$ such that $Bu|_{\partial G} = \varphi$.

Definition 1.2.2. *The problem*

$$l^+ v(x) = g(x), \qquad x \in G, \qquad (1.2.18)$$

$$b'v|_{\partial G} = \varphi, \qquad (1.2.19)$$

$$\langle Mv, \mathfrak{N}_\Lambda\rangle = 0 \qquad (1.2.20)$$

is called a formally adjoint to problem (1.1.1)-(1.1.2) with respect to Green's formula (1.2.17).

Let

$$\mathfrak{N}^+ = \{v \in (C^\infty(\overline{G}))^N : l^+ v = 0, \, b'v|_{\partial G} = 0, \, \langle Mv, \mathfrak{N}_\Lambda\rangle = 0\}. \qquad (1.2.21)$$

Theorem 1.2.3. *Under the conditions of Theorem 1.2.2, let $\mathfrak{N}_{\Lambda^+} = 0$, and let $s \in \mathbf{R}$, $p \in (1, +\infty)$. Then problem (1.1.10), with $F \in K_{s,p}$, possesses a solution $u \in \widetilde{H}^{T+s,p,(\tau)}$ if and only if the relation*

$$(f_0, v) + \sum_{j=1}^{N} \sum_{r=1}^{æ-s_j} \langle f_{jr}, e'_{kr} v\rangle + \sum_{h=1}^{m} \langle \varphi_h, c'_h v\rangle = 0 \quad (\forall v \in \mathfrak{N}^+). \qquad (1.2.22)$$

holds. In addition, $\mathfrak{N}^+$ is finite-dimensional.

Proof. By passing to the limit we verify that Green's formula (1.2.17) is true for $u \in \widetilde{H}^{T+s,p,(\tau)}$ and $v \in (C^\infty(\overline{G}))^N$. Then, the *necessity* directly follows from obtained Green's formula.

Let us prove the *sufficiency*. Problem (1.1.10) is solvable if and only if conditions (1.1.11) hold. If in (1.1.11) we substitute $l_j u|_{\overline{G}}$ in place of f_{j0}, $D_\nu^{k-1} l_r u|_{\partial G}$ in place of f_{rk}, and $b_h u|_{\partial G}$ in place of φ_h, then, by using Green's formula (1.2.17), we obtain that

$$(u, l^+ V_0) + \sum_{r=1}^{N} \sum_{k=1}^{æ-s_r} \langle D_\nu^{k-1} l_r u, V_{rk} - e'_{kr} V_0 \rangle + \sum_{h=1}^{m} \langle b_h u|_{\partial G}, \psi_h - c'_h V_0 \rangle$$
$$+ \sum_{h=1}^{m} \langle c_h u, b'_h V_0 \rangle + \langle \eta'', M V_0 \rangle = 0 \tag{1.2.23}$$
$$\left(\forall V \in \mathfrak{N}^*, \ u \in (C^\infty(\overline{G}))^N\right).$$

Relation (1.2.23) with $u \in (C_0^\infty(G))^N$ directly implies that

$$(u, l^+ V_0) = 0, \qquad \forall u \in (C_0^\infty(G))^N.$$

Therefore, $l^+ V_0 = 0$ ($\forall V \in \mathfrak{N}^*$). Now let $\eta = \eta'' \in \mathfrak{N}_\Lambda$. Then $Bu|_{\partial G} = \Lambda \eta = 0$, and, it follows from (1.2.23) that $\langle \eta'', M V_0 \rangle = 0$ ($\forall V \in \mathfrak{N}^*$). Therefore, $\langle M V_0, \mathfrak{N}_\Lambda \rangle = 0$.

Since $\mathfrak{N}_{\Lambda+} = 0$ and matrix (1.2.16) is τ-complete, the range $\{Bu|_{\partial G} : u \in \widetilde{H}^{T+s,p,(\tau)}\}$ coincides with $\mathcal{R}\Lambda_q = V^{q,\sigma,p}$ ($\forall q \in \mathbf{R}$). Therefore, $\{Bu|_{\partial G} : u \in C^\infty(\overline{G})\}$ is dense in $V^{q,\sigma,p}$ for any $q \in \mathbf{R}$. Then, it follows from (1.2.23) that

$$b'_h V_0|_{\partial G} = 0, \quad c'_h V_0|_{\partial G} = \psi_h \qquad (h = 1, \dots, m),$$
$$l'_{kr} V_0|_{\partial G} = v_{rk} \qquad (r = 1, \dots, N, \ k = 1, \dots, æ - s_r).$$

Then, $V_0 \in \mathfrak{N}^+$ ($\forall V \in \mathfrak{N}^*$), and conditions (1.1.11) coincide with conditions (1.2.22). This completes the proof of the theorem. □

1.2.5. Let us note that if $l(x, D)$ is a (T, S)-order matrix expression,

$$t_1 \geq \cdots \geq t_N \geq 0 = s_1 \geq \cdots \geq s_N,$$

then $l^+(x, D)$ is a (T', S')-order expression, where

$$t'_j = t_1 + s_j, \quad s'_j = t_j - t_1 \quad (j = 1, \dots, N).$$

Let

$$H^{T,p}_{(b)} = \{u \in H^{T,p}(G) : bu|_{\partial G} = 0\} \tag{1.2.24}$$

denote a subspace of the space $H^{T,p}(G)$, and let $H^{T',p'}_{(b)+}$ denote the space of all vectors $v \in H^{T',p'}(G)$ such that

$$(lu, v) = (u, l^+ v), \tag{1.2.25}$$

$$\forall u : u \in H^{T,p}_{(b)},\ D_\nu^{k-1} L_r u|_{\partial G} = 0,\ r = 1, \dots, N,\ k = 1, \dots, -s_r + æ.$$

It is clear that $H^{T',p'}_{(b)+}$ is a subspace of the space $H^{T',p'}(G)$.
We set also that

$$H^{T+k,p}_{(b)} = H^{T,p}_{(b)} \bigcap H^{T+k,p}(G),\ H^{T'+k,p'}_{(b)+} = H^{T',p'}_{(b)+} \bigcap H^{T'+k,p'}(G) \quad (k \geq 0).$$

Lemma 1.2.3. *Under the conditions of Theorem 1.2.2 let $\mathfrak{N}_{\Lambda+} = 0$. Then the space $H^{T',p'}_{(b)+}$ consists of those and only those elements $v \in H^{T',p'}$ such that*

$$b'v|_{\partial G} = 0, \qquad \langle Nv, \mathfrak{N}_\Lambda \rangle = 0. \tag{1.2.26}$$

Proof. The proof follows from both the Green's formula and Lemma 1.2.1. Indeed, Green's formula (1.2.17) and relation (1.2.25) directly imply that

$$\sum_{h=1}^{m} \langle c_h u, b'_h v \rangle + \langle \eta'', Mv \rangle = 0 \quad \left(u \in H^{T,p}_{(b)}\right).$$

Choosing here the element u such that $\eta'' = 0$, by virtue of Lemma 1.2.1, we get that

$$\sum_{h=1}^{m} \langle \varphi_h, b'_h v \rangle = 0 \qquad (\varphi_h \in C^\infty(\partial G),\ h = 1, \dots, m).$$

Therefore, $b'v|_{\partial G} = 0$, and relations (1.2.26) are, thus, established. □

Now consider the problems

$$lu = f, \qquad u \in H^{T,2}_{(b)} \tag{1.2.27}$$

and

$$l^+ v = g, \qquad v \in H^{T',2}_{(b)+} \tag{1.2.28}$$

with homogeneous boundary conditions.

Lemma 1.2.4. *Under the conditions of Lemma 1.2.3, let $g \in H^{-S',2}(G)$ and $(g, \mathfrak{N}) = 0$. Then there exists a solution $v \in H^{T',2}(G)$ of problem (1.2.28).*

Proof. Since problem (1.1.1)-(1.1.2) is elliptic, the a priori estimate

$$\|u\|^2_{H^{T+æ,2}(G)} \le c(\|lu\|^2_{H^{æ-S,2}(G)} + \sum_{j=1}^{m} \langle\langle b_j u\rangle\rangle^2_{æ-\sigma_h-\frac{1}{2},2} + \sum_{j=1}^{N} \|u_j\|^2_{L_2(G)})$$

is valid ([R12], [KosR], [RR3]). In addition, if $(u, \mathfrak{N}) = 0$, then the last term of the right-hand side of this inequality is missing, i.e.,

$$\|u\|^2_{H^{T+æ,2}(G)} \le c(\|lu\|^2_{H^{æ-S,2}(G)} + \sum_{j=1}^{m} \langle\langle b_j u\rangle\rangle^2_{æ-\sigma_h-\frac{1}{2},2}), \qquad (1.2.29)$$

$$u \in PH^{T+æ,2}(G) = \{w \in H^{T+æ,2}(G) : (w, \mathfrak{N}) = 0\}.$$

Estimate (1.2.29) gives us an opportunity to set that the expression

$$[u, v] = (lu, lv)_{H^{æ-S,2}(G)} + \sum_{j=1}^{m} \langle B_j u, B_j v\rangle_{B^{æ-\sigma_h-1/2,2}(\partial G)}$$

defines new scalar product in the subspace $PH^{T+æ,2}(G)$ of the space $H^{T+æ,2}(G)$.

It is clear that the expression (u, g) is a linear continuous functional of $u \in PH^{T+æ,2}(G)$. Therefore, there exists an element $w \in PH^{T+æ,2}(G)$ such that

$$[u, w] = (u, g) \qquad (\forall u \in PH^{T+æ,2}(G)). \qquad (1.2.30)$$

However, if $u' \in \mathfrak{N}$, then $[u', w] = 0$, and $(u', g) = 0$. Therefore, equality (1.2.30) is true for $u + u'$ in place of u. Thus,

$$[u, w] = (u, g) \qquad (\forall u \in H^{T+æ,2}(G)),$$

or, in detail,

$$\sum_{h=1}^{N} \sum_{|\alpha| \le æ - s_h} (D^\alpha l_h u, D^\alpha l_h w) + \sum_{h=1}^{m} \langle b_h u, b_h w\rangle_{B^{æ-\sigma_h-\frac{1}{2},2}(\partial G)} = (u, g), \quad (1.2.31)$$

$$u \in H^{æ+T,2}(G).$$

It follows from lemma ([RSh3, p. 469]) on increasing of smoothness that if

$$g \in H^{k-\tau,2}(G) = \prod_{j=1}^{N} H^{k-\tau_j,2}, \qquad \tau_j = t_j + æ, k \ge \tau_1,$$

then

$$w \in H^{k+\tau,2}(G) = \prod_{j=1}^{N} H^{k+\tau_j,2}(G).$$

Hence, by integration by parts, we get from (1.2.31) that

$$\sum_{h=1}^{N}(l_h u, \sum_{|\alpha|\leq æ-s_h} D^{2\alpha} l_h w) + \sum_{h=1}^{N}\sum_{j=1}^{æ-s_h}\langle D_\nu^{j-1} l_h u, T_{jh} w\rangle$$

$$+\sum_{h=1}^{m}\langle b_h u, b_h w\rangle_{B^{æ-\sigma_h-\frac{1}{2},2}(\partial G)} = (u, g), \qquad (1.2.32)$$

$$(\forall u \in H^{æ+T,2}(G)).$$

We set

$$\sum_{|\alpha|\leq æ-s_h} D^{2\alpha} l_h w = v_h \in H^{t_1+s_h,2}(G), \qquad h = 1, \dots, N.$$

Then the equality

$$(lu, v) = (u, g),$$

holds for all elements $u \in H^{æ+T,2}(G)$ such that $bu|_{\partial G} = 0$, $D_\nu^{j-1} l_h u|_{\partial G} = 0$, $h = 1, \dots, N$, $j = 1, \dots, æ - s_j$, and the vector v is a solution of problem (1.2.28). □

In view of Theorem 1.2.3, the kernel $\mathfrak{N}^+$ of the problem (1.2.18)-(1.2.20) is finite-dimensional. Therefore, every element $v \in H^{T',p'}(G)$ can be represented in the form

$$v = v' + v'', \quad v' \in \mathfrak{N}^+, \quad (v'', \mathfrak{N}^+) = 0,$$

and the operator $P' : v \to v''$ is continuous in $H^{T',p}(G)$. Hence, Lemma 1.2.4 yields that the operator

$$v \mapsto L^+ v : P^+ H^{T',2}_{(b)+} \to PH^{-S',2}(G)$$

is a mutually continuous one-to-one mapping.

By virtue of Banach theorem on inverse operator, this mapping is an isomorphism. Therefore, the estimate

$$\|v\|_{H^{T',2}(G)} \leq \|l^+ v\|_{H^{-S',2}(G)} \quad (\forall v \in P^+ H^{T',2}_{(b)+})$$

is valid. Then, in view of finite dimensionality of $\mathfrak{N}^+$, the estimate

$$\|v\|_{H^{T',2}(G)} \leq c(\|L^+ v\|_{H^{-s',2}(G)} + \|v\|_{(L_2(G))^N}), \qquad v \in H^{T',2}_{(b)+} \qquad (1.2.33)$$

is true.

Now consider problem (1.2.18)-(1.2.20), formally adjoint to problem (1.1.1)-(1.1.2). Assume that, for any $\psi \in B^{T'-\sigma'-1/2,2}(\partial G)$, there exists an element $v_0 \in H^{T',2}(G)$ such that $b'v_0|_{\partial G} = \psi$ and $\langle Mv_0, \mathfrak{N}_\Lambda\rangle = 0$, and the extension operator

$$T : \psi \mapsto v_0 : B^{T'-\sigma'-1/2,2}(\partial G) \to H^{T',2}(G) \tag{1.2.34}$$

is continuous.

Lemma 1.2.5. *Under the conditions of Lemma 1.2.3, let extension operator (1.2.34) be continuous. Then the a priori estimate*

$$\|v\|_{H^{T',2}(G)} \leq c(\|L^+v\|_{H^{-s',2}(G)} + \sum_{h=1}^{m}\langle\langle b'_h v\rangle\rangle_{t'-\sigma'_h-1/2,2} + \|v\|_{(L_2(G))^N}) \tag{1.2.35}$$

$$(v \in H^{T',2}).$$

holds.

Proof. Let $v \in H^{T',2}(G)$, $(Mv, \mathfrak{N}_\Lambda) = 0$, and let $b'v|_{\partial G} = \psi$. Assume that $T\psi = v_0 \in H^{T',2}(G)$. Then $v - v_0 \in H^{T',2}_{(b)+}$, and, in view of estimate (1.2.33), we get

$$\|v - v_0\|_{H^{T',2}(G)} \leq c\,(l^+(v - v_0)\|_{-S',2} + \|v - v_0\|_{(L_2(G))^N}).$$

Therefore

$$\begin{aligned}
\|v\|_{H^{T',2}(G)} &\leq \|v - v_0\|_{H^{T',2}(G)} + \|v_0\|_{H^{T',2}(G)} \\
&\leq c\,(\|l^+v\|_{H^{-S',2}(G)} + \|l^+v_0\|_{H^{-S',2}(G)} \\
&\qquad + \|v_0\|_{H^{T',2}(G)} + \|v\|_{(L_2(G))^N}) \\
&\leq c_1(\|L^+v\|_{H^{-S',2}(G)} + \|v_0\|_{H^{T',2}(G)} + \|v\|_{(L_2(G))^N}).
\end{aligned}$$

On the other hand,

$$\begin{aligned}
\|v_0\|_{H^{T',2}(G)} &= \|T\psi\|_{H^{T',2}(G)} \\
&\leq c_2\langle\langle\psi\rangle\rangle_{t'-\sigma'_h-1/2,2} = c_2\langle\langle b'v\rangle\rangle_{t'-\sigma'_h-1/2,2},
\end{aligned}$$

and required estimate (1.2.35) is thus established for the vectors $v \in H^{T',2(G)}$ satisfying the condition $(Mv, \mathfrak{N}_\Lambda) = 0$.

Let $\{e_1, \ldots, e_q\}$ be a basis in $\mathfrak{N}_\Lambda$. The expression $\langle Mv, e_i\rangle$ is a continuous functional of $v \in H^{T',2}(G)$. Therefore there exists the element $e_i' \in H^{T',2}(G)$ such that $\langle Mv, e_i\rangle = (v, e_i')_{H^{T',2}(G)}$, $i = 1, \ldots, q$. The condition $\langle Mv, \mathfrak{N}_\Lambda\rangle = 0$ is equivalent to the relation

$$(v, e_i')_{H^{T',2}(G)} = 0, \qquad i = 1, \ldots, q.$$

Let $L = L(e_1', \ldots, e_q')$ denote a linear hull spanned on the elements $e_1', \ldots, e_q'$. Since L is finite-dimensional, every element $v \in H^{T',2}(G)$ can be represented in the form of the sum

$$v = v' + v'', \qquad v' \in L, \qquad (v'', L) = 0.$$

Estimate (1.2.35) is already established for the element v''. It follows from the finite dimensionality of L that it is true also for $v = v' + v'' \in H^{T',2}(G)$. This copletes the proof of the lemma. □

Theorem 1.2.4. *Let problem (1.1.1)–(1.1.2) be elliptic, and let* $\mathfrak{N}_{\Lambda^+} = 0$. *Assume that there exists extension operator (1.2.34). Then problem (1.2.18)–(1.2.19) formally adjoint to problem (1.1.1)–(1.1.2) with respect to Green's formula is also elliptic.*

Proof. Because of Lemma 1.2.5 the a priori estimate (1.2.35) is valid. The existence of this estimates is the necessary and sufficient condition for the ellipticity of problem (1.2.18)–(1.2.19) (see, for example, [ADN], [Vol], [Sol]). □

1.2.6. In what follows we assume that

$$\mathfrak{N}_\Lambda = 0, \qquad \mathfrak{N}_{\Lambda^+} = 0. \tag{1.2.36}$$

It will be, for example, in the case where the matrix B is elliptic with a parameter, or where B is Dirichlet matrix ([RSh3]). Under this assumption the term $\langle \eta'', Mv\rangle$ is missing in formulas (1.2.12), (1.2.17), (1.2.21), and (1.2.26). Then the formally adjoint problem is defined by equalities (1.2.18)–(1.2.19), and all the considerations are rather simplified.

The case where $\mathfrak{N}_\Lambda \neq 0$ and $\mathfrak{N}_{\Lambda^+} \neq 0$ is considered in [R2] for one equation and in [RSh4] for Petrovskii elliptic system. It is based on the stadying of the set of various projection operators onto finite-dimensional subspaces.

1.3. On Various Theorems on Isomorphisms for General Elliptic Boundary Value Problems for Systems of Equations

1.3.1. Let us describe two methods those permit us to obtain new theorems on isomorphisms from known isomorphism theorems.

Let B_1 and B_2 be Banach spaces and let T be a linear operator that isomorphically maps the space B_1 onto the space B_2. Let E_1 be a subspace of B_1, and let $E_2 = TE_1$. Then it is clear that the operator T naturally defines a linear operator T_1 that isomorphically maps the factor space B_1/E_1 onto the factor space B_2/E_2.

Further, let Q_2 be a Banach space, and let $Q_2 \subset B_2$ (the imbedding is algebraic and topological). Then $Q_1 = T^{-1}Q_2$ is a linear (generally speaking, nonclosed) subset of the space B_1. However, the space Q_1 becomes a Banach space (denoted by Q_{1T}) with respect to the graph norm

$$\|x\|_{Q_{1T}} = \|x\|_{B_1} + \|Tx\|_{Q_2} \quad (x \in Q_1).$$

The restriction of the operator T onto Q_1 establishes the isomorphism $Q_{1T} \to Q_2$.

These simple procedures those are called **'Pasting' (or factorization) method** and **graph method** give a possibility to obtain in [R12] and [KosR] all known theorems and a number of new theorems on isomorphisms for the case of one equation from the isomorphisms theorems from [R10] and [R2].

The results on Green's formula obtained above enable us to use these two methods for the case of general boundary value problems for Douglis - Nirenberg systems.

1.3.2. The norm $|||\cdot|||$ of the space $\widetilde{H}^{s,p,(r)}$ is defined by formula (1.1.4) for $s \neq k+1/p$, $k = 0,\dots,r-1$. Therefore in what follows we shall consider such spaces for these values of $s \in \mathbf{R}$. For values

$$s = k + 1/p \qquad (k = 0,\dots,r-1)$$

the theorems on isomorphisms are obtained by interpolation theorem.
It was mentioned in 1.1.3 that the closure $A = A_{s,p}$ of the mapping

$$u \mapsto (lu, bu|_{\partial G}), \qquad u \in (C^\infty(\overline{G}))^N,$$

acts continuously in the pair of spaces (1.1.9). In addition, the following theorem is true.

Theorem 1.3.1. *([R11], [R1, Ch.X]) The operator $A = A_{s,p}$ is Noetherian. The restriction $\widehat{A} = \widehat{A}_{s,p}$ of the operator A realises the isomorphism*

$$P\widetilde{H}^{T+s,p,(\tau)} \to Q^+ K_{s,p}. \tag{1.3.1}$$

If the defect is missing (i.e., $\mathfrak{N} = 0$ and $\mathfrak{N}^* = 0$), then the operator $A = A_{s,p}$ realizes the isomorphism

$$\widetilde{H}^{T+s,p,(\tau)} \to K_{s,p}. \tag{1.3.2}$$

For simplicity, in what follows we shall assume that the defekt is absent, i.e., $\mathfrak{N} = 0$ and $\mathfrak{N}^* = 0$. Otherwise, it will be necessary to substitute the spaces of form (1.3.2) in place of the spaces of form (1.3.1).

1.3.3. Relations (1.2.36) imply that the operator $\Lambda = \Lambda_q$ realizes an isomorphism between spaces (1.2.6). Therefore, since the matrix B is defined by formula (1.2.16), the norm

$$\begin{aligned} \|u\|_{\widetilde{H}^{T+s,p,(\tau)}} &= \left(\sum_{j=1}^{N} |||u_j|||^p_{t_j+s,p,(\tau_j)}\right)^{1/p} \\ &= \sum_{j=1}^{N}\left(\|u_j\|^p_{t_j+s,p} + \sum_{k=1}^{\tau_j} \langle\langle D_\nu^{k-1}u_j, \partial G\rangle\rangle^p_{t_j+s-k+1-1/p,p}\right)^{1/p} \end{aligned} \tag{1.3.3}$$

is equivalent to the norm

$$\left(\sum_{j=1}^{N}\|u_j\|_{t_j+s,p} + \sum_{r=1}^{N}\sum_{k=1}^{-s_r+æ} \langle\langle D_\nu^{k-1} l_r u\rangle\rangle_{s-s_r-k+1-1/p,p} \right.$$
$$\left. + \sum_{j=1}^{m}\langle\langle b_j u\rangle\rangle_{s-\sigma_j-1/p,p} + \sum_{h=1}^{m}\langle\langle c_h u\rangle\rangle_{s-\sigma_h^c-1/p,p}\right)^{1/p}, \tag{1.3.4}$$

and the space $\widetilde{H}^{T+s,p,(\tau)}$ coincides with the completion $\widetilde{H}_B^{T+s,p,(\tau)}$ of the space $(C^\infty(\overline{G}))^N$ in the norm (1.3.4). If $s \geq æ$ (1.1.3), then norms (1.3.3) and (1.3.4) are equivalent to the norm $\sum_{j=1}^N \|u_j\|_{t_j+s,p}$, and the spaces $\widetilde{H}^{T+s,p,(\tau)}$ and $\widetilde{H}_B^{T+s,p,(\tau)}$ coincides with the space

$$\prod_{j=1}^{N} H^{t_j+s,p}(G) := H^{T+s,p}.$$

If $s < æ$, then norms (1.3.3) and (1.3.4) are equivalent to the norms

$$\|u\|_{H^{T+s,p}} + \sum_{k,j:t_j+s-k+1-1/p<0} \langle\langle D_\nu^{k-1} u_j \rangle\rangle_{t_j+s-k+1-1/p,p} \tag{1.3.3$'$}$$

and

$$\|u\|_{H^{T+s,p}} + \sum_{r,k:s-s_r-k+1-1/p<0} \langle\langle D_\nu^{k-1} l_r u \rangle\rangle_{s-s_r-k+1-1/p,p}$$
$$+ \sum_{j:s-s_j-1/p<0} \langle\langle b_j u \rangle\rangle_{s-s_j-1/p,p} + \sum_{h:s-\sigma_h^c-1/p<0} \langle\langle c_h u \rangle\rangle_{s-\sigma_h^c-1/p,p}, \tag{1.3.4$'$}$$

respectively.

The closure S of the mapping

$$u \mapsto (u|_{\overline{G}}, \{D_\nu^{j-1} u_k | j,k : t_j + s - k + 1 - 1/p < 0\}),\ u \in (C^\infty(\overline{G}))^N,$$

is an isometry between the space $\widetilde{H}^{T+s,p,(\tau)}$ with the metric (1.3.3)$'$ and the space

$$\mathcal{H}^{T+s,p,(\tau)} := H^{T+s,p} \times \prod_{k,j:t_j+s-k+1-1/p<0} B^{t_j+s-k+1-1/p,p}(\partial G) \tag{1.3.5}$$

(see [R11], [R1]). Analogously, the closure S_B of the mapping

$$u \mapsto (u|_{\overline{G}}, \{D_\nu^{k-1} l_r u|_{\partial G} | r,k : s - s_r - k + 1 - 1/p < 0\},$$
$$\{b_j u|_{\partial G} : s - \sigma_j - 1/p < 0\}, \{c_h u|_{\partial G} : s - \sigma_h^c - 1/p < 0\}),\quad u \in (C^\infty(\overline{G}))^N$$

is an isometry between the space $\widetilde{H}^{T+s,p,(\tau)}$ with metric (1.3.4)$'$ and the space

$$\mathcal{H}_B^{T+s,p,(\tau)}$$
$$:= H^{T+s,p} \times \prod_{r,k:s-s_r-k+1-1/p<0} B^{s-s_r-k+1-1/p,p}(\partial G)$$
$$\times \prod_{j:s-\sigma_j-1/p<0} B^{s-\sigma_j-1/p,p}(\partial G) \times \prod_{h:s-\sigma_h^c-1/p<0} B^{s-\sigma_h^c-1/p,p}(\partial G). \tag{1.3.6}$$

This gives us a possibility to identify every element $u \in \widetilde{H}_B^{T+s,p,(\tau)}$ with the element

$$S_B u = \Big(u_0, \{u_{rk} : s - s_r - k + 1 - 1/p < 0\},$$
$$\{u_j^b : s - \sigma_j - 1/p < 0\}, \{u_h^c : s - \sigma_h^c - 1/p < 0\} \Big) \in \mathcal{H}_B^{T+s,p,(\tau)}. \tag{1.3.7}$$

We shall write $u = S_B u$ for any $u \in \widetilde{H}_B^{T+s,p(\tau)}$.

For any sequence $u_i \in (C^\infty(\overline{G}))^N$ that converges to u in $\widetilde{H}_B^{T+s,p(\tau)}$ as $i \to \infty$
the sequence $D_\nu^{k-1} l_r u_i|_{\partial G}$ converges to u_{rk} in $B^{s-s_r-k+1-1/p,p}(\partial G)$ for r,k : $s - s_r - k + 1 - 1/p < 0$,
the sequence $b_j u_i|_{\partial G}$ converges to u_j^b in $B^{s-\sigma_j-1/p,p}(\partial G)$, $s - \sigma_j^b - 1/p < 0$,
the sequence $c_h u_i|_{\partial G}$ converges to u_h^c, in the space $B^{s-\sigma_h^c-1/p,p}(\partial G)$, $s - \sigma_h^c - 1/p < 0$.
In this sence,

$$u_{rk} = D_\nu^{k-1} l_r u|_{\partial G}, \quad u_j^b = b_j u|_{\partial G}, \quad u_h^c = c_h u|_{\partial G}.$$

Otherwise, if $s - s_r - k + 1 - 1/p > 0$, $u_i \in (C^\infty(\overline{G}))^N$ such that $u_i \to u$ in $\widetilde{H}^{T+s,p,(\tau)}$, then $D_\nu^{k-1} l_r u_i|_{\partial G} \to D_\nu^{k-1} l_r u_0|_{\partial G}$. If $s - \sigma_j - 1/p > 0$, then $b_j u_i|_{\partial G} \to b_j u_0|_{\partial G}$ in $B^{s-\sigma_j-1/p,p}(\partial G)$. If $s - \sigma_h^c - 1/p > 0$, then $c_h u_i|_{\partial G} \to c_h u_0|_{\partial G}$.

Thus, the expressions $D_\nu^{k-1} l_r u|_{\partial G}$, $b_j u|_{\partial G}$ and $c_h u|_{\partial G}$ are defined for every element $u \in \mathcal{H}_B^{T+s,p,(\tau)} \simeq \widetilde{H}^{T+s,p,(\tau)}$.

Green's formula (1.2.17) and equations (1.1.10) easily imply that element $u = Su \in \widetilde{H}^{T+s,p,(\tau)}$ is a generalized solution of problem (1.1.1)–(1.1.2) if and only if

$$(u_0, l^+ v) + \sum_{h=1}^{m} \langle c_h u, b_h' v\rangle = (f_0, v) + \sum_{h=1}^{m} \langle \varphi_h, c_h' v\rangle + \sum_{r=1}^{N} \sum_{k=1}^{-s_r+æ} \langle f_{rk}, e_{kr}' v\rangle, \tag{1.3.8}$$
$$\forall v \in (C^\infty(\overline{G}))^N.$$

1.3.4. It may be that the various elements of the space $\widetilde{H}_B^{T+s,p,(\tau)}$ have the same components

$$(u_0, c_1 u|_{\partial G}, \ldots, c_m u|_{\partial G}).$$

'Pasting' these elements and doing the corresponding factorization in the space of images, we obtain new statement on isomorphisms from the isomorphism $\widetilde{H}_B^{T+s,p,(\tau)} \to K^{s,p}$ Formula (1.3.6) easily implies that after this pasting the space of pre-images

$$H^{T+s,p} \times \prod_{h: s-\sigma_h^c-1/p<0} B^{s-\sigma_h^c-1/p,p}(\partial G) \tag{1.3.9}$$

is the completion of $(C^\infty(\overline{G}))^N$ in the norm

$$\|u\|_{T+s,p} + \sum_{h=1}^{m} \langle\langle c_h u\rangle\rangle_{s-\sigma_h^c-1/p,p},$$

and the space of images (see (1.1.9) and (1.3.8)) is the direct product

$$\Big(\prod_{j=1}^{N} H^{s-s_j,p}(G) \times \prod_{h:s-\sigma_h-1/p>0} B^{s-\sigma_h-1/p,p}(\partial G)\Big) \Big/ M^1_{s,p}, \qquad (1.3.10)$$

where

$$M^1_{s,p} = \Big\{(f,\{\varphi_h\}) \in \prod_{j=1}^{N} H^{s-s_j,p}(G) \times \prod_{h:s-\sigma_h-1/p>0} B^{s-\sigma_h-1/p,p}(\partial G) :$$

$$(f,v) + \sum_{h:s-\sigma_h-1/p>0} \langle \varphi_h, c'_h v\rangle + \sum_{k,r:s-s_r-k+1-1/p>0} \langle D_\nu^{k-1} f_r, e'_{kr} v\rangle = 0,$$

$$\forall v \in (C^\infty(\overline{G}))^N : c'_h v|_{\partial G} = 0,\ e'_{kr} v|_{\partial G} = 0,\ h : s - \sigma^c_h - 1/p < 0,$$

$$k, r : s - s_r - k + 1 - 1/p < 0\Big\}$$

is a subspace of the direct product

$$\prod_{j=1}^{N} H^{s-s_j,p}(G) \times \prod_{h:s-\sigma_h-1/p>0} B^{s-\sigma_h-1/p,p}(\partial G).$$

As a result, we obtain the following statement.

Theorem 1.3.2. *Let problem (1.1.1)–(1.1.2) be elliptic, and let relations (1.2.36) hold. For simplicity, assume that* $\mathfrak{N} = 0$ *and* $\mathfrak{N}^* = 0$. *Then the closure* $A_1 = A_{1sp}$ *of the mapping*

$$u \mapsto \Big(lu,\ \{b_h u|_{\partial G} : s - \sigma_h - 1/p > 0\}\Big), \qquad u \in (C^\infty(\overline{G}))^N,$$

realizes an isomorphism from space (1.3.9) into space (1.3.10).

Corollary 1.3.1. *The estimate*

$$\|u\|_{T+s,p} + \sum_{j=1}^{m} \langle\langle c_j u\rangle\rangle_{s-\sigma^c_j-1/p,p}$$

$$\le c_1\Big(\|lu\|_{s-S,p} + \sum_{h:s-\sigma_h-1/p>0} \langle\langle b_h u\rangle\rangle_{s-\sigma_h-1/p,p} + \|u\|_{T+s-k,p}\Big), \qquad (1.3.11)$$

$$\forall u \in (C^\infty(\overline{G}))^N,$$

is valid. Here the number $k > 0$ may be choosen arbitrary large. Note that if $\mathfrak{N} = 0$, then one can take, the term $\|u\|_{T+s-k,p}$ is absent.

1.3.5. The various elements of the space $\widetilde{H}^{T+s,p,(\tau)}$ may have the same components

$$(u_0, c_1 u|_{\partial G}, \ldots, c_m u|_{\partial G}, b_1 u|_{\partial G}, \ldots, b_m u|_{\partial G}).$$

Pasting them and pasting the corresponding elements in the space of images we get new theorem on isomorphisms.
To formulate it let us denote by $\widetilde{H}^{T+s,p,(\tau)}_{c,b}$ the completion of the space $C^\infty(\overline{G}))^N$ in the norm

$$|||u|||_{T+s,p,(c,b)}$$

$$:= \left(\|u\|^p_{T+s,p} + \sum_{j=1}^{m} \langle\langle c_j u \rangle\rangle^p_{s-\sigma^c_h-1/p,p} + \sum_{j=1}^{m} \langle\langle b_j u \rangle\rangle^p_{s-\sigma_j-1/p,p} \right)^{1/p}. \quad (1.3.12)$$

It follows from the consideration of Subsection 1.3.3 that the space $\widetilde{H}^{T+s,p,(\tau)}_{c,b}$, that is the space of pre-images obtained by this pasting, is isomorphic to the direct product

$$H^{T+s,p} \times \prod_{h:s-\sigma^c_h-1/p<0} B^{s-\sigma^c_h-1/p,p}(\partial G)$$
$$\times \prod_{h:s-\sigma_h-1/p<0} B^{s-\sigma_h-1/p,p}(\partial G); \quad (1.3.13)$$

The space of images is the factor-space

$$\left(H^{s-S,p} \times \prod_{h=1}^{m} B^{s-\sigma_h-1/p,p}(\partial G) \right) \Big/ M^2_{s,p} := K^2_{s,p}, \quad (1.3.14)$$

where

$$H^{s-S,p} = \prod_{j=1}^{N} H^{s-s_j,p}(G),$$

and

$$M^2_{s,p} = \Big\{ (f, \varphi_1, \ldots, \varphi_m) \in H^{s-S,p} \times \prod_{h=1}^{m} B^{s-\sigma_h-1/p,p}(\partial G) :$$

$$(f, v) + \sum_{h=1}^{m} \langle \varphi_h, c'_h v \rangle + \sum_{k,r:s-s_r-k+1-1/p>0} \langle D^{k-1}_\nu f_r, e'_{kr} v \rangle = 0,$$

$$\forall v \in C^\infty(\overline{G}))^N : e'_{k_r} v|_{\partial G} = 0,\ k, r : s - s_r - k + 1 - 1/p < 0)\Big\} \tag{1.3.15}$$

(see (1.1.9) and (1.3.7)).

Thus, we establish the validity of the following statement.

Theorem 1.3.3. *Under the conditions of Theorem 1.3.2, the closure $A_2 = A_{2sp}$ of the mapping*

$$u \mapsto (lu, bu), \qquad u \in (C^\infty(\overline{G}))^N,$$

realizes an isomorphism

$$\widetilde{H}_{c,b}^{T+s,p,(\tau)} \to K^2_{s,p}. \tag{1.3.16}$$

Corollary 1.3.2. *The estimate*

$$|||u|||_{T+s,p,(c,b)} \le c_1\Big(\|lu\|_{s-S,p} + \sum_{j=1}^{m} \langle\langle b_j u\rangle\rangle_{s-\sigma_j-1/p,p} + \|u\|_{T+s-k,p}\Big), \tag{1.3.17}$$

$$u \in (C^\infty(\overline{G}))^N,$$

is valid. Here the number $k > 0$ may be choosen arbitrary large.

Note that if $\mathfrak{N} = 0$, then one can take, the term $\|u\|_{T+s-k,p}$ is absent.

1.3.6. The various elements of the space $\widetilde{H}_B^{T+s,p,(\tau)}$ may have the same components

$$(u_0, b_1 u\big|_{\partial G}, \dots, b_m u\big|_{\partial G}).$$

Paste them and make a corresponding factorization in the space of images. Denote by $\widetilde{H}_b^{T+s,p,(\tau)}$ the completion of $(C^\infty(\overline{G}))^N$ in the norm

$$|||u|||_{T+s,p,(b)} = \left(\|u\|^p_{T+s,p} + \sum_{j=1}^{m} \langle\langle b_j u\rangle\rangle^p_{s-\sigma_j-1/p,p}\right)^{1/p}. \tag{1.3.18}$$

It is clear (see Subsection 1.3.3) that the obtained space $\widetilde{H}_b^{T+s,p,(\tau)}$ of pre-images is isomorphic to the direct product

$$\widetilde{H}_b^{T+s,p,(\tau)} \simeq H^{T+s,p} \times \prod_{h:s-\sigma_h-1/p<0} B^{s-\sigma_h-1/p,p}(\partial G) \tag{1.3.19}$$

The space of images is described by the formula

$$\left(H^{s-S,p} \times \prod_{h=1}^{m} B^{s-\sigma_h-1/p,p}(\partial G)\right) \Big/ M^3_{s,p} = K^3_{s,p}, \tag{1.3.20}$$

where

$$M^3_{s,p} = \Big\{(f,\varphi_1,\ldots,\varphi_m) \in H^{s-S,p} \times \prod_{h=1}^{m} B^{s-\sigma_h-1/p,p}(\partial G) :$$

$$(f,v) + \sum_{h=1}^{m} \langle \varphi_h, c'_h v\rangle + \sum_{k,r:s-s_r-k+1-1/p>0} \langle D_\nu^{k-1} f_r, e'_{kr} v\rangle = 0,$$

$$\forall v \in (C^\infty(\overline{G}))^N : e'_{kr} v|_{\partial G} = 0,\ b'_h v|_{\partial G} = 0,$$

$$k, r : s - s_r - k + 1 - 1/p < 0,\ h = 1,\ldots,m\Big\} \tag{1.3.21}$$

is the subspace of the direct product

$$H^{s-S,p} \times \prod_{h=1}^{m} B^{s-\sigma_h-1/p,p}(\partial G).$$

Theorem 1.3.4. *Under the assumption of Theorem 1.3.2, the closure* $A_3 = A_{3sp}$ *of the mapping*

$$u \mapsto (lu, bu), \qquad u \in (C^\infty(\overline{G}))^N,$$

realizes an isomorphism

$$\widetilde{H}_b^{T+s,p,(r)} \to K^3_{s,p}. \tag{1.3.22}$$

Corollary 1.3.3. *The estimate*

$$|||u|||_{T+s,p,(b)} \leq c(\|lu\|_{s-S,p} + \sum_{j=1}^{m} \langle\langle b_j u\rangle\rangle_{s-\sigma_j-1/p,p} + \|u\|_{T+s-k,p}), \tag{1.3.23}$$

is valid. Here the number $k > 0$ *may be choosen arbitrary large.*

Note that if $\mathfrak{N} = 0$*, then one can take, the term* $\|u\|_{T+s-k,p}$ *is absent.*

Remark 1.3.1. If the matrix

$$\begin{pmatrix} c'(x,D) \\ b'(x,D) \\ e'(x,D) \end{pmatrix}$$

is τ'-complete then it follows from definition (1.3.21) of the space $M^3_{s,p}$ that

$$M^3_{s,p} = \Big\{(f,0,\dots,0) \in H^{s-S,p} \times \prod_{j=1}^{m} B^{s-\sigma_j-1/p,p}(\partial G):$$

$$D_\nu^{k-1} f_r|_{\partial G} = 0 \; (k,r: s - s_r - k + 1 - 1/p > 0), \; (f,v) = 0,$$

$$\forall v \in (C^\infty(\overline{G}))^N : e'_{kr} v = 0, \; k,r: s - s_r - k + 1 - 1/p < 0,$$

$$b'_h v|_{\partial G} = 0, \; h = 1,\dots,m\Big\}.$$

If we denote by $\widetilde{M}^3_{s,p}$ the subspace of the space $H^{s-S,p}$ consisting of the elements f such that $(f,0,\dots,0) \in M^3_{s,p}$, then

$$K^3_{s,p} = H^{s-S,p} \Big/ \widetilde{M}^3_{s,p} \times \prod_{j=1}^{m} B^{s-\sigma_j-1/p,p}(\partial G) \tag{1.3.24}$$

1.3.7.
Let

$$C^\infty_{(b)} = \{u \in (C^\infty(\overline{G}))^N : bu|_{\partial G} = 0\},$$

and let

$$\begin{aligned} H^{T+s}_{(b)} &= \{u \in \widetilde{H}_b^{T+s,p,(\tau)} : bu|_{\partial G} = 0\} \\ &= \Big\{u \in H^{T+s,p}(G) : b_h u|_{\partial G} = 0, \; h : s - \sigma_h - 1/p > 0\Big\} \end{aligned}$$

denote the subspace of $\widetilde{H}_b^{T+s,p,(\tau)}$. The space $H^{T+s,p}_{(b)}$ is the closure of $C^\infty_{(b)}$ in the space $H^{t+s,p}$. Theorem 1.3.4 directly implies the following assertion.

Theorem 1.3.5. *Under the condition of Theorem 1.3.2, the closure $A_{3(b)}$ of the mapping*

$$u \mapsto lu, \qquad u \in C^\infty_{(b)},$$

realizes an isomorphism

$$H^{T+s,p}_{(b)} \to H^{s-S,p} \Big/ M^4_{s,p}, \tag{1.3.25}$$

where

$$M^4_{s,p} = \Big\{ f \in H^{s-S,p} : (f,v) + \sum_{k,r:s-s_r-k+1-1/p>0} \langle D_\nu^{k-1} f_r, e'_{kr} v\rangle = 0,$$

$$v \in (C^\infty(\overline{G}))^N : e'_{kr} v\big|_{\partial G} = 0,\ b'v|_{\partial G} = 0$$

$$k,r : s - s_r - k + 1 - 1/p < 0, \Big\}$$

is the subspace of the space $H^{s-S,p}$.

In the special case of one equation with normal boundary conditions, isomorphism (1.3.25) was established in [BKR] (see also [Ber, Ch. III, §6, Subsec. 10], [RS1, §5.5]).

1.3.8. Let us show an example illustrated the using of the graph method. Consider isomorphism (1.3.22) with the space $K^3_{s,p}$ of the form (1.3.24). Let $Y_{s,p}$ be a functional Banach space such that

$$(C^\infty(\overline{G}))^N \subset Y_{s,p} \subset H^{s-S,p} \Big/ \widetilde{M}^3_{s,p},$$

and let the space $(C^\infty(\overline{G}))^N$ is densein Y. For example,

$$Y_{s,p} = \prod_{r=1}^{N} H^{s-s_r+t,p}(G) \Big/ \widetilde{M}^3_{s+t,p}, \quad t > 0.$$

We obtain that the set

$$Q_{s,p} = (A_{3sp})^{-1} \left(Y_{s,p} \times \prod_{h=1}^{m} B^{s-\sigma_h-1/p,p}(\partial G) \right)$$

is a linear, generally speaking, nonclosed subset of the set $\widetilde{H}^{T+s,p,(\tau)}$. However, in the graph morm

$$\|u\|_{\widetilde{H}_b^{T+s,p,(\tau)}} + \|lu\|_{Y_{s,p}} = \|u\|_{Q_{s,p}}$$

the space $Q_{s,p}$ is a complete Banach space, and the operator A_{3sp} naturally establishes an isomorphism

$$Q_{s,p} \to Y_{s,p} \times \prod_{h=1}^{m} B^{s-\sigma_h-1/p,p}(\partial G). \qquad (1.3.26)$$

In the special case of one equation with normal boundary conditions, isomorphism (1.3.26) was obtained by Lions and Magenes for $s - s_N < 0$ and $Y_{s,p} = L_p(G)$ (see, for details, [R12], [R1]).

1.4. Addition. Proof of Lemma 1.2.2

Now let Ω denote a half-space $\mathbf{R}^n_+$ with the boundary

$$\partial\Omega = \mathbf{R}'_{n-1} = \{x \in \mathbf{R}^n : x_n = 0\}.$$

Consider problem (1.1.1)–(1.1.2) in the case where the coefficients are constant and all the expressions are homogeneous with respect to the derivatives $(D_1, \ldots, D_n)$.

Let us study the problem

$$l_0(D)u = \left(l^0_{rj}(D)\right)_{r,j=1,\ldots,N} u(x) = 0, \quad x \in \mathbf{R}^n_+, \tag{1.4.1}$$

$$b_0(D)u = \left(b^0_{hj}(D)\right)_{\substack{h=1,\ldots,m\\ j=1,\ldots,N}} u(x) = \varphi(x), \quad x \in \mathbf{R}'_{n-1}, \tag{1.4.2}$$

where

$$l^0_{rj}(D) = \begin{cases} \sum_{|\mu|=s_r+t_j} a^{rj}_\mu D^\mu, & s_r + t_j \geq 0, \\ 0, & s_r + t_j < 0, \end{cases}$$

$$b^0_{hj}(D) = \begin{cases} \sum_{|\mu|=\sigma_h+t_j} b^{hj}_\mu D^\mu, & \sigma_h + t_j \geq 0, \\ 0, & \sigma_h + t_j < 0. \end{cases}$$

The coefficients a^{rj}_μ and b^{hj}_μ are given complex numbers.

Using the partial Fourier transform $F' : x' \mapsto \xi'$ and setting $F'(u(x', x_n)) = \widehat{u}(\xi', x_n)$, we obtain from (1.4.1)–(1.4.2) that

$$l_0(\xi', D_n)\widehat{u}(\xi', x_n) = 0, \quad x_n > 0, \tag{1.4.3}$$

$$b_0(\xi', D_n)\widehat{u}(\xi', x_n)\big|_{x_n=0} = \widehat{\varphi}(\xi'). \tag{1.4.4}$$

Note that problem (1.4.3)–(1.4.4) is a problem for the system of ordinary equations with constant coefficients on the semiaxis. Rewrite this problem in the form

$$l_0(\xi', D_n)V(x_n) = 0, \tag{1.4.5}$$

$$b_0(\xi', D_n)V(x_n)\big|_{x_n=0} = h, \quad h = (h_1, \ldots, h_m)'. \tag{1.4.6}$$

Let $\mathcal{M}_+ = \mathcal{M}_+(\xi')$, $\xi' \neq 0$, denote the space of stable smooth solutions of equation (1.4.5) (stable means function tending to zero as $x_n \to \infty$).

Since l_0 is properely elliptic expression, the space $\mathcal{M}_+$ is m-dimensional (see [R1, Ch.X]).

Let us give various forms of the Lopatinskii condition, equivalent to each other:

L1. *For any $\xi' \neq 0$ the problem (1.4.5)–(1.4.6) has not more than one solution in $\mathcal{M}_+$.*

L2. *For any $\xi' \neq 0$ and any set of numbers $h_1, \dots, h_m$ problem (1.4.5)–(1.4.6) has a solution in $\mathcal{M}_+$.*

If we substitute the operator

$$\widehat{D}_j = F'^{-1} \frac{\xi_j}{|\xi'|}(1 + |\xi'|)F'$$

in place of the operator $D_j = F'^{-1}\xi_j F'$ for every $j = 1, \dots, n-1$ to problem (1.4.1)–(1.4.2), then we obtain a model problem in the half-space:

$$\widehat{l}_0(D)u(x) = f(x), \qquad x \in \mathbf{R}^n_+, \tag{1.4.7}$$

$$\widehat{b}_0(D)u(x) = \varphi(x), \qquad x \in \mathbf{R}'_{n-1} \tag{1.4.8}$$

(sf. [R1, Sec. 4.2]). This problem is elliptic if and only if problem (1.4.1)–(1.4.2) is elliptic.

We represent the expressions l_{rj} and b_{hj} in the form

$$l^0_{rj}(D) = \sum_{k=1}^{s_r+t_j+1} l^{rj}_k(D')D^{k-1}_n, \qquad r, j : s_r + t_j \geq 0,$$

and

$$b^0_{hj}(D) = \sum_{k=1}^{\sigma_h+t_j+1} b^{hj}_k(D')D^{k-1}_n, \qquad h, j : \sigma_h + t_j \geq 0,$$

respectively. Here $l^{rj}_k(D')$ and $b^{hj}_k(D')$ are the differential expressions with respect to the variables $x_1, \dots, x_{n-1}$ of corresponding orders $s_r + t_j - k + 1$ and $\sigma_h + t_j - k + 1$. According to formulas (1.1.7), (1.1.7′), (1.1.8) we have that the equality $\widehat{l}_0 u = f$ is valid if and only if both the relations

$$\begin{aligned} \left(\widehat{l}_{0r}u\right)_+ &= \sum_{j=1}^{N} \widehat{l}^0_{rj}(D)u_{j0+} - i \sum_{j:s_r+t_j\geq 1} \sum_{k=1}^{s_r+t_j} \left(J^k \widehat{l}^0_{rj}(D)\right)\left(u_{jk}(x') \times \delta(x_n)\right) \\ &= f_{r0+} \qquad (r = 1, \dots, N), \end{aligned} \tag{1.4.9}$$

and

$$D_n^{h-1}\widehat{l}_{0r}u\big|_{x_n=0} = \sum_{j=1}^{N}\sum_{k=1}^{s_r+t_j+1} \widehat{l}_k^{rj}(D')u_{j,k+h-1}(x') = f_{rh} \tag{1.4.10}$$

$$(h = 1,\dots,æ - s_r, \quad r : æ - s_r \geq 1)$$

hold. Here u_{j0+} and f_{r0+} are the continuations by zero of the elements u_{j0} and f_{r0} onto the half-space $x_n < 0$, and $\widehat{l}_{0r}$ is the line of number r of the matrix $\widehat{l}_0$. Moreover, equality (1.4.8) holds if and only if

$$\widehat{b}_h u \equiv \sum_{j=1}^{N}\sum_{k=1}^{\sigma_h+t_j+1} b_k^{hj}(\widehat{D}')u_{jk}(x') = \varphi_h(x'), \qquad h = 1,\dots,m. \tag{1.4.11}$$

Since equation (1.4.9) is considered in the whole space $\mathbf{R}^n$, it should be study the problem

$$\widehat{l}_0 v = \Phi \in \prod_{r=1}^{N} H^{s-s_r,p}(\mathbf{R}^n), \quad s \in \mathbf{R},\, p \in (1,+\infty) \tag{1.4.12}$$

in $\mathbf{R}^n$.

It turns out ([R1, Lemma 10.2.1]) that for any vector

$$\Phi = (\Phi_1,\dots,\Phi_N), \qquad \Phi_j \in H^{s-s_j,p}(\mathbf{R}^n),$$

problem (1.4.12) has one and only one solution $v = (v_1,\dots,v_N)$ such that

$$v_j \in H^{t_j+s,p}(\mathbf{R}^n),$$

and

$$\sum_{j=1}^{N} \|v_j, \mathbf{R}^n\|_{t_j+s,p} \leq c\sum_{j=1}^{N} \|\Phi_j, \mathbf{R}^n\|_{s-s_j,p}, \tag{1.4.13}$$

where the constant $c > 0$ does not depend on V and Φ. In addition, let $\operatorname{supp}\Phi \subset \overline{\mathbf{R}^n_+}$ and

$$\Phi \subset \prod_{r=1}^{N} H_+^{s-s_r,p}(\mathbf{R}^n) \bigcap \prod_{r=1}^{N} H_+^{s-s_r,2}(\mathbf{R}^n),$$

where

$$H_+^{t,p}(\mathbf{R}^n) = \left\{ f \in H^{t,p}(\mathbf{R}^n) : \operatorname{supp} f \subset \overline{\mathbf{R}^n_+} \right\}$$

is a subspace of $H^{t,p}(\mathbf{R}^n)$. Then ([R1, Lemma 10.2.2]) the inclusion supp $V \subset \overline{\mathbf{R}}^n_+$ is true if and only if the equalities

$$\int_{-\infty}^{+\infty} \widehat{L}^{-1}(\xi',\xi_n)(\xi_n + i\sqrt{1+|\xi|^2})^{s+t_k-m} \sum_{r=1}^{N} \widehat{L}_{rk}(\xi',\xi_n)\widetilde{\Phi}_r(\xi',\xi_n)\xi_n^j \, d\xi_n = 0 \tag{1.4.14}$$

$$\left(k = 1,\dots,N, \quad j = 0,\dots,m-1\right),$$

are valid for almost all $\xi' \neq 0$. Here

$$\widehat{L}(\xi',\xi_n) = \det(\widehat{l}_0(\xi',\xi_n)),$$

and $\widehat{L}_{rk}(\xi',\xi_n)$ is a cofactor of the element $\widehat{l}^0_{rj}(\xi)$ of the matrix $\widehat{l}^0(\xi)$.

If Φ is a sufficiently smooth function, then equalities (1.4.14) hold for any $\xi' \neq 0$.

Taking into account the facts that in relation (1.4.9) the term

$$f_{r0+} - i \sum_{j:s_r+t_j\geq 1} \sum_{k=1}^{s_r+t_j} \left(J^k \widehat{l}^0_{rj}(D)\right)(u_{jk}(x') \times \delta(x_n)) =: \Phi_r$$

belongs to the space $H_+^{s-s_r,p}(\mathbf{R}^n)$ and

$$\|\Phi_r, \mathbf{R}^n\|_{s-s_r,p}$$

$$\leq c\left(\|f_{r0+}, \mathbf{R}^n\|_{s-s_r,p} + \sum_{j:s_r+t_j\geq 1} \sum_{k=1}^{s_r+t_j} \langle\langle u_{jk}, \mathbf{R}'_{n-1}\rangle\rangle_{t_j+s-k+1-1/p,p}\right),$$

we obtain from equalities (1.4.14) that system (1.4.9) has a solution

$$u_{0+} = (u_{1\,0+},\dots,u_{N0+}), \quad u_{j0+} \in H_+^{t_j+s,p}(\mathbf{R}^n), \tag{1.4.15}$$

if and only if the relation

$$-i\int_{-\infty}^{+\infty} \widehat{L}^{-1}(\xi',\xi_n)(\xi_n + i\sqrt{1+|\xi'|^2})^{s+t_k-m}$$

$$\times \sum_{r=1}^{N} \widehat{L}_{rk}(\xi',\xi_n)\xi_n^j \sum_{\alpha:s_r+t_\alpha\geq 1} \sum_{\beta=1}^{s_r+t_\alpha} (J^\beta \widehat{l}^0_{r\alpha}(\xi)\widehat{u}_{\alpha\beta}(\xi'))\, d\xi_n$$

$$= \int_{-\infty}^{+\infty} \widehat{L}^{-1}(\xi',\xi_n)(\xi_n + i\sqrt{1+|\xi'|^2})^{s+t_k-m} \sum_{r=1}^{N} \widehat{L}_{rk}(\xi',\xi_n)\xi_n^j \widetilde{f}_{r0+}(\xi',\xi_n)\, d\xi_n \tag{1.4.16}$$

$$(k = 1, \ldots, N, \qquad j = 0, \ldots, m-1)$$

holds.

Let $\widehat{g}_{kj}(\xi')$ denote the right-hand side of (1.4.16), and let $\widehat{\widehat{c}}_{kj\alpha\beta}(\xi')$ denote the coefficients of $\widehat{u}_{\alpha\beta}(\xi')$ in the left-hand side of (1.4.16).

Note that if $\beta > s_r + t_\alpha$, then $J^\beta \widehat{l}^0_{r\alpha}(0) \equiv 0$. Therefore one can change the summation over β from 1 to $s_r + t_\alpha$ by the summation from 1 to $t_\alpha = \max\{s_1 + t_\alpha, \ldots, s_N + t_\alpha\}$. We obtain

$$\sum_{\alpha=1}^{N_1} \sum_{\beta=1}^{t_\alpha} \widehat{\widehat{c}}_{kj\alpha\beta} \widehat{u}_{\alpha\beta}(\xi') = \widehat{g}_{kj}(\xi') \tag{1.4.17}$$

$$(k = 1, \ldots, N, \qquad j = 0, \ldots, m-1).$$

Here $N_1 \leq N$ is a number of the indexes j such that $t_j \geq 1$.

Conditions (1.4.17) are necessary and sufficient for the existence of solution (1.4.15) of system (1.4.9). These conditions forms a linear system of Nm equations with respect to $t_1 + \cdots + t_N$ variables $\{\widehat{u}_{\alpha\beta}\}$. It turns out that ([R1, Lemma 10.2.3]) among the Nm equations (1.4.17) for any ξ' there exist m linearly independent equations such that the rest of the equations are expressed in terms of these m equations.

Therefore in what follows we mean that in the system (1.4.17) for any point $\xi' \neq 0$ only m linearly independent equations are remainded and the rest of them are rejected.

In addition, note that every mth-order minor of matrix (1.4.17) is a cotinuous function of ξ'. Thus, if some m conditions are linearly independent at the point $\xi' \neq 0$, then they are linearly independent in a neighborhood of the point.

By passing to the Fourier images in (1.4.10) and (1.4.11) we obtain

$$\sum_{j=1}^{N} \sum_{k=1}^{s_r+t_j+1} \widehat{l}_k^{rj}(\xi') \widehat{u}_{j,k+h-1}(\xi') = \widehat{f}_{rh}(\xi') \tag{1.4.18}$$

$$\Big(h = 1, \ldots, æ - s_r, \qquad r : æ - s_r \geq 1\Big),$$

$$\sum_{j=1}^{N} \sum_{k=1}^{\sigma_h} \widehat{b}_k^{hj}(\xi') \widehat{u}_{jk}(\xi') = \widehat{\varphi}_h(\xi'), \quad h = 1, \ldots, m. \tag{1.4.19}$$

System (1.4.17)–(1.4.19) is a linear system of $|\tau| = \tau_1 + \cdots + \tau_N$ equations in $|\tau|$ unknowns

$$\Big\{\widehat{u}_{jk}(\xi') \,\big|\, j : \tau_j \geq 1,\ k = 1, \ldots, \tau_j\Big\}.$$

It turns out that ([R1, Sec. 10.2]) the conditions of the ellipticity of problem (1.4.7)–(1.4.8) are equivalent to the following condition:
for any $\xi' \neq 0$ the determinant $\widehat{\Delta}(\xi')$ of system (1.4.17)–(1.4.19) does not equal to zero.

Let us denote by $\widehat{c}(\xi')$, $\widehat{e}(\xi')$ and $\widehat{b}(\xi')$ the matrixes of systems (1.4.17), (1.4.18) and (1.4.19) respectively, and let

$$\Lambda(D') = \begin{pmatrix} e(D') \\ b(D') \\ c(D') \end{pmatrix}.$$

Then the mapping

$$\Lambda : \eta \mapsto F$$

is mutually continuous one-to-one correspondence from the space of the Cauchy data

$$\prod_{j=1}^{N} \prod_{k=1}^{\tau_j} B^{t_j-k+1-1/p,p}(\mathbf{R}'_{n-1})$$

into the corresponding soace of the right-hand sides. Then $B : Bu = \Lambda\eta$ is a τ-complete matrix, and Lemma 1.2.2 is proved. □

1.5. Elliptic Boundary Value Problems with a Parameter for General Systems of Equations

1.5.1. In a bounded domain $G \in \mathbf{R}^n$ with the boundary $\partial G \in C^\infty$ we consider an parameter-elliptic boundary value problem for a system of order $(T, S) = (t_1, \ldots, t_N, s_1, \ldots, s_N)$ of Douglis-Nirenberg structure (see, for example, [RS1], [RS2], [R1])

$$l(x, D, q_1)u(x) = f(x) \qquad (x \in G), \tag{1.5.1}$$

$$b(x, D, q_1)u(x) = \varphi(x) \qquad (x \in \partial G). \tag{1.5.2}$$

Here $q_1 = qe^{i\theta}$, $q \in \mathbf{R}$, $\theta \in [\theta_1, \theta_2]$ (the case where $\theta = \theta_1 = \theta_2$ is not excluded), the system

$$l = l(x, D, q_1) = \left(l_{rj}(x, D, q_1)\right)_{r,j=1,\ldots,N}$$

and the boundary expressions

$$b(x, D, q_1) := \left(b_{hj}(x, D, q_1)\right)_{\substack{h=1,\ldots,m \\ j=1,\ldots,N}}$$

are described by the following relations:

$$\operatorname{ord} l_{rj} \leq s_r + t_j \text{ for } s_r + t_j \geq 0,\ l_{rj} = 0 \text{ for } s_r + t_j < 0\ (r, j = 1, \ldots, N);$$

$$\operatorname{ord} b_{hj} \leq \sigma_h + t_j \text{ for } \sigma_h + t_j \geq 0,\ b_{hj} = 0 \text{ for } \sigma_h + t_j < 0$$
$$(j = 1, \dots, N, \quad h = 1, \dots, m).$$

Here $t_1, \dots, t_N, s_1, \dots, s_N, \sigma_1, \dots, \sigma_m$ are given integer numbers such that

$$s_1 + \cdots + s_N + t_1 + \cdots + t_N = 2m,$$

$$t_1 \geq \cdots \geq t_N \geq 0 = s_1 \geq \ldots \geq s_N, \qquad \sigma_1 \geq \ldots \geq \sigma_m.$$

Recall that $u(x) = (u_1(x), \dots, u_N(x))$, $x \in \overline{G}$. Moreover,

$$l_{rj}(x, D, q_1) = \begin{cases} \sum_{|\mu|+k \leq s_r + t_j} a^{rj}_{\mu k}(x) q_1^k D^\mu & \text{for } r, j : s_r + t_j \geq 0, \\ 0 & \text{for } r, j : s_r + t_j < 0 \end{cases} \tag{1.5.3}$$

$$(\mu = (\mu_1, \dots, \mu_n),\ |\mu| = \mu_1 + \cdots + \mu_n),$$

$$b_{hj}(x, D, q_1) = \begin{cases} \sum_{|\alpha|+k \leq \sigma_h + t_j} b^{hj}_{\alpha k}(x) q_1^k D^\alpha, & \text{for } h, j : \sigma_h + t_j \geq 0, \\ 0, & \text{for } h, j : \sigma_h + t_j < 0 \end{cases} \tag{1.5.4}$$

$$(j = 1, \dots, N),$$

Here and in what follows we assume that the coefficients $a^{rj}_{\mu k}(x)$ and $b^{hj}_{\alpha k}(x)$ of all differential expressions are infinitely smooth in G and ∂G, respectively. Recall that system (1.5.1) is called elliptic with a parameter (or, shortly, parameter-elliptic system) if for every point $x \in \overline{G}$, and any $\xi \in \mathbf{R}^n$, $q \in \mathbf{R}$ $(|\xi| + |q| > 0)$, and $\theta \in [\theta_1, \theta_2]$, we have

$$L(x, \xi, q_1) = \det\left(l^0_{rj}(x, \xi, q_1)\right)_{r,j=1,\dots,N} \neq 0, \tag{1.5.5}$$

where

$$l^0_{rj}(x, \xi, q_1) = \begin{cases} \sum_{|\mu|+k = s_r + t_j} a^{rj}_{\mu k}(x) q_1^k \xi^\mu & \text{for } r, j : s_r + t_j \geq 0, \\ 0 & \text{for } r, j : s_r + t_j < 0 \end{cases}$$

is a principal symbol of the expression l_{rj}. Parameter-elliptic system (1.5.1) is called properly elliptic with a paremeter (or, shortly, properly parameter-elliptic system) if, for every point $x \in \overline{G}$, $q \in \mathbf{R}$ $(|\xi| + |q| > 0)$, $\theta \in [\theta_1, \theta_2]$, and every vector $\tau \in \mathbf{R}^n$ tangential to ∂G at the point x, the polynomial $L(\eta) = L(x, \tau + \eta\nu, q_1)$ (where ν is a unit vector normal to ∂G at the point x), is a polynomial of even order $2m = |S| + |T| = \sum_{j=1}^N (s_j + t_j)$, and accurately m of its roots have positive (negative) imaginary parts.

Then $L(\eta) = L_+(\eta)L_-(\eta)$, where L_+ (L_-) is an mth-order polynomial whose all roots lie in the upper (lower) half-plane.

Definition 1.5.1. *Boundary value problem (1.5.1)–(1.5.2) is called parameter-elliptic problem if system (1.5.1) is properly parameter-elliptic, and the Lopatinskii condition is sutisfied, i.e., for every point $x \in \partial G$ and any $\theta \in [\theta_1, \theta_2]$, the rows of the matrix*

$$\left(\sum_{j=1}^{N} b_{hj}^{0}(x, \tau + \eta\nu, q_1) L_{rj}(x, \tau + \eta\nu, q_1)\right)_{\substack{h=1,\dots,m\\ r=1,\dots,N}},$$

whose elements are considered as polynomials of η, are linearly independent modulo $L_+(\eta)$. Here $|\tau| + |q| > 0$, the expression $b_{hj}^0(x, \xi, q_1)$ denotes the principal symbol of b_{hj}, the expression L_{rj} denotes the cofactor of the element l_{rj}^0 of the determinant $L(\eta)$.

Let us note ([RS1], [RS2], [R1]) that problem (1.5.1)–(1.5.2) is parameter-elliptic if the matrix $l(x, D_x, e^{i\theta}D_t)$ is properly elliptic in the cylinder $\overline{G} \times \mathbf{R}$, and matrix $b(x, D_x, e^{i\theta}D_t)$ satisfies the Lopatinskii condition on the boundary $\partial G \times \mathbf{R}$, i.e., the problem

$$l(x, D_x, e^{i\theta}D_t)u = f \quad (\text{in } G \times \mathbf{R}),$$

$$b(x, D_x, e^{i\theta}D_t)u = \varphi \quad (\text{in } \partial G \times \mathbf{R}),$$

where $\theta \in [\theta_1, \theta_2]$, is elliptic in the cylinder $\overline{G} \times \mathbf{R}$.

1.5.2. To study elliptic problems with a parameter, it is convenient to introduce relevant spaces and norms which depend on the parameter $q \in \mathbf{R}$.

Let $s, q \in \mathbf{R}$, $p \in]1, +\infty[$. We denote by $H^{s,p}(\mathbf{R}^n, q)$ the space of distributions such that

$$\|u, \mathbf{R}^n, q\|_{s,p} = \|F^{-1}(1 + |\xi|^2 + |q|^2)^{s/2} F u\|_{L_p(\mathbf{R}^n)} < \infty. \tag{1.5.6}$$

Here $F_{x\mapsto\xi}$ denotes the Fourier transformation, and $F_{x\mapsto\xi}^{-1}$ denotes the inverse transformation. It is easy to see ([R1, Sec. 1.13]) that, for each fixed $q \in \mathbf{R}$, norm (1.5.6) is equivalent to the norm $\|u, \mathbf{R}^n, 0\|$, and, therefore, the set $H^{s,p}(\mathbf{R}^n, q)$ coincides with the set $H^{s,p}(\mathbf{R}^n, 0) = H^{s,p}(\mathbf{R}^n)$. However, it is now more convenient to consider only those norm equivalent to (1.5.6) for which the corresponding bilateral estimates can be written with constants that do not depend on q. For example, if $s \geq 0$, then, in this sense, norm (1.5.6) is equivalent to the norm

$$\|F^{-1}((1+|\xi|^2)^{s/2}+|q|^s)Fu\|_{L_p(\mathbf{R}^n)},$$

and if $|q| \geq q_0 > 0$, then (1.5.6) is equivalent to the norm

$$\|F^{-1}(|\xi|^2+|q|^2)^{s/2}Fu\|_{L_p(\mathbf{R}^n)}. \tag{1.5.6'}$$

Note that

$$\|u,\mathbf{R}^n,q\|_{s,p} \geq c_p|q|^t\|u,\mathbf{R}^n,q\|_{s-t,p} \tag{1.5.7}$$

$$\Big(t \geq 0,\ \ s \in \mathbf{R},\ \ q \in \mathbf{R},\ \ u \in H^{s,p}(\mathbf{R}^n,q)\Big),$$

where the constant c_p does not depend on u and q.

Let $s \geq 0$, $p, p' \in]1,+\infty[$, $1/p+1/p' = 1$. For any domain $\Omega \subset \mathbf{R}^n$ whose boundary $\partial\Omega$ is infinitely smooth, we denote by $H^{s,p}(\Omega,q)$ the space of restrictions of elements of $H^{s,p}(\mathbf{R}^n,q)$ to Ω with quotient space topology. The norm in the space $H^{s,p}(\Omega,q)$ $(s \geq 0, 1 < p < \infty)$ is defined by the relation

$$\|u,\Omega,q\|_{s,p} = \inf \|v,\mathbf{R}^n,q\|_{s,p},$$

where the infimum is taken over all elements $v \in H^{s,p}(\mathbf{R}^n,q)$ which are equal to u in $\overline{\Omega}$.

We denote by $H^{-s,p}(\Omega,q)$ $(s \geq 0)$ the space dual to $H^{s,p'}(\Omega,q)$ $(1/p+1/p' = 1)$ with respect to the extension $(\cdot,\cdot) = (\cdot,\cdot)_\Omega$ of the scalar product in $L_2(\Omega)$. For any $s \in \mathbf{R}$ and $p,p' \in]1,+\infty[$, the norm in $H^{s,p}(\Omega,q)$ is denoted by $\|u,\Omega,q\|_{s,p}$.

Futher on we consider only the cases where either $\Omega = G$ is a bounded domain, $\partial\Omega = \partial G$, or

$$\Omega = \mathbf{R}^n_\pm = \{x = (x',x_n) \in \mathbf{R}^n : x_n \gtrless 0\}, \quad \partial\Omega = \mathbf{R}'_{n-1} = \{x \in \mathbf{R}^n : x_n = 0\}.$$

Inequality (1.5.7) implies the following estimate ([R1, Sec. 9.1])

$$\|u,\Omega,q\|_{s,p} \geq c_p|q|^t\|u,\Omega,q\|_{s-t,p} \tag{1.5.8}$$

$$(t \geq 0,\ s \in \mathbf{R},\ q \in \mathbf{R},\ u \in H^{s,p}(\Omega,q)),$$

where the constant c_p does not depend on u and q.

For $s \in \mathbf{R}$ and $p \in]1,+\infty[$ we consider a Besov space $B^{s,p}(\mathbf{R}'_{n-1})$ with the norm $\langle\langle u,\mathbf{R}'_{n-1}\rangle\rangle_{s,p}$. In this space we introduce the norm which depends on the parameter $q \in \mathbf{R}$. We set:

$$\langle\langle \varphi,\mathbf{R}^{n-1},q\rangle\rangle_{s,p} = \langle\langle F'^{-1}(1+|\xi'|^2+q^2)^{s/2}F'\varphi\rangle\rangle_{B^{0,p}(\mathbf{R}'_{n-1})}. \tag{1.5.9}$$

We denote by $B^{s,p}(\mathbf{R}'_{n-1},q)$ the space $B^{s,p}(\mathbf{R}'_{n-1})$ with the norm defined by (1.5.9). If $s>0$, then norm (1.5.9) is equivalent to the norm

$$\inf_{\substack{u\in H^{s+1/p,p}(\mathbf{R}^n)\\ u|_{x_n=0}=\varphi}} \|u,\mathbf{R}^n,q\|_{s+1/p,p}. \tag{1.5.10}$$

The spaces $B^{s,p}(\mathbf{R}'_{n-1},q)$ and $B^{-s,p'}(\mathbf{R}'_{n-1},q)$ are dual to each other with respect to the extension $\langle\cdot,\cdot\rangle=\langle\cdot,\cdot\rangle_{\mathbf{R}^{n-1}}$ of the scalar product in $L_2(\mathbf{R}'_{n-1})$.

By using the procedure of decomposition of unity, norm (1.5.9) defines the norm $\langle\langle\varphi,\partial G,q\rangle\rangle_{s,p}$ and the space $B^{s,p}(\partial G,q)$ with this norm ($s\in\mathbf{R}$ and $p\in]1,+\infty[$). The spaces $B^{s,p}(\partial G,q)$ and $B^{-s,p'}(\partial G,q)$ are dual to each other with respect to the extension $\langle\cdot,\cdot\rangle=\langle\cdot,\cdot\rangle_{\partial G}$ of the scalar product in $L_2(\partial G)$.

If $s>0$, then $B^{s,p}(\partial\Omega,q)$ coincides with the space of restrictions of elements of $H^{s+1/p,p}(\Omega,q)$ to $\partial\Omega$ with factor-topology.

Let us now introduce the norm depending on the parameter $q\in\mathbf{R}$ in the space $\widetilde{H}^{s,p,(r)}(\Omega)$ (see Subsection 1.1.2). Let r be a natural number, $p\in]1,+\infty[$, $s\in\mathbf{R}$, $s\neq k+1/p$ $(k=0,\ldots,r-1)$. Assume that $C_0^\infty(\overline{\Omega})$ denotes the space of the restrictions of functions of $C_0^\infty(\mathbf{R}^n)$ to Ω. We denote by $\widetilde{H}^{s,p,(r)}(\Omega,q)$ the completion of $C_0^\infty(\overline{\Omega})$ in the norm

$$|||u,\Omega,q|||_{s,p,r}=\left(\|u,\Omega,q\|^p_{s,p}+\sum_{j=1}^{r}\langle\langle D_\nu^{j-1}u,\partial\Omega,q\rangle\rangle^p_{s-j+1-1/p}\right)^{1/p}. \tag{1.5.11}$$

Norm (1.5.11) coincides with the norm of an element

$$\tilde{u}=(u_0,\ldots,u_r),\quad u_0\in C_0^\infty(\overline{\Omega}),\quad u_j=D_\nu^{j-1}u_0\big|_{\partial\Omega}\quad (j=1,\ldots,r)$$

in the space

$$\widetilde{\mathcal{H}}^{s,p,(r)}(\Omega,q)=H^{s,p}(\Omega,q)\times\prod_{j=1}^{r}B^{s-j+1-1/p,p}(\partial\Omega,q).$$

Therefore the space $\widetilde{H}^{s,p,(r)}(\Omega,q)$ is isometric to the closure $\widetilde{\mathcal{H}}_0^{s,p,(r)}(\Omega,q)$ of the set $\{\tilde{u}=(u_0,u_1,\ldots,u_r):u_0\in C_0^\infty(\overline{\Omega})\}$ in $\widetilde{\mathcal{H}}^{s,p,(r)}(\Omega,q)$. This gives us a possibility to identify the space $\widetilde{H}^{s,p,(r)}(\Omega,q)$ with the space $\widetilde{\mathcal{H}}_0^{s,p,(r)}(\Omega,q)$. Furtermore, an element

$$u=(u_0,u_1,\ldots,u_r)\in\widetilde{\mathcal{H}}^{s,p,(r)}(\Omega,q)$$

belongs to

$$\widetilde{\mathcal{H}}_0^{s,p,(r)}(\Omega,q)=\widetilde{H}^{s,p,(r)}(\Omega,q)$$

if and only if

$$u_j = D_\nu^{j-1} u_0\big|_{\partial\Omega} \quad (\forall j : s - j + 1 - 1/p > 0).$$

Therefore, the space $\widetilde{H}^{s,p,(r)}(\Omega, q)$ consists of vectors $(u_0, u_1, \ldots, u_r)$,

$$u_0 \in H^{s,p}(\Omega, q), \quad u_j \in B^{s-j+1-1/p,p}(\partial\Omega, q) \quad (j = 1, \ldots, r),$$

such that, if $s-j+1-1/p > 0$, then $u_j = D_\nu^{j-1} u_0\big|_{\partial\Omega}$, and if $s-j+1-1/p < 0$, then u_j dos not depend on u_0.

For $s = k + 1/p$ $(k = 0, \ldots, r-1)$ we define the space $\widetilde{H}^{s,p,(r)}(\Omega, q)$ and norm (1.5.11) by the method of complex interpolation.

Finally, for the case $r = 0$ we set:

$$\widetilde{H}^{s,p,(0)}(\Omega, q) := H^{s,p}(\Omega, q), \quad |||u, \Omega, q|||_{s,p,(0)} := \|u, \Omega, q\|_{s,p}.$$

1.5.3. Denote:

$$M = M(x, D, e^{i\theta}q) = \sum_{|\mu|+k\leq l} a_{\mu k}(x)(e^{i\theta}q)^k D^\mu$$

$$(x \in \Omega, \quad l \leq r),$$

and

$$N = N(x, D, e^{i\theta}q) = \sum_{|\mu|+k\leq l_1} b_{\mu k}(x)(e^{i\theta}q)^k D^\mu$$

$$(x \in \partial\Omega, \quad l_1 \leq r - 1),$$

and suppose that the coefficients $a_{\mu k}$ and $b_{\mu k}$ are infinitely smooth functions whose all derivatives are bounded. Then (see [R1, Sec.9.1]) there exists a constant $c > 0$ independent of u, q and θ, such that, for $u \in C_0^\infty(\overline{\Omega})$, the estimates

$$\|Mu, \Omega, q\|_{s-l,p} \leq c|||u, \Omega, q|||_{s,p,(r)},$$

$$\langle\langle Nu, \partial\Omega, q\rangle\rangle_{s-l-1/p,p} \leq c|||u, \Omega, q|||_{s,p,(r)},$$

$$|||Mu, \Omega, q|||_{s-l,p,(r-l)} \leq c|||u, \Omega, q|||_{s,p,(r)}$$

are valid. This implies that the closure $A = A(q) = A_{s,p}(q)$ of the mapping

$$u \mapsto (l\,(x, D, q)\,u, b\,(x, D, q)\,u) \qquad \left(u \in \left(C^\infty(\overline{G})\right)^N\right)$$

acts continuously in the pair of spaces

$$\widetilde{H}^{T+s,p,(\tau)}(G,q) \to K_{s,p}, \tag{1.5.12}$$

$$\widetilde{H}^{T+s,p,(\tau)}(G,q) := \prod_{j=1}^{N} \widetilde{H}^{t_j+s,p,(\tau_j)}(G,q),$$

$$K_{s,p} := \prod_{r=1}^{N} \widetilde{H}^{s-s_r,p,(æ-s_r)}(G,q) \times \prod_{h=1}^{m} B^{s-\sigma_h-1/p,p}(\partial G,q).$$

Here

$$\tau_j = t_j + æ \; (j = 1,\dots,N), \quad æ = \max\{0, \sigma_1+1,\dots,\sigma_m+1\}. \tag{1.5.13}$$

Note that, if

$$A_{s,p}(q)u = F = (f,\varphi)$$

$$\begin{gathered} u = (u_1,\dots,u_N), \quad u_j = (u_{j0},\dots,u_{j\tau_j}) \in \widetilde{H}^{t_j+s,p,(\tau_j)}(G,q), \\ f = (f_1,\dots,f_N), f_j = (f_{j0},\dots,f_{j,æ-s_j}) \in \widetilde{H}^{s-s_j,p,(æ-s_j)}(G,q), \\ \varphi = (\varphi_1,\dots,\varphi_m), \varphi_j \in B^{s-\sigma_j-1/p,p}(\partial G,q) \quad (j=1,\dots,m), \end{gathered} \tag{1.5.14}$$

then

$$\begin{gathered} l_j u|_{\overline{G}} = f_{j0} \qquad (j=1,\dots,N), \\ D_\nu^{k-1} l_j u\big|_{\partial G} = f_{jk} \qquad (j=1,\dots,N,\ 1 \le k \le æ - s_j), \\ bu\big|_{\partial G} = \varphi = (\varphi_1,\dots,\varphi_m). \end{gathered} \tag{1.5.15}$$

The following assertion is true.

Theorem 1.5.1. *Let $p \in]1,\infty[$, $s \in \mathbf{R}$, and let problem (1.5.1)–(1.5.2) be parameter-elliptic. Assume that both the coefficients and the boundary ∂G are infinitely smooth. Then there exists a number $q_0 > 0$ such that, for $q \ge q_0$ and $\theta \in [\theta_1, \theta_2]$, the operator $A_{s,p} = A_{s,p}(q)$, acting continuously in the pair of spaces (1.5.12), realizes an isomorphism between these spaces. Moreover, there exists a constant $c_s > 0$ independent of u, q $(|q| \ge q_0)$ and $\theta \in [\theta_1, \theta_2]$, such that the following estimate is valid:*

$$c_s^{-1}\|u\|_{\widetilde{H}^{T+s,p,(\tau)}(G,q)} \le \|A_{s,p}u\|_{\widetilde{K}^{s,p,(æ-S)}} \le c_s\|u\|_{\widetilde{H}^{s,p,(\tau)}(G,q)} \tag{1.5.16}$$

In addition, the function $s \mapsto c_s$ $(s \in \mathbf{R})$ is bounded for every compact set.

For the case where $p = 2$ these theorem is proved in [RS1] and [RS2], the proof for the general case is presented in [R1, Sec. 9, 10].

Consider equation (1.5.1) in $\mathbf{R}^n$ and suppose that the coefficients $a_{\mu k}(x)$ are infinitely smooth functions whose all derivatives are bounded. Then, of course, the closure $l = l_{s,p} = l(x, D, q)$ of the mapping

$$u \mapsto l(x, D, q)u \qquad \left(u \in (C_0^\infty(\mathbf{R}^n))^N\right)$$

acts continuously in the pair of spaces

$$\prod_{j=1}^{N} H^{t_j+s,p}(\mathbf{R}^n, q) := H^{T+s,p}(\mathbf{R}^n, q) \to \prod_{r=1}^{N} H^{s-s_r,p}(\mathbf{R}^n, q) := H^{s-S,p}(\mathbf{R}^n, q).$$

Furtermore, the following estimate

$$\|l(x, D, q)u\|_{H^{s-S,p}(\mathbf{R}^n,q)} \le c\|u\|_{H^{T+s,p}(\mathbf{R}^n,q)}$$

holds with the constant $c > 0$ independent of u and q.

1.5.4. We mention here the several lemmas which are used in the proof of the Theorem 1.5.1

Let us fix the point x in (1.5.1), and consider the parameter-elliptic system $l(D, q_1)u = f$. Let $l_0(D, q_1)$ be the principal part of the matrix $l(D, q_1)$. There hold the following assertions (see [R1, Sec. 10.2]).

Lemma 1.5.1. *Let $q \ge q_0 > 0$, $p \in]1, +\infty[$, and $s \in \mathbf{R}$. The closure l_0 of the mapping*

$$u \mapsto l_0(D, q_1)u \qquad \left(u \in (C_0^\infty(\mathbf{R}^n))^N\right)$$

establishes an isomorphism

$$H^{T+s,p}(\mathbf{R}^n, q) \to H^{s-S,p}(\mathbf{R}^n, q).$$

Lemma 1.5.2. *In order that problem*

$$l_0(D, q_1)u = \Phi \in H_+^{s-S,p}(\mathbf{R}^n, q) \qquad (q \ge q_0 > 0)$$

be solvable in

$$H_+^{T+s,p}(\mathbf{R}^n, q) := \left\{u \in H^{T+s,p}(\mathbf{R}^n, q) : \operatorname{supp} u \subset \overline{\mathbf{R}}_+{}^n\right\}$$

it is necessary and sufficient that the equalities

$$\int_{-\infty}^{+\infty} L^{-1}(\xi', \xi_n, q)\left(\xi_n + i\sqrt{q^2 + |\xi'|^2}\right)^{s+t_k-m} \sum_{r=1}^{N} L_{rk}(\xi', \xi_n, q)\tilde{\Phi}_r(\xi', \xi_n)\xi_n^j d\xi_n = 0$$

$$(k = 1, \ldots, N,\ j = 0, \ldots, m-1)$$

hold for almost all $\xi' \neq 0$. Here

$$\widetilde{\Phi}_r(\xi', \xi_n) = (F\Phi_r)(\xi', \xi_n)$$

is a Fourier transform of the element Φ_r, and L_{rk} is the cofactor of the element l_{rk} of the matrix $l_0(\xi, q)$.

Note that, under the conditions of this lemma, Φ is a sufficiently smooth function, then the mentioned equalities hold for all $\xi' \neq 0$.

Lemma 1.5.3. *For every $\xi' \neq 0$, in the set of Nm equalities of Lemma 1.5.6, there are m linearly independent conditions, and the other equalities can be represented as their linear combination.*

In the half-space $\mathbf{R}^n_+$ consider a parameter-elliptic model problem

$$l_0(D, q_1)u(x) = f(x) \quad (x \in \mathbf{R}^n_+), \qquad b_0(D, q_1)u(x)\big|_{x_n=0} = \varphi(x), \quad (1.5.17)$$

where $b_0(D, q_1)$ is the principal part of the matrix $b(D, q_1)$ of the form (1.5.4) whose coefficients are complex constant numbers.

Let us represent the expressions $l^0_{rj}(D, q_1)$ and $b^0_{hj}(D, q_1)$ in the following form:

$$l^0_{rj}(D, q_1) = \sum_{k=1}^{s_r+t_j+1} l^{rj}_k(D', q)D_n^{k-1} \quad (\forall r, j :\ s_r + t_j \geq 0),$$

$$b^0_{hj}(D, q_1) = \sum_{k=1}^{\sigma_h+t_j+1} b^{hj}_k(D', q)D_n^{k-1} \quad (\forall h, j :\ \sigma_h + t_j \geq 0),$$

where $l^{rj}_k(D', q)$ and $b^{hj}_k(D'$ are expressions of orders $s_r + t_j - k + 1$ and $\sigma_h + t_j - k + 1$, respectively, over the variables $x_1, \ldots, x_n, q$.

Denote by $N_1 \leq N$ the number of the subscripts j such that $t_j \geq 1$.

If

$$u = (u_1, \ldots, u_N) \in \widetilde{H}^{T+s,p,(\tau)}(\mathbf{R}^n_+, q),$$

$$u_j = (u_{j0}, \ldots, u_{j\tau_j}) \in \widetilde{H}^{t_j+s,p,(\tau_j)}(G, q) \quad (j = 1, \ldots, N),$$

$$F = (f, \varphi) = (f_1, \ldots, f_N, \varphi_1, \ldots, \varphi_m) \in K_{s,p}(\mathbf{R}^n_+, q),$$

$$f_j = (f_{j0}, \ldots, f_{j,æ-s_j}) \in \widetilde{H}^{s-s_j,p,(æ-s_j)}(G, q), \quad (j = 1, \ldots, m),$$

then, from the formulas of Section 1.1, the equality $l_0 u = f$ is valid if and only if there hold the relations

$$(l_{0r}u)_+ = \sum_{j=1}^{N} l^0_{rj}(D,q)u_{j0+} - i \sum_{j:s_r+t_j\geq 1} \sum_{k=1}^{s_r+t_j} \left(J^k l^0_{rj}(D,q)\left(u_{jk}(x') \times \delta(x_n)\right)\right)$$

$$= f_{r0+} \qquad (r = 1,\ldots,N), \tag{1.5.18}$$

$$D_n^{h-1} l_{0r} u\big|_{x_n=0} = \sum_{j=1}^{N} \sum_{k=1}^{s_r+t_j+1} l_k^{rj}(D',q_1)u_{j,k+h-1}(x') = f_{rh} \tag{1.5.19}$$

$$(h = 1,\ldots,æ - s_r,\ r : æ - s_r \geq 1),$$

where u_{j0+} and f_{r0+} are the continuation of the elements u_{j0} and f_{r0} to the half-space $x_n < 0$ by zero, and, furthermore, the equality $b_0 u\big|_{x_n=0} = \varphi$ is valid if and only if there hold the relations

$$b_h u \equiv \sum_{j=1}^{N} \sum_{k=1}^{\sigma_h+t_j+1} b_k^{hj}(D',q)u_{jk}(x') = \varphi_h(x') \tag{1.5.20}$$

$$(h = 1,\ldots,m).$$

Since equations (1.5.18) are considered in whole $\mathbf{R}^n$, one can use here the results of Lemmas 1.5.6 and 1.5.3 with

$$\Phi_r = f_{r0+} + i \sum_{j:s_r+t_j\geq 1} \sum_{k=1}^{s_r+t_j} \left(J^k l^0_{rj}(D,q)u_{jk}(x') \times \delta(x_n)\right) \in H_+^{s-s_r,p}(\mathbf{R}^n) \tag{1.5.21}$$

$$(r = 1,\ldots,N).$$

Lemma 1.5.6 implies that system (1.5.18) possesses a solution

$$u_{0+} = (u_{10+},\ldots,u_{N0+}), \quad u_{j0+} \in H_+^{t_j+s,p}(\mathbf{R}^n) \tag{1.5.22}$$

if and only if

$$-i \int_{-\infty}^{+\infty} L^{-1}(\xi',\xi_n,q_1)\left(\xi_n + i\sqrt{q^2+|\xi'|^2}\right)^{s+t_k-m} \sum_{r=1}^{N} L_{rk}(\xi',\xi_n,q_1)\xi_n^j$$

$$\times \sum_{\alpha:s_r+t_\alpha\geq 1} \sum_{\beta=1}^{s_r+t_\alpha} J^\beta l^0_{r\alpha}(\xi,q_1)\widehat{u}_{\alpha\beta}(\xi')d\xi_n$$

$$= \int_{-\infty}^{+\infty} L^{-1}(\xi', \xi_n, q_1)\Big(\xi_n + i\sqrt{q^2 + |\xi'|^2}\Big)^{s+t_k-m} \sum_{r=1}^{N} L_{rk}(\xi', \xi_n, q_1)\xi_n^j \widetilde{f}_{r0+}(\xi', \xi_n)d\xi_n \tag{1.5.23}$$

$$(k = 1, \ldots, N,\ j = 0, \ldots, m-1).$$

Denote by $\widehat{g}_{kj}(\xi')$ the right-hand side of (1.5.23), and by $\widetilde{c}_{kj\alpha\beta}(\xi', q_1)$ the coefficients of $\widehat{u}_{\alpha\beta}(\xi')$ in the left-hand side of (1.5.23). Taking into account the fact that $J^\beta l^0_{r\alpha}(D, q_1) \equiv 0$ for $\beta > s_r + t_\alpha$, one can change the summation over β from 1 to $s_r + t_\alpha$ in the right-hand side of (1.5.23) by the summation over β from 1 to $t_\alpha = \max\{s_1 + t_\alpha, \ldots, s_N + t_\alpha\}$. We obtain the equations

$$\sum_{j=1}^{N_1} \sum_{\beta=1}^{t_\alpha} \widetilde{c}_{kj\alpha\beta}(\xi', q_1)\widehat{u}_{\alpha\beta}(\xi') = \widehat{g}_{kj}(\xi') \tag{1.5.24}$$

$$(k = 1, \ldots, N, \quad j = 0, \ldots, m-1).$$

In view of Lemma 1.5.3, here, for every $\xi' \neq 0$, there are m linearly independent equations, and the other equations are expressed as their linear combinations. Therefore, in what follows, speaking about system (1.5.24), we always assume that, for every $\xi' \neq 0$, this system contains only m linearly independent equations, and the others are thrown away.

Passing to the Fourier images in (1.5.19) and (1.5.20), we obtain

$$\sum_{j=1}^{N} \sum_{k=1}^{s_r+t_j+1} l_k^{rj}(\xi', q_1)\widehat{u}_{j,k+h-1}(\xi') = \widehat{f}_{r\alpha}(\xi') \tag{1.5.25}$$

$$(h = 1, \ldots, æ - s_r, \qquad r : æ - s_r \geq 1),$$

$$\sum_{j=1}^{N} \sum_{k=1}^{\sigma_h+t_j+1} b_k^{hj}(\xi', q)\widehat{u}_{jk}(\xi') = \widehat{\varphi}_h(\xi') \qquad (h = 1, \ldots, m). \tag{1.5.26}$$

As we agreed above, assume that, for every $\xi' \neq 0$, system (1.5.24) contains only m linearly independent equations, and other equations are thrown away. Then system (1.5.23)–(1.5.25) is a system of $|\tau| = \tau_1 + \cdots + \tau_N$ linear equations in $|\tau|$ unknown functions. Let $\Delta(\xi', q_1)$ denote the determinant of this system. Note also that each mth-order minor of system (1.5.24) is a continuous function of ξ'. Therefore, if certain m equations of (1.5.24) are linearly independent at the point $\xi' \neq 0$, then these equations are also linearly independent in a neighborhood of this point.

The following assertion is true (see [R1, Sec. 10.2]).

Lemma 1.5.4. *Model problem (1.5.17) is elliptic with a parameter if and only if*

$$\Delta(\xi', q_1) \neq 0 \qquad (\forall(\xi', q) : |\xi'| + |q| > 0). \tag{1.5.27}$$

This lemma enables us to solve linear system (1.5.24)–(1.5.26), find the vector

$$U = \left(u_{jk} : j = 1, \ldots, N,\ k = 1, \ldots, \tau_j\right) \in \prod_{j=1}^{N} \prod_{k=1}^{\tau_j} B^{t_j+s-k+1-1/p,p}(\mathbf{R}^{n-1}, q) \tag{1.5.28}$$

of the Cauchy data of the solution

$$u = (u_0, U) \in \widetilde{H}^{T+s,p,(\tau)}(\mathbf{R}^n_+, q)$$

of problem (1.5.17), and estimate this vector.

This gives us a possibility to prove the theorem on complete collection of isomorphisms for the model problem in the half-space $\mathbf{R}^n_+$, similar to Theorem 1.5.1. Then, Theorem 1.5.1 on complete collection of isomorphisms follows from this theorem by using the standard procedure (see [R1, Ch. IX, X]).

Note that, if the element

$$u = (u_0, U) \in \widetilde{H}^{T+s,p,(\tau)}(\mathbf{R}^n_+, q)$$

is a solution of the problem

$$l_0(D, q)u = 0, \qquad b_0(D, q)u\big|_{\partial G} = \varphi,$$

then the Fourier image $\widehat{U}$ of the Cauchy data vector is a solution of the following system:

$$F'_{x'\mapsto\xi'}\left(D_n^{n-1} l_{0r} u\big|_{x_n=0}\right)(\xi') := \sum_{j=1}^{N} \sum_{k=1}^{s_r+t_j+1} l_k^{rj}(\xi', q_1)\widehat{u}_{j,k+h-1}(\xi') = 0$$

$$(h = 1, \ldots, æ - s_r, \quad r : æ - s_r \geq 1),$$

$$F'_{x'\mapsto\xi'}\left(b_h u\big|_{x_n=0}\right)(\xi') := \sum_{j=1}^{N} \sum_{k=1}^{\sigma_h+t_j+1} b_k^{hj}(\xi', q_1)\widehat{u}_{j,k}(\xi') = \widehat{\varphi}_h(\xi') \tag{1.5.29}$$

$$(h = 1, \ldots, m),$$

$$\sum_{j=1}^{N_1} \sum_{\beta=1}^{t_\alpha} \widetilde{c}_{kj\alpha\beta}(\xi', q)\widehat{u}_{\alpha\beta}(\xi') = 0,$$

where $g = 0$ because $f_0 = 0$, and, therefore, $f_0 = 0$ and $f_{j,k} = 0$. This system is considered to be including only m linearly independent equalities, the other equalities are expressed as their linear combination. The determinant of system (1.5.29) is nonzero.

Rewrite equalities (1.5.29) in the form

$$\widehat{E}\widehat{U} = \begin{pmatrix} \widehat{e}(\xi', q) \\ \widehat{b}(\xi', q) \\ \widehat{\widetilde{c}}(\xi', q) \end{pmatrix} \widehat{U}(\xi') = \begin{pmatrix} 0 \\ \widehat{\varphi}(\xi') \\ 0 \end{pmatrix}. \tag{1.5.30}$$

Here $\widehat{e}(\xi', q)$ is a $|\tau| - 2m \times |\tau|$-matrix, and $b(\xi', q)$ and $\widehat{\widetilde{c}}(\xi', q)$ are $m \times |\tau|$-matrixes.

It follows from (1.5.29) that the element $\widehat{U}(\xi')$ runs an m-dimensional linear space: if we denote by $\widehat{U}^j(\xi')$ a solution of system (1.5.29) with

$$\widehat{\varphi}_h = \delta_{jh} = \begin{cases} 0 & \text{for } h \neq j \\ 1 & \text{for } h = j \end{cases}$$

then each solution of problem (1.5.29) is equal to

$$\widehat{U} = \sum_{h=1}^{m} \widehat{\varphi}_h(\xi')\widehat{U}^h(\xi').$$

1.5.5. We now consider in $\mathbf{R}^n_\pm = \{x \in \mathbf{R}^n : x_n \gtrless 0\}$ the Cauchy problem

$$l_0(D, q)u^\pm(x) = 0 \quad \text{in } \mathbf{R}^n_\pm, \qquad u^\pm = (u_0^\pm, U^\pm) \in \widetilde{H}^{T+s,p,(\tau)}(\mathbf{R}^n_\pm, q),$$

where $U^\pm$ is the Cauchy data vector. If $b^\pm(D', q)U^\pm = \varphi^\pm$, then (1.5.30) yields that

$$\widehat{E}^\pm\widehat{U}^\pm(\xi') = \begin{pmatrix} 0 \\ \widehat{\varphi}^\pm(\xi') \\ 0 \end{pmatrix},$$

and, for various $\varphi_\pm$ the sets $\{\widehat{U}^\pm(\xi')\}$ run m-dimensional spaces. The intersection of these spaces consist of only zero point. Indeed, in the contrary case, the elements $u^\pm \in \widetilde{H}^{T+s,p,(\tau)}(\mathbf{R}^n_+, q)$ have the same vectors of the Cauchy data, and $l_0^\mp u^\pm = 0$. Then, the element $u_0 = \begin{cases} u_0^+, & x_n > 0 \\ u_0^-, & x_n < 0 \end{cases}$ is a solution in $\mathbf{R}^n$ of the problem $lu_0 = 0$ $(u_0 \in H^{T+s,p}(\mathbf{R}^n, q))$, and $u_0 \equiv 0$ (see the end of Subsection 1.5.3). Therefore, in $\mathbf{R}^n$, $u_0^\pm = 0$ and $lu_0^\pm = 0$, and, thus, $U^\pm = 0$. Hence, the Cauchy data vectors $\{\widehat{U}^+(\xi') + \widehat{U}^-(\xi')\}$ run a $2m$-dimensional space.

Let

$$B^{T+s,p,(\tau)}(\mathbf{R}^{n-1}, q) = \prod_{k=1}^{N}\prod_{j=1}^{\tau_k} B^{t_k+s-j+1-1/p,p}(\mathbf{R}^{n-1}, q)$$

denote the space of the Cauchy data, and let

$$\mathcal{U}_{s,p} = \left\{ U \in B^{T+s,p,(\tau)}(\mathbf{R}^{n-1}, q) : e(D,q)U = 0 \right\}$$

be a subspace of $B^{T+s,p,(\tau)}(\mathbf{R}^{n-1}, q)$. Assume that $\widehat{B}^{T+s,p,(\tau)}$ and $\widehat{U}_{s,p}$ denote the Fourier images of these spaces, respectively. In addition, $\widehat{U}_{s,p}$ is a $2m$-dimensional subspace of $\widehat{B}^{T+s,p,(\tau)}$.

Analogously, let

$$\mathcal{U}^{\pm}_{s,p} = \left\{ U^{\pm} \in B^{T+s,p,(\tau)}(\mathbf{R}^{n-1}, q) : e(D,q)U^{\pm} = 0 \,\&\, \tilde{c}(D,q)U^{\pm} = 0 \right\},$$

and let $\widehat{\mathcal{U}}^{\pm}_{s,p}$ be the Fourier images of $\mathcal{U}^{\pm}_{s,p}$. In addition, $\widehat{\mathcal{U}}^{\pm}_{s,p}$ is an m-dimensional subspace of the space $\mathcal{U}_{s,p}$, and, futhermore, $\widehat{\mathcal{U}}^{+}_{s,p} \cap \widehat{\mathcal{U}}^{-}_{s,p} = \{0\}$. This implies that

$$\mathcal{U}_{s,p} = \mathcal{U}^{+}_{s,p} \dot{+} \mathcal{U}^{-}_{s,p}.$$

The mapping

$$\begin{pmatrix} \varphi^{+} \\ \varphi^{-} \end{pmatrix} \mapsto \begin{pmatrix} U^{+} \\ U^{-} \end{pmatrix} = U \in \mathcal{U}_{s,p}$$

is a continuous one-to-one mapping, and therefore, it is mutually continuous mapping in the corresponding spaces. This yields that the projection operators $P_{\pm} : U \mapsto U^{\pm}$ act continuously in $\mathcal{U}_{s,p}$ $(s \in \mathbf{R},\ p \in]1, +\infty[)$.

1.5.6. Green's formula

Consider the matrix $B(x, D, qe^{i\theta})$ of the system of $|\tau| = \tau_1 + \cdots + \tau_N$ boundary conditions which contain polynomially a parameter $q_1 = qe^{i\theta}$:

$$B_h u(x) = \sum_{j=1}^{N} B_{hj}(x, D, qe^{i\theta}) u_j(x), \qquad (h = 1, \ldots, |\tau|), \tag{1.5.31}$$

$$B_{hj}(x, D, qe^{i\theta}) = \sum_{k=1}^{\tau_j} \Lambda_{hjk}(x, D', qe^{i\theta}) D_{\nu}^{k-1} \quad (x \in \partial G,\ j = 1, \ldots, N),$$

$$\Lambda_{hjk}(x, D', qe^{i\theta}) = \sum_{l=0}^{\tau_j+\sigma_h-k+1} \Lambda_{hjkl}(x, D')(qe^{i\theta})^{k},$$

where $\Lambda_{hjkl}(x, D')$ are tangential, generally speaking, pseudo-differential operators. Moreover, we assume that there exist integer numbers $\sigma_h < 0$ $(h = 1, \ldots, |\tau|)$ such that the order of the operator Λ_{hjkl} with respect to (D', q) is: $\operatorname{ord}_{(D',q)} \Lambda_{hjkl} \le \sigma_h + \tau_j - k + 1 - l$ for $\sigma_h + \tau_j - k + 1 - l \ge 0$, and $\Lambda_{hjkl} \equiv 0$ for $\sigma_h + \tau_j - k + 1 - l < 0$.

The matrix (1.5.31) defines on ∂G the system of the equations

$$\sum_{j=1}^{N}\sum_{k=1}^{\tau_j} \Lambda_{hjk}(x, D', q_1)\eta_{jk} = \varphi_h \tag{1.5.32}$$

$$(h = 1, \ldots, |\tau|,\, x \in \partial G,\, q_1 = qe^{i\theta},\, \eta_{jk} = D_\nu^{k-1}u_j\big|_{\partial G}),$$

or,shortly,

$$\Lambda(q)\eta = \varphi$$

with the polynomially contained parameter q_1.

If we associate the number $\tau_{js} = \tau_j - k + 1$ with the column with index jk in the system (1.5.32), and the number σ_h with the equation with index h, then we obtain that $\operatorname{ord}_{(D',q)} \Lambda_{hjk}(x, D', q_1) \leq \sigma_h + \tau_{jk}$ for $\sigma_h + \tau_{jk} \geq 0$, and $\Lambda_{hjk} \equiv 0$ for $\sigma_h + \tau_{jk} < 0$, in other words, system (1.5.31) is a Douglis-Nirenberg parameter-elliptic system of order $((\tau_{jk}), (\sigma_1, \ldots, \sigma_{|\tau|}))$. Therefore, for any $s \in \mathbf{R}$ and $p \in]1, +\infty[$, the closure $\Lambda(q) = \Lambda_{s,p}(q)$ of the mapping

$$\eta \mapsto \Lambda(q)\eta \qquad \left(\eta \in (C^\infty(\partial G))^{|\tau|}\right)$$

acts continuously in the pair of spaces

$$U^{s,\tau,p}(q) := \prod_{j=1}^{N}\prod_{k=1}^{\tau_j} B^{s+\tau_j-k,p}(\partial G, q) \to V^{s,\sigma,p}(q) := \prod_{h=1}^{|\tau|} B^{s-\sigma_h-1,p}(\partial G, q), \tag{1.5.33}$$

and there holds the estimate

$$||\Lambda(q)\eta||_{V^{s,\sigma,p}(q)} \leq c||\eta||_{U^{s,\tau,p}(q)}$$

with the constant $c > 0$ independent of η and q.

Definition 1.5.2. *Matrix (1.5.31) is called τ-complete if (1.5.32) is a parameter-elliptic system of order*

$$\left((\tau_{jk}), (\sigma_1, \ldots, \sigma_{|\tau|})\right)$$
$$= \left(\tau_{11}, \tau_{12}, \ldots, \tau_{1\tau_1}, \tau_{21}, \ldots, \tau_{2\tau_2}, \ldots, \tau_{N1}, \ldots, \tau_{N\tau_N}, \sigma_1, \ldots, \sigma_{|\tau|}\right)$$

on ∂G.

Futher on matrix (1.5.31) is assumed to be τ-complete. It is clear, that the system

$$\sum_{h=1}^{|\tau|} \Lambda^+_{hjk}(x, D', q)\zeta_h = \psi_{jk} \;\; (k = 1, \ldots, \tau_j,\, j = 1, \ldots, N,\, x \in \partial G), \tag{1.5.34}$$

or, shortly,

$$\Lambda^+\zeta = \psi,$$

which is formally adjoint to system (1.5.32), is also parameter-elliptic. In addition,

$$\langle \Lambda\eta, \zeta\rangle = \langle \eta, \Lambda^+\zeta\rangle \qquad \left(\eta, \zeta \in (C^\infty(\partial G))^{|\tau|}\right),$$

where $\langle\cdot,\cdot\rangle$ denotes the scalar product in $(L_2(\partial G))^{|\tau|}$.

Since system (1.5.32) is a Douglis-Nirenberg parameter-elliptic system, there exists a constant $q_0 > 0$ such that, for $|q| \geq q_0$, the operator $\Lambda(q) = \Lambda_{s,p}(q)$ realizes an isomorphism between spaces (1.5.33), and there exists a constant $c > 0$ independent of η and q $(|q| \geq q_0 > 0)$ such that

$$c^{-1}\|\eta\|_{U^{s,\tau,p}(q)} \leq \|\Lambda(q)\eta\|_{V^{s,\sigma,p}(q)} \leq c\|\eta\|_{U^{s,\tau,p}(q)}. \qquad (1.5.35)$$

The analogous statement is true also for the operator $\Lambda^+(q)$.

Let us now deduce the Green's formula. Our reasoning will be quite analogous to the reasoning in Subsection 1.2.2.

Assume that, in the certain neighborhood of the boundary ∂G in G, we have that

$$l_{rj}(x, D, q, \theta) = \sum_{k=0}^{s_r+t_j} l_k^{rj}(x, D', q_1)D_\nu^k \qquad (\forall r, j : s_r + t_j \geq 0,\ q_1 = qe^{i\theta}),$$

where $l_k^{rj}(x, D', q_1)$ is a tangential operator of order

$$\operatorname*{ord}_{(D',q_1)} l_k^{rj}(x, D', q_1) \leq s_r + t_j - k.$$

Using the integration by parts and the interchange of the order of summation, we find

$$(lu, v) - (u, l^+, v) = \sum_{j:t_j\geq 1}\sum_{s=1}^{t_j}\langle D_\nu^{s-1}u_j, M_j^s v\rangle, \qquad (1.5.36)$$

$$M_j^s v = -i \sum_{r:s_r+t_j\geq s}\ \sum_{k=s}^{s_r+t_j} D_\nu^{k-j}\left(l_k^{rj}(x, D', q_1)\right)^+ v_r.$$

If we associate the vector

$$\eta = (\eta_1, \ldots, \eta_N),$$

$$\eta_j = (\eta_{j1}, \ldots, \eta_{j,\tau_j}), \quad \eta_{jk} = D_\nu^{k-1}u_j\big|_{\partial G} \qquad (k = 1, \ldots, \tau_j,\ j = 1, \ldots, N),$$

with the element $u = (u_1, \dots, u_N)$, and the vector

$$Mv = (\zeta_1, \dots, \zeta_N),$$

$$\zeta_j = (\zeta_{j1}, \dots, \zeta_{j,\tau_j}), \quad \zeta_{jk} = \begin{cases} M_j^k v|_{\partial G} & \text{for } k = 1, \dots, t_j \\ 0 & \text{for } t_j < k \le \tau_j \end{cases},$$

with the element $v = (v_1, \dots, v_N)$, then we obtain that $Bu|_{\partial G} = \Lambda(q)\eta$, and we can rewrite formula (1.5.36) in the form

$$(lu, v) - (u, l^+, v) = \langle \eta, Mv \rangle \qquad \left(u, v \in (C^\infty(\overline{G}))^N\right).$$

From the other hand, for $|q| \ge q_0$,

$$\eta = (\Lambda(q))^{-1}\Lambda(q)\eta = (\Lambda(q))^{-1} Bu|_{\partial G}.$$

Therefore,

$$(lu, v) - (u, l^+, v) = \langle Bu, B'v \rangle \qquad \left(u, v \in (C^\infty(\overline{G}))^N\right), \tag{1.5.37}$$

where

$$B'v = (\Lambda(q)^{-1})^+ Mv. \tag{1.5.38}$$

Thus, we have the following result.

Theorem 1.5.2. *Let matrix (1.5.31) be τ-complete. Then, there exists a number $q_0 > 0$ such that, for $|q| \ge q_0$,*
i) the operator $\Lambda(q)$ realizes an isomorphism (1.5.33), and
ii) there holds Green's formula (1.5.37).

Let us rewrite the Green's formula in more convenient form. We complement the matrix $b(x, D, q_1)$ by the new rows so that the obtained system $B(x, D, q_1)$ is τ-complete.

Let

$$c(x, D, q_1) = (c_{hj}(x, D, q_1))_{\substack{h=1,\dots,m, \\ j=1,\dots,N}} \qquad q_1 = qe^{i\theta}, \tag{1.5.39}$$

$\operatorname{ord} c_{hj} \le \sigma_h^c + t_j$ for $\sigma_h^c + t_j \ge 0$, and $c_{hj} \equiv 0$ for $\sigma_h^c + t_j < 0$. The expression

$$c_h(x, D, q_1) = (c_{h1}(x, D, q_1), \dots, c_{hN}(x, D, q_1)), \quad (h = 1, \dots, m),$$

is the row with index h of the matrix $c(x, D, q_1)$. Let

$$e(x, D, q_1) = (D_\nu^{k-1} l_r(x, D, q_1))_{\substack{r=1,\dots,N \\ k:1 \le k \le -s_r + \varkappa}}. \tag{1.5.40}$$

The expression

$$l_r(x, D, q_1) = (l_{r1}(x, D, q_1), \dots, l_{rN}(x, D, q_1))$$

is the row with index r of the matrix $l(x, D, q_1)$. The matrix $c(x, D, q_1)$ contains m rows, the matrix $e(x, D, q_1)$ contains $|\tau| - 2m$ rows.

It turns out that, if the problem (1.5.1)–(1.5.2) is parameter-elliptic, then there exists a matrix $c(x, D, q_1)$ with $\sigma_h^c < 0$, such that the $|\tau| \times N$-matrix

$$B(x, D, q_1) = \begin{pmatrix} e(x, D, q_1) \\ b(x, D, q_1) \\ c(x, D, q_1) \end{pmatrix} \tag{1.5.41}$$

is τ-complete. One can verify this statement by the repeating of the reasonings of Section 1.4 (see also [RS1] and [RS2]) with the understandable changes: insted of the problem (1.4.7)–(1.4.8) we have to consider a problem with a parameter

$$l_0(D, q)u(x) = f(x) \quad (x \in \mathbf{R}_+^n), \qquad b_0(D, q)u(x) = \varphi(x) \quad (x \in \mathbf{R}'_{n-1}).$$

Then, Theorem 1.5.2 directly implies the following assertion.

Theorem 1.5.3. *Let the problem (1.5.1)–(1.5.2) be parameter-elliptic, and let matrixes (1.5.39) and (1.5.40) complement the matrix $b(x, D, q_1)$ to τ-complete matrix (1.5.41). Then there exists a number $q_0 > 0$ such that, for $q \geq q_0$, there holds the Green's formula*

$$(lu, v) + \sum_{j=1}^{m} \langle b_h u, c'_h v\rangle + \sum_{r=1}^{N} \sum_{k=1}^{-s_r+æ} \langle D_\nu^{k-1} l_r u, e'_{kr} v\rangle = (u, l^+ v) + \sum_{h=1}^{m} \langle c_h u, b'_h v\rangle \tag{1.5.42}$$

$$\left(u, v \in \left(C^\infty(\overline{G})\right)^N\right).$$

To obtain formula (1.5.42), it is necessary to represent matrix (1.5.38) in the form of the matrix transposed to the matrix $\begin{pmatrix} -c'(x, D, q_1) \\ -e'(x, D, q_1) \\ b'(x, D, q_1) \end{pmatrix}$.

1.5.7.

Definition 1.5.3. *The problem*

$$l^+ v(x) = g(x), \qquad x \in G, \tag{1.5.43}$$

$$b' v\big|_{\partial G} = \psi, \tag{1.5.44}$$

is called a formally adjoint to problem (1.5.1)–(1.5.2) with respect to Green's formula (1.5.42).

Theorem 1.5.4. *The problem (1.5.43)–(1.5.44), formally adjoint to problem (1.5.1)–(1.5.2) with respect to Green's formula (1.5.42), is elliptic with a parameter if and only if the problem (1.5.1)–(1.5.2) is elliptic with a parameter.*

Proof. We prove the theorem in the same way as we proved Theorem 1.2.4.

Let the problem (1.5.1)–(1.5.2) be parameter-elliptic, and let $q \geq q_0$. Denote:

$$\mathfrak{N}^+ = \left\{ v \in \left(C^\infty(\overline{G})\right)^N : l^+ v = 0 \ \& \ b'v\big|_{\partial G} = 0, \right\}.$$

Let us show that $\mathfrak{N}^+ \equiv 0$. Indeed, by passing to the limit, we make sure that Green's formula (1.5.42) is valid for the elements

$$u \in \widetilde{H}^{T+s,p,(\tau)}(G,q), \quad v \in \left(C^\infty(\overline{G})\right)^N.$$

Then, for the solvability of problem (1.5.14) with $F \in \widetilde{K}^{s,p}$ it is necessary that

$$(f_0, v) + \sum_{j=1}^{N} \sum_{r=1}^{æ - s_j} \langle f_{jr}, e'_{kr} v \rangle + \sum_{h=1}^{m} \langle \varphi_h, c'_h v \rangle = 0 \qquad (\forall v \in \mathfrak{N}^+). \quad (1.5.45)$$

But, in view of the fact that, for $q \geq q_0$, the problem (1.5.14) is solvable (uniquely) for any $F \in \widetilde{K}^{s,p}$, we have that (1.5.45) yields the relation

$$v \equiv 0 \qquad (\forall v \in \mathfrak{N}^+).$$

Note that, if $l(x, D, q_1)$ is a matrix expression of order (T, S) with

$$t_1 \geq \ldots \geq t_N \geq 0 = s_1 \geq \ldots \geq s_N,$$

then $l^+(x, D, q_1)$ is a matrix expression of order (T', S') with

$$t'_j = t_1 + s_j, \quad s'_j = t_j - t_1 \quad (j = 1, \ldots, N).$$

Furtermore, l is a properly elliptic expression with a parameter if and only if l^+ is a properly elliptic expression with a parameter.

Let

$$H^{T,p}_{(b)} = \left\{ u \in H^{T,p}(G,q) : bu\big|_{\partial G} = 0 \right\} \qquad (1.5.46)$$

denote a subspace of the space $H^{T,p}(G,q)$, and let $H^{T',p'}_{(b)+}$ denote a set of elements v such that

$$(lu, v) = (u, l^+ v) \qquad (1.5.47)$$

$$\left(\forall u \in H^{T,p}_{(b)} : D^{k-1}_{\nu} L_r u\big|_{\partial G} = 0 \quad (r = 1, \ldots, N,\, k = 1, \ldots, -s_r + æ)\right).$$

It is clear that $H^{T',p'}_{(b)+}$ is a subspace of the space $H^{T',p'}(G,q)$.

In addition, we denote, for $k \geq 0$,

$$H^{T+k,p}_{(b)} = H^{T,p}_{(b)} \bigcap H^{T+k,p}(G,q),$$

$$H^{T'+k,p'}_{(b)+} = H^{T',p'}_{(b)+} \bigcap H^{T'+k,p'}(G,q).$$

Lemma 1.5.5. *Let the conditions of Theorem 1.5.1 be valid and $q \geq q_0$. Then, the space $H^{T',p'}_{(b)+}$ consists of elements $v \in H^{T',p'}(G,q)$ such that*

$$b'v\big|_{\partial G} = 0, \tag{1.5.48}$$

and does not include other elements.

Indeed, Green's formula (1.5.42) and equality (1.5.47) imply that the element v belongs to $H^{T',p'}_{(b)+}$ if and only if $v \in H^{T',p'}(G,q)$ and the equality

$$\sum_{h=1}^{m} \langle c_h u, b'_h v \rangle = 0 \tag{1.5.49}$$

$$\left(\forall u \in H^{T,p}_{(b)} \;\&\; D^{k-1}_{\nu} L_r u\big|_{\partial G} = 0 \quad (r = 1, \ldots, N,\, k = 1, \ldots, -s_r + æ)\right)$$

holds.

Since the operator $\Lambda(q)$ realizes isomorphism (1.5.33), we obtain that (1.5.49) is equivalent to equalities (1.5.48), and Lemma 1.5.5 is thus established. □

Now consider the problems

$$lu = f \qquad \left(u \in H^{T,2}_{(b)}\right) \tag{1.5.50}$$

and

$$l^{+}v = g \qquad \left(v \in H^{T',2}_{(b)+}\right) \tag{1.5.51}$$

with homogeneous boundary conditions.

Lemma 1.5.6. *Under the conditions of Theorem 1.5.1, let $q \geq q_0$. Then, for each $g \in H^{-S',2}(G,q)$, there exists one and only one solution $v \in H^{T',2}(G,q)$ of problem (1.5.51).*

Proof. In view of Theorem 1.5.1, the a priori estimate

$$\|u\|^2_{H^{T+æ,2}(G,q)} \le c\left(\|lu\|^2_{H^{æ-S,2}(G,q)} + \sum_{j=1}^{m}\langle\langle b_j u, \partial G, q\rangle\rangle^2_{æ-\sigma_h-\frac{1}{2},2}\right) \quad (1.5.52)$$

is valid.

Estimate (1.5.52) gives us an opportunity to suppose that the expression

$$[u,v] = (lu, lv)_{H^{æ-S,2}(G,q)} + \sum_{j=1}^{m}\langle B_j u, B_j v\rangle_{B^{æ-\sigma_h-1/2,2}(\partial G,q)}$$

defines new scalar product in the space $H^{T+æ,2}(G,q)$. It is clear that the expression (u,g) is a linear continuous functional of $u \in H^{T+æ,2}(G,q)$. Therefore there exists an element $w \in H^{T+æ,2}(G,q)$ such that

$$[u,w] = (u,g) \qquad (\forall u \in H^{T+æ,2}(G,q)),$$

or, in detail,

$$\sum_{h=1}^{N}\sum_{|\alpha|\le æ-s_h}(D^\alpha l_h u, D^\alpha l_h w) + \sum_{h=1}^{m}\langle b_h u, b_h w\rangle_{B^{æ-\sigma_h-\frac{1}{2},2}(\partial G,q)} = (u,g) \quad (1.5.53)$$

$$(u \in H^{T+æ,2}(G,q)).$$

It follows from the lemma on increasing of smoothness ([RSh3, p. 469]), that, if

$$g \in H^{k-\tau,2}(G) = \prod_{j=1}^{N} H^{k-\tau_j,2} \qquad (\tau_j = t_j + æ,\ k \ge \tau_1),$$

then

$$w \in H^{k+\tau,2}(G) = \prod_{j=1}^{N} H^{k+\tau_j,2}(G).$$

Hence, by integration by parts we obtain from (1.2.31) that

$$\sum_{h=1}^{N}\left(l_h u, \sum_{|\alpha|\le æ-s_h} D^{2\alpha} l_h w\right) + \sum_{h=1}^{N}\sum_{j=1}^{æ-s_h}\langle D_\nu^{j-1} l_h u, T_{jh} w\rangle$$

$$+\sum_{h=1}^{m}\langle b_h u, b_h w\rangle_{B^{æ-\sigma_h-\frac{1}{2},2}(\partial G)} = (u,g) \quad \left(\forall u \in H^{k+T,2}(G)\right). \quad (1.5.54)$$

We set

$$\sum_{|\alpha|\le æ-s_h} D^{2\alpha} l_h w = v_h \in H^{t_1+s_h,2}(G), \qquad h=1,\dots,N.$$

Then the equality

$$(lu, v) = (u, g)$$

holds for any element $u \in H^{k+T,2}(G)$ such that $bu|_{\partial G} = 0$ and

$$D_\nu^{j-1} l_h u\big|_{\partial G} = 0 \qquad (h=1,\dots,N,\ j=1,\dots,æ-s_j),$$

and the vector v is a solution of problem (1.5.51).

The uniqueness follows easily from Theorem 1.5.1: if there exists the another solution v', then

$$(lu, v-v') = 0, \quad \forall u \in H^{k+T,2}(G) : bu|_{\partial G} = 0 \ \& \ D_\nu^{j-1} l_h u|_{\partial G} = 0$$

$$(h=1,\dots,N,\ 1\le j\le æ-s_j),$$

and, hence, $v = v'$. □

The fact that the operator

$$v \mapsto l^+ v : H^{T',2}_{(b)+}(G,q) \to H^{-S',2}(G,q)$$

is continuous one-to-one mapping yields that the inverse operator is also continuous. This implies the validity of the estimate

$$\|v\|_{H^{T',2}(G,q)} \le c\|l^+ v\|_{H^{-S',2}(G,q)} \qquad (\forall v \in H^{T',2}_{(b)+}). \tag{1.5.55}$$

We next consider the problem (1.5.43)–(1.5.44), formally adjoint to the problem (1.5.1)–(1.5.2).

For any element $\psi \in B^{T'-\sigma'-1/2,2}(\partial G, q)$ there exists an element $v_0 \in H^{T',2}(G,q)$ such that $b'v_0|_{\partial G} = \psi$, and the continuation operator

$$T : \psi \mapsto v_0 : \ B^{T'-\sigma'-1/2,2}(\partial G, q) \to H^{T',2}(G,q) \tag{1.5.56}$$

is continuous (see Subsection 1.5.1). Thus, if $v \in H^{T',2}(G,q)$ is a solution of the problem (1.5.43)–(1.5.44) then, in view of Lemma 1.5.6, $v - v_0 \in H^{T',2}_{(b)+}$. Therefore, (1.5.55) and (1.5.56) imply that

$$\|v - v_0\|_{H^{T',2}(G,q)} \le c\|l^+(v-v_0)\|_{-S',2},$$

and, further,

$$\begin{aligned}
\|v\|_{H^{T',2}(G,q)} &\leq \|v - v_0\|_{H^{T',2}(G,q)} + \|v_0\|_{H^{T',2}(G,q)} \\
&\leq c\left(\|l^+ v\|_{H^{-S',2}(G,q)} + \|l^+ v_0\|_{H^{-S',2}(G,q)} + \|v_0\|_{H^{T',2}(G,q)}\right) \\
&\leq c_1\left(\|l^+ v\|_{H^{-S',2}(G,q)} + \|v_0\|_{H^{T',2}(G,q)}\right) \\
&= c_2\left(\|l^+ v\|_{H^{-S',2}(G,q)} + \|T\psi\|_{H^{T',2}(G,q)}\right) \\
&\leq c_3\left(\|l^+ v\|_{H^{-S',2}(G,q)} + \langle\langle \psi, \partial G, q\rangle\rangle_{T'-\sigma'-1/2,2}\right).
\end{aligned}$$

As a result, we obtain the a priori estimate

$$\|v\|_{H^{T',2}(G,q)} \leq \text{const}\left(\|l^+ v\|_{H^{-S',2}(G,q)} + \langle\langle b'v, \partial G, q\rangle\rangle_{T'-\sigma'-1/2,2}\right)$$

which implies the parameter-ellipticity of the problem (1.5.43)–(1.5.44).

Thus, the parameter-ellipticity of the problem (1.5.43)–(1.5.44) follows from the parameter-ellipticity of the problem (1.5.1)–(1.5.2). By using these reasonings in the inverse direction, we obtain that the parameter-ellipticity of the problem (1.5.1)–(1.5.2) follows from the parameter-ellipticity of the problem (1.5.43)–(1.5.44). Theorem 1.5.4 is now completely proved. □

Remark 1.5.1. All mentioned results with the same proofs remain true in the case where

$$q = (q^1, \quad, q^k) \in \mathbf{R}^k, \quad q_1 = \left(q^1 e^{i\theta_1}, \ldots, q^k e^{i\theta_k}\right),$$

$$\theta_0^i \leq \theta_i \leq \theta_1^0, \quad |q|^2 = \left(q^1\right)^2 + \cdots + \left(q^k\right)^2.$$

1.6. On Various Theorems on Complete Collection of Isomorphisms for Parameter-Elliptic Boundary Value Problems for Systems of Equations

1.6.1. Recall that the norm of the space $\widetilde{H}^{s,p,(r)}(G,q)$ is defined by formula (1.5.11) for the case where $s \neq k + 1/p$ $(k = 0, \ldots, r-1)$. For these reason, in what follows, we consider such spaces only for $s \in \mathbf{R}$, $s \neq k + 1/p$ $(k = 0, \ldots, r-1)$. By using the interpolation theorem, theorems on isomorphisms will be odtained also for the case $s = k+1/p\,(k = 0,\dots,r-1)$.

1.6.2. Since matrix (1.5.39) is τ-complete, estimates (1.5.35) directly imply that the norm

$$\|u\|_{\widetilde{H}^{T+s,p,(\tau)}} = \left(\sum_{j=1}^{N} |||u_j, G, q|||^p_{t_j+s,p,(\tau_j)}\right)^{1/p} \tag{1.6.1}$$

is equivalent to the norm

$$\left(\sum_{j=1}^{N} \|u_j, G, q\|^p_{t_j+s,p} + \sum_{r=1}^{N} \sum_{k=1}^{-s_r+æ} \langle\langle D_\nu^{k-1} l_r u, \partial G, q\rangle\rangle^p_{s-s_r-k+1-1/p,p} \right.$$

$$\left. + \sum_{j=1}^{m} \langle\langle b_j u, \partial G, q\rangle\rangle^p_{s-\sigma_j-1/p,p} + \sum_{h=1}^{m} \langle\langle c_h u, \partial G, q\rangle\rangle^p_{s-\sigma_h^c-1/p,p}\right)^{1/p}, \tag{1.6.2}$$

and the space $\widetilde{H}^{T+s,p,(\tau)}(G,q)$ coincides with the completion $\widetilde{H}_B^{T+s,p,(\tau)}$ of the set $\left(C^\infty(\overline{G})\right)^N$ in the norm (1.6.2). For $s > æ$ (1.5.13), norms (1.6.1) and (1.6.2) are equivalent to the norm $\left(\sum_{j=1}^N \|u_j, G, q\|^p_{t_j+s,p}\right)^{1/p}$, and the spaces $\widetilde{H}^{T+s,p,(\tau)}(G,q)$ and $\widetilde{H}_B^{T+s,p,(\tau)}(G,q)$ coincide with the space

$$\prod_{j=1}^{N} H^{t_j+s,p}(G,q) =: H^{T+s,p}(G,q).$$

For $s < æ$, norms (1.6.1) and (1.6.2) are equivalent to the norms

$$\left(\|u\|^p_{H^{T+s,p}(G,q)} + \sum_{k,j:t_j+s-k+1-1/p<0} \langle\langle D_\nu^{k-1} u_j, \partial G, q\rangle\rangle^p_{t_j+s-k+1-1/p,p}\right)^{1/p} \tag{1.6.3}$$

and

$$\|u\|^p_{H^{T+s,p}(G,q)} + \left(\sum_{r,k:s-s_r-k+1-1/p<0} \langle\langle D_\nu^{k-1} l_r u, \partial G, q\rangle\rangle_{s-s_r-k+1-1/p,p}\right.$$

$$+ \sum_{j:s-s_j-1/p<0} \langle\langle b_j u, \partial G, q\rangle\rangle^p_{s-s_j-1/p,p}$$

$$\left. + \sum_{h:s-\sigma_h^c-1/p<0} \langle\langle c_h u, \partial G, q\rangle\rangle^p_{s-\sigma_h^c-1/p,p}\right)^{1/p}, \tag{1.6.4}$$

respectively.

The closure S of the mapping

$$u \mapsto \left(u\big|_{\overline{G}}, \left\{ D_\nu^{j-1} u_k \,\middle|\, j,k : t_j + s - k + 1 - 1/p < 0 \right\} \right) \quad \left(u \in \left(C^\infty(\overline{G})\right)^N \right)$$

is an isometry between the spaces $\widetilde{H}^{T+s,p,(\tau)}(G,q)$ with metric (1.6.3), and

$$\mathcal{H}^{T+s,p,(\tau)}(G,q) = H^{T+s,p}(G,q) \times \prod_{k,j: t_j+s-k+1-1/p<0} B^{t_j+s-k+1-1/p,p}(\partial G, q) \tag{1.6.5}$$

(see Subsection 1.1.2, and [R1, Ch.II]).

Analogously, the closure S_B of the mapping

$$u \mapsto \Big(u\big|_{\overline{G}}, \quad \left\{ D_\nu^{k-1} l_r u\big|_{\partial G} \,\middle|\, r,k : s - s_r - k + 1 - 1/p < 0 \right\},$$

$$\left\{ b_j u\big|_{\partial G} : s - \sigma_j - 1/p < 0 \right\}, \quad \left\{ c_h u\big|_{\partial G} : s - \sigma_h^c - 1/p < 0 \right\} \Big)$$

$$\left(u \in \left(C^\infty(\overline{G})\right)^N \right)$$

is an isometry between the space $\widetilde{H}^{T+s,p,(\tau)}(G,q)$ with metric (1.6.4), and

$$\mathcal{H}_B^{T+s,p,(\tau)}(G,q)$$

$$= H^{T+s,p}(G,q) \times \prod_{r,k: s-s_r-k+1-1/p<0} B^{s-s_r-k+1-1/p,p}(\partial G)$$

$$\times \prod_{j: s-\sigma_j-1/p<0} B^{s-\sigma_j-1/p,p}(\partial G) \times \prod_{h: s-\sigma_h^c-1/p<0} B^{s-\sigma_h^c-1/p,p}(\partial G). \tag{1.6.6}$$

This gives us a possibility to identify every element $u \in \widetilde{H}_B^{T+s,p,(\tau)}(G,q)$ with the element

$$S_B = \Big(u_0, \quad \{ u_{rk} : s - s_r - k + 1 - 1/p < 0 \}, \quad \{ u_j^b : s - \sigma_j - 1/p < 0 \},$$

$$\{ u_h^c : s - \sigma_h^c - 1/p < 0 \} \Big) \in \mathcal{H}_B^{T+s,p,(\tau)}(G,q). \tag{1.6.7}$$

For any element $u \in \widetilde{H}_B^{T+s,p,(\tau)}(G,q) \simeq \mathcal{H}_B^{T+s,p,(\tau)}(G,q)$, we write: $u = S_B u$.

For each sequence $u_j \in \left(C^\infty(\overline{G})\right)^N$ that converges to u in $\widetilde{H}_B^{T+s,p(\tau)}$ as $j \to \infty$, we have that the sequence $D_\nu^{k-1} l_r u_j\big|_{\partial G}$ converges to u_{rk} $(r,k : s - s_r - k + 1 - 1/p < 0)$ in $B^{s-s_r-k+1-1/p,p}(\partial G, q)$, the sequence $b_k u_j\big|_{\partial G}$

converges to u_k^b $(s-\sigma_j^b-1/p<0)$ in $B^{s-\sigma_k-1/p,p}(\partial G,q)$, the sequence $c_h u_j\big|_{\partial G}$ converges to u_h^c $(s-\sigma_h^c-1/p<0)$ in the space $B^{s-\sigma_h^c-1/p,p}(\partial G,q)$. In this sence,

$$u_{rk}=D_\nu^{k-1}l_r u\big|_{\partial G},\qquad u_k^b=b_k u\big|_{\partial G},\qquad u_h^c=c_h u\big|_{\partial G}.$$

Otherwise, if $s-s_r-k+1-1/p>0$, $u_j\in\left(C^\infty(\overline{G})\right)^N$, and $u_j\to u$ in $\widetilde{H}^{T+s,p,(\tau)}(G,q)$, then $D_\nu^{k-1}l_r u_j\big|_{\partial G}\to D_\nu^{k-1}l_r u_0\big|_{\partial G}$, if $s-\sigma_k-1/p>0$, then $b_k u_j\big|_{\partial G}\to b_k u_0\big|_{\partial G}$, and, if $s-\sigma_h^c-1/p>0$, then $c_h u_j\big|_{\partial G}\to c_h u_0\big|_{\partial G}$. In this way we obtain that the expressions $D_\nu^{k-1}l_r u\big|_{\partial G}$, $b_j u\big|_{\partial G}$, and $c_h u\big|_{\partial G}$ are defined for every element $u\in\mathcal{H}_B^{T+s,p,(\tau)}(G,q)\simeq\widetilde{H}^{T+s,p,(\tau)}(G,q)$.

Green's formula (1.5.42) and relations (1.5.13) easily imply that element $u=Su\in\widetilde{H}^{T+s,p,(\tau)}(G,q)$ is a generalized solution of the problem (1.5.1)–(1.5.2) if and only if

$$(u_0,l^+v)+\sum_{h=1}^{m}\langle c_h u,b_h' v\rangle=(f_0,v)+\sum_{h=1}^{m}\langle\varphi_h,c_h' v\rangle+\sum_{r=1}^{N}\sum_{k=1}^{-s_r+æ}\langle f_{rk},e_{kr}' v\rangle \tag{1.6.8}$$

$$\left(\forall v\in\left(C^\infty(\overline{G})\right)^N\right).$$

If $u=S_B u\in\mathcal{H}_B^{T+s,p,(\tau)}(G,q)\simeq\widetilde{H}^{T+s,p,(\tau)}(G,q)$, then equality (1.6.8) takes the form

$$\begin{aligned}(u_0,l^+v)&+\sum_{h:s-\sigma_h^c-1/p>0}\langle c_h u_0,b_h' v\rangle+\sum_{h:s-\sigma_h^c-1/p<0}\langle u_h^c,b_h' v\rangle\\ &=(f_0,v)+\sum_{h:s-\sigma_h-1/p>0}\langle\varphi_h,c_h' v\rangle+\sum_{h:s-\sigma_h-1/p<0}\langle u_h^b,c_h' v\rangle\\ &\quad+\sum_{r,k:s-s_r-k+1-1/p>0}\langle f_{rk},e_{kr}' v\rangle\\ &\quad+\sum_{r,k:s-s_r-k+1-1/p<0}\langle u_{rk},e_{kr}' v\rangle\quad\left(\forall v\in\left(C^\infty(\overline{G})\right)^N\right),\end{aligned} \tag{1.6.9}$$

where

$$\varphi_h=b_h u_0\big|_{\partial G}\qquad(s-\sigma_h-1/p>0),$$
$$f_{rk}=D_\nu^{k-1}l_r u_0\big|_{\partial G}\qquad(s-s_r-k+1-1/p>0).$$

1.6.3. It may be that the various elements of the space $\widetilde{H}_B^{T+s,p,(\tau)}(G,q)$ have the same components $(u_0,c_1u\big|_{\partial G},\dots,c_m u\big|_{\partial G})$. 'Pasting' these elements and doing the corresponding factorization in the space of images (see Section 1.3), we obtain new statement on isomorphisms from the isomorphism $\widetilde{H}_B^{T+s,p,(\tau)}(G,q) \to \widetilde{K}^{s,p}$. Since

$$E_1 = E_{1\,s,p} = \{S_B u : u_0 = 0,\ u_h^c = 0\,(h : s - \sigma_h^c - 1/p < 0)\}$$
$$\subset \mathcal{H}_B^{T+s,p,(\tau)}(G,q),$$

formula (1.6.6) easily implies that, after this pasting, the space of preimages

$$H^{T+s,p}(G,q) \times \prod_{h:s-\sigma_h^c-1/p<0} B^{s-\sigma_h^c-1/p,p}(\partial G, q) \tag{1.6.10}$$

is a completion of $\left(C^\infty(\overline{G})\right)^N$ in the norm

$$\left(\|u, G, q\|_{T+s,p}^p + \sum_{h=1}^{m} \langle\langle c_h u, \partial G, q\rangle\rangle_{s-\sigma_h^c-1/p,p}^p \right)^{1/p},$$

and the space of images (see (1.5.12) and (1.6.8)) is the space

$$K_{s,p}^1 := \left(\prod_{j=1}^{N} H^{s-s_j,p}(G,q) \times \prod_{h:s-\sigma_h-1/p>0} B^{s-\sigma_h-1/p,p}(\partial G, q) \right) \Big/ M_{s,p}^1, \tag{1.6.11}$$

where

$$M_{s,p}^1 = \Bigg\{ \Big(f_0, \{\varphi_h\} \Big) \in \prod_{j=1}^{N} H^{s-s_j,p}(G) \times \prod_{h:s-\sigma_h-1/p>0} B^{s-\sigma_h-1/p,p}(\partial G) :$$
$$(f_0, v) + \sum_{h:s-\sigma_h-1/p>0} \langle \varphi_h, c_h' v\rangle + \sum_{k,r:s-s_r-k+1-1/p>0} \langle D_\nu^{k-1} f_r, e_{kr}' v\rangle = 0,$$
$$\forall v \in \left(C^\infty(\overline{G})\right)^N : c_h' v\big|_{\partial G} = 0, \quad e_{kr}' v|_{\partial G} = 0,$$
$$h : s - \sigma_h^c - 1/p < 0, \quad k, r : s - s_r - k + 1 - 1/p < 0 \Bigg\}$$

is a subspace of the direct product

$$\prod_{j=1}^{N} H^{s-s_j,p}(G,q) \times \prod_{h:s-\sigma_h-1/p>0} B^{s-\sigma_h-1/p,p}(\partial G, q).$$

As a result, we obtain the following statement.

Theorem 1.6.1. *Under the conditions of Theorem 1.5.1, let $q \geq q_0$. Then the closure $A_1 = A_{1\,s,p}$ $(s \in \mathbf{R},\, p \in]1,\infty[)$ of the mapping*

$$u \mapsto \left(lu,\ \{b_h u|_{\partial G} : s - \sigma_h - 1/p > 0\}\right) \qquad \left(u \in (C^\infty(\overline{G}))^N\right)$$

realizes an isomorphism from space (1.6.10) into space (1.6.11).

Corollary 1.6.1. *There holds the estimate*

$$\|u, G, q\|_{T+s,p} + \sum_{j=1}^{m} \langle\langle c_j u, \partial G, q\rangle\rangle_{s-\sigma_j^c-1/p,p}$$

$$\leq c_1\left(|||lu, G, q|||_{s-S,p} + \sum_{h:s-\sigma_h-1/p>0} \langle\langle b_h u, \partial G, q\rangle\rangle_{s-\sigma_h-1/p,p}\right) \tag{1.6.12}$$

$$\left(\forall u \in (C^\infty(\overline{G}))^N,\quad q \geq q_0\right).$$

Here the constant c_1 does not depend on u and q.

1.6.4. The various elements of the space $\widetilde{H}^{T+s,p,(\tau)}(G,q)$ can have the same components

$$(u_0, c_1 u|_{\partial G}, \ldots, c_m u|_{\partial G}, b_1 u|_{\partial G}, \ldots, b_m u|_{\partial G}).$$

'Pasting' them and pasting the corresponding elements in the space of images, we obtain new theorem on isomorphisms.
We have now that (see (1.6.7))

$$E_1 = E_{1\,s,p} = \Big\{S_B u : u_0 = 0,\quad u_h^c = 0\ (h : s - \sigma_h^c - 1/p < 0),$$

$$u_j^b = 0\ (j : s - \sigma_j - 1/p < 0)\Big\} \subset \mathcal{H}_B^{T+s,p,(\tau)}(G,q).$$

Denote by $\widetilde{H}_{c,b}^{T+s,p,(\tau)}(G,q)$ a completion of $(C^\infty(\overline{G}))^N$ in the norm

$$|||u, G, q|||_{T+s,p,(c,b)} := \left(\|u, G, q\|_{T+s,p}^p + \sum_{j=1}^{m} \langle\langle c_j u, \partial G, q\rangle\rangle_{s-\sigma_j^c-1/p,p}^p\right.$$

$$\left. + \sum_{j=1}^{m} \langle\langle b_j u, \partial G, q\rangle\rangle_{s-\sigma_j-1/p,p}^p\right)^{1/p}. \tag{1.6.13}$$

The considerations of Subsection 1.6.2 imply that the space $\widetilde{H}_{c,b}^{T+s,p,(\tau)}(G,q)$ is the space of pre-images obtained by this pasting; it is isomorphic to the direct product

$$\begin{aligned}\widetilde{H}_{c,b}^{T+s,p,(\tau)}(G,q) &\simeq H^{T+s,p} \times \prod_{h:s-\sigma_h^c-1/p<0} B^{s-\sigma_h^c-1/p,p}(\partial G,q) \\ &\quad \times \prod_{h:s-\sigma_h-1/p<0} B^{s-\sigma_h-1/p,p}(\partial G,q) \\ &\simeq \mathcal{H}_B^{T+s,p,(\tau)}(G,q)\Big/E_{1\,s,p}; \end{aligned} \tag{1.6.14}$$

The space of images is the quotient space

$$\left(H^{s-S,p}(G,q)\times\prod_{h=1}^{m} B^{s-\sigma_h-1/p,p}(\partial G,q)\right)\Big/ M_{s,p}^2 := K_{s,p}^2, \tag{1.6.15}$$

where $H^{s-S,p}(G,q) := \prod_{j=1}^{N} H^{s-s_j,p}(G,q)$, and

$$M_{s,p}^2 = \Big\{(f_0,\varphi_1,\dots,\varphi_m) \in H^{s-S,p}(G,q)\times\prod_{h=1}^{m} B^{s-\sigma_h-1/p,p}(\partial G,q) :$$

$$(f_0,v)+\sum_{h=1}^{m}\langle\varphi_h, c_h' v\rangle + \sum_{k,r:s-s_r-k+1-1/p>0}\langle D_\nu^{k-1} f_{r0}, e_{kr}' v\rangle = 0$$

$$\forall v\in C^\infty(\overline{G}))^N : e_{kr}' v|_{\partial G} = 0, (k,r: s-s_r-k+1-1/p<0)\Big\}. \tag{1.6.16}$$

Thus, we establish the validity of the following statement.

Theorem 1.6.2. *Under the conditions of Theorem 1.5.1, let $q \geq q_0$. Then the closure*

$$A_2 = A_{2\,s,p} \qquad (s\in\mathbf{R},\ p\in]1,\infty[)$$

of the mapping

$$u \mapsto (lu, bu) \qquad \left(u\in(C^\infty(\overline{G}))^N\right)$$

realizes an isomorphism

$$\widetilde{H}_{c,b}^{T+s,p,(\tau)}(G,q) \to K_{s,p}^2. \tag{1.6.17}$$

Corollary 1.6.2. *There holds the estimate*

$$|||u,G,q|||_{T+s,p,(c,b)} \leq c\Big(\|lu,G,q\|_{s-S,p} + \sum_{j=1}^{m} \langle\langle b_j u, \partial G, q\rangle\rangle_{s-\sigma_j-1/p,p}\Big) \tag{1.6.18}$$

$$\Big(u \in \left(C^\infty(\overline{G})\right)^N\Big)$$

with the constant $c > 0$ independent of u and q $(q \geq q_0)$.

1.6.5. It is possible that the various elements of the space $\widetilde{H}_B^{T+s,p,(\tau)}(G,q)$ have the same components $(u_0, b_1u|_{\partial G}, \dots, b_m u|_{\partial G})$. 'Paste' them and make a corresponding factorization in the space of images.

Denote by $\widetilde{H}_b^{T+s,p,(\tau)}$ a completion of $\left(C^\infty(\overline{G})\right)^N$ in the norm

$$|||u,G,q|||_{T+s,p,(b)} = \left(\|u,G,q\|^p_{T+s,p} + \sum_{j=1}^{m} \langle\langle b_j u, \partial G, q\rangle\rangle^p_{s-\sigma_j-1/p,p}\right)^{1/p}. \tag{1.6.19}$$

It is clear (see Subsection 1.6.2) that $\widetilde{H}_b^{T+s,p,(\tau)}$ is the space of pre-images; it is isomorphic to the direct product

$$\widetilde{H}_b^{T+s,p,(\tau)}(G,q) \simeq H^{T+s,p}(G,q) \times \prod_{h:s-\sigma_h-1/p<0} B^{s-\sigma_h-1/p,p}(\partial G,q). \tag{1.6.20}$$

The space of images is described by following:

$$\Big(H^{s-S,p}(G,q) \times \prod_{h=1}^{m} B^{s-\sigma_h-1/p,p}(\partial G,q)\Big)\Big/ M^3_{s,p} = K^3_{s,p}, \tag{1.6.21}$$

where

$$M^3_{s,p} = \Big\{(f,\varphi_1,\dots,\varphi_m) \in H^{s-S,p}(G,q) \times \prod_{h=1}^{m} B^{s-\sigma_h-1/p,p}(\partial G) :$$

$$(f,v) + \sum_{h=1}^{m} \langle\varphi_h, c'_h v\rangle + \sum_{k,r:s-s_r-k+1-1/p>0} \langle D_\nu^{k-1} f_r, e'_{kr} v\rangle = 0,$$

$$\forall v \in \left(C^\infty(\overline{G})\right)^N : e'_{kr} v|_{\partial G} = 0 \qquad (k,r:\ s-s_r-k+1-1/p<0),$$

$$b'_h v|_{\partial G} = 0 \qquad (h = 1,\dots,m)\Big\} \tag{1.6.22}$$

is a subspace of the direct product

$$H^{s-S,p}(G,q) \times \prod_{h=1}^{m} B^{s-\sigma_h-1/p,p}(\partial G,q).$$

Since, for $q \geq q_0$, the matrix $\begin{pmatrix} c'(x,D,q) \\ b'(x,D,q) \\ e'(x,D,q) \end{pmatrix}$ is a τ'-complete matrix, it follows from (1.6.22) that

$$M^3_{s,p} = \Big\{ (f,\varphi) \in H^{s-S,p}(G,q) \times \prod_{j=1}^{m} B^{s-\sigma_j-1/p,p}(\partial G,q) :$$

$$\varphi = 0,\ D_\nu^{k-1} f_r\big|_{\partial G} = 0\ (k,r:\ s - s_r - k + 1 - 1/p > 0),\ (f,v) = 0$$

$$\forall v \in \left(C^\infty(\overline{G})\right)^N :\ e'_{kr} v = 0\ (k,r: s - s_r - k + 1 - 1/p < 0),$$

$$b'_h v|_{\partial G} = 0\ (h = 1,\dots,m) \Big\}. \tag{1.6.23}$$

Denote by $\widetilde{M}^3_{s,p}$ a subspace of the space $H^{s-S,p}(G,q)$, which consists of elements f such that $(f,0,\dots,0) \in M^3_{s,p}$. Then,

$$K^3_{s,p} = H^{s-S,p}(G,q) \Big/ \widetilde{M}^3_{s,p} \times \prod_{j=1}^{m} B^{s-\sigma_j-1/p,p}(\partial G,q). \tag{1.6.24}$$

Theorem 1.6.3. *Under the assumption of Theorem 1.5.1, let $q \geq q_0$. Then, the closure $A_3 = A_{3\,s,p}$ $(s \in \mathbf{R},\ p \in]1,\infty[)$ of the mapping*

$$u \mapsto (lu, bu) \qquad \left(u \in (C^\infty(\overline{G}))^N\right)$$

realizes an isomorphism

$$\widetilde{H}_b^{T+s,p,(\tau)}(G,q) \to K^3_{s,p}. \tag{1.6.25}$$

Corollary 1.6.3. *There holds the estimate*

$$|||u,G,q|||_{T+s,p,(b)} \leq c\Big(\|lu,G,q\|_{s-S,p} + \sum_{j=1}^{m} \langle\langle b_j u, \partial G, q\rangle\rangle_{s-\sigma_j-1/p,p} \Big), \tag{1.6.26}$$

with the constant $c > 0$ independent of u and q $\ (|q| \geq q_0)$.

1.6.6. Let

$$C^\infty_{(b)} = \left\{ u \in \left(C^\infty(\overline{G})\right)^N : bu\big|_{\partial G} = 0 \right\}.$$

Denote by

$$H^{T+s}_{(b)} = \left\{ u \in \widetilde{H}^{T+s,p,(\tau)}_b(G,q) : bu\big|_{\partial G} = 0 \right\}$$

$$= \left\{ u \in H^{T+s,p}(G) : b_h u = 0 \ (h : s - \sigma_h - 1/p > 0) \right\}$$

a subspace of $\widetilde{H}^{T+s,p,(\tau)}_b(G,q)$. The space $H^{T+s,p}_{(b)}$ is the closure of $C^\infty_{(b)}$ in the space $H^{T+s,p}(G,q)$. Theorem 1.6.3 directly implies the validity of the following assertion.

Theorem 1.6.4. *Under the condition of Theorem 1.5.1, let* $q \geq q_0$. *Then, the closure* $A_{3(b)}$ *of the mapping*

$$u \mapsto lu \qquad \left(u \in C^\infty_{(b)}\right)$$

realizes an isomorphism

$$H^{T+s,p}_{(b)}(G,q) \to H^{s-S,p}(G,q) \Big/ M^3_{s,p}. \tag{1.6.27}$$

In the special case of one equation with normal boundary conditions, isomorphism (1.6.27) has been established in [R4].

Theorem 1.6.4 directly implies the validity of the estimate

$$\|u, G, q\|_{T+s,p} \leq c\|lu\|_{H^{s-S,p}(G,q)} \qquad \left(u \in C^\infty_{(b)}\right) \tag{1.6.28}$$

with the constant $c > 0$ independent of u and q $(|q| \geq q_0)$.

Estimate (1.6.28) yields the estimate of the resolvent which is important for the spectral theory.

1.6.7. Let us show an example illustrated the using of the graph method. Consider isomorphism (1.6.25) and denote by $Y_{s,p}$ a functional Banach space such that

$$\left(C^\infty(\overline{G})\right)^N \subset Y_{s,p} \subset H^{s-S,p}(G,q) \Big/ \widetilde{M}^3_{s,p},$$

and $\left(C^\infty(\overline{G})\right)^N$ is dense in $Y_{s,p}$. For example, let

$$Y_{s,p} = \prod_{r=1}^{N} H^{s-s_r+t,p}(G,q) \Big/ \widetilde{M}^3_{s+t,p}, \quad t > 0.$$

Then, the set

$$Q_{s,p} = (A_{3\,s,p})^{-1}\left(Y_{s,p} \times \prod_{h=1}^{m} B^{s-\sigma_h-1/p,p}(\partial G, q)\right)$$

is a linear, generally speaking, nonclosed subset of the set $\widetilde{H}^{T+s,p,(\tau)}(G,q)$. However, in the graph morm

$$\|u\|_{\widetilde{H}^{T+s,p,(\tau)}(G,q)} + \|lu\|_{Y_{s,p}} = \|u\|_{Q_{s,p}}$$

the space $Q_{s,p}$ is a complete Banach space, and the operator $A_{3\,s,p}$ naturally establishes an isomorphism

$$Q_{s,p} \to Y_{s,p} \times \prod_{h=1}^{m} B^{s-\sigma_h-1/p,p}(\partial G, q). \tag{1.6.29}$$

In the special case of one equation with normal boundary conditions, isomorphism (1.6.29) has been obtained by Lions and Magenes for the case where $s - s_N < 0$ and $Y_{s,p} = L_p(G)$ (see [R13]).

1.7. Cauchy Problem and Calderón Projections for General Parameter Elliptic Systems

1.7.1. Let $\Omega \subset \mathbf{R}^n$ be a domain with the boundary $\partial\Omega \in C^\infty$. We consider only the cases where either $\Omega = G$ is a bounded domain, $\partial\Omega = \partial G$, or

$$\Omega = \mathbf{R}^n_\pm = \{(x', x_n) \in \mathbf{R}^n : x_n \gtrless 0\},$$

$$\partial\Omega = \mathbf{R}'_{n-1} = \{(x', x_n) \in \mathbf{R}^n : x_n = 0\}.$$

In Ω we consider a parameter elliptic boundary value problem

$$l(x, D, q_1)u(x) = f(x), \qquad x \in \Omega, \tag{1.7.1}$$

$$b(x, D, q_1)u(x) = \varphi(x), \qquad x \in \partial\Omega, \tag{1.7.2}$$

for the system of order $(T, S) = (t_1, \dots, t_N, s_1, \dots, s_N)$ of Douglis–Nirenberg structure (see 1.5.1). Here $q_1 = qe^{i\theta}$, $q \in \mathbf{R}$, $\theta \in [\theta_1, \theta_2]$ (the case where $\theta_1 = \theta = \theta_2$ is not excluded),

$$l(x, D, q_1) = \left(l_{rj}(x, D, q_1)\right)_{r,j=1,\dots,N}, b(x, D, q_1) = \left(b_{hj}(x, D, q_1)\right)_{\substack{h=1,\dots,m\\ j=1,\dots,N}},$$

the expressions l_{rj} and b_{hj} are defined by relations (1.5.3) and (1.5.4). Suppose that the coefficients $a^{rj}_{\mu k}$ and $b^{hj}_{\alpha k}$ are infinitely smooth complex-valued functions whose all derivatives are bounded.

We complement the matrix $b(x, D, q_1)$ to τ-complete matrix (1.5.27). The matrix $e(x, D, q_1)$ is expressed by formula (1.5.26). The matrix $c(x, D, q_1)$ (and, hence, the matrixes c' and b' from Green's formula (1.5.28)) is not uniquely defined. Therefore the following question arises:
Is it possible to choose the matrix $c(x, D, q_1)$ with $\sigma_h^c < 0$ so that the matrix (1.2.24) is τ-complete and the problem

$$l(x, D, q_1)u(x) = f(x), \qquad x \in \Omega, \tag{1.7.3}$$

$$c(x, D, q_1)u(x) = \psi(x), \qquad x \in \partial\Omega \tag{1.7.4}$$

is also elliptic with a parameter?
The positive answer is contained in the following lemma, the proof of this assertion will be given below in 1.7.4

Lemma 1.7.1. *Let problem (1.7.1)–(1.7.2) be parameter elliptic. Then there exists the matrix $c(x, D, q_1)$ with $\sigma_h^c < 0$ such that the matrix (1.5.27) is τ-complete and the problem (1.7.3)–(1.7.4) is parameter elliptic.*

In what follows we will choose the matrix $c(x, D, q_1)$ so that the statement of this Lemma holds. Thus, if problem (1.7.1)–(1.7.2) is parameter elliptic, then problem (1.7.3)–(1.7.4) is parameter elliptic, and the problem

$$E(x, D_x, q)\, U = \begin{pmatrix} e(x, D, q_1) \\ b(x, D, q_1) \\ c(x, D, q_1) \end{pmatrix} U = \Phi \tag{1.7.5}$$

is parameter elliptic on $\partial\Omega$. Therefore there exist a number $q_0 > 0$ such that for $q \geq q_0$ all these problems are unique solvable and the correspondent operators isomorphically map the spaces of solutions onto correspondent spaces of the right-hand sides.

Let

$$\begin{gathered} æ = \max\{0, \sigma_1 + 1, \ldots, \sigma_m + 1\}, \quad \tau_j = t_j + æ, \\ \tau = (\tau_1, \ldots, \tau_N), \quad |\tau| = \tau_1 + \cdots + \tau_N. \end{gathered} \tag{1.7.6}$$

For any $s \in \mathbf{R}$ and any $p \in (1, +\infty)$ the closure $A_B = A_{B,s,p}$ of the mapping

$$u \mapsto (lu|_{\overline{\Omega}}, \{D_\nu^{k-1} l_j u|_{\partial\Omega} : j = 1, \ldots, N, k = 1, \ldots, æ - s_j\}, bu|_{\partial\Omega}),$$

$$u \in (C_0^\infty(\overline{\Omega}))^N, \; l_j u = \sum_{k=1}^{N} l_{jk} u_k,$$

acts continuously in the pair of spaces

$$A_B : \widetilde{H}^{T+s,p,(\tau)}(\Omega, q) \to K^{s,p,(\sigma)}(\Omega, q), \tag{1.7.7}$$

where

$$\widetilde{H}^{T+s,p,(\tau)}(\Omega, q) := \prod_{j=1}^{N} \widetilde{H}^{t_j+s,p,(\tau_j)}(\Omega, q),$$

$$K^{s,p,(\sigma)}(\Omega, q) \quad := \prod_{k=1}^{N} \widetilde{H}^{s-s_r,p,(\text{æ}-s_r)}(\Omega, q) \times \prod_{h=1}^{m} B^{s-\sigma_h-1/p,p}(\partial\Omega, q).$$

In addition

$$\|A_B u\|_{K^{s,p,(\sigma)}(\Omega,q)} \leq c\|u\|_{\widetilde{H}^{T+s,p,(\tau)}(\Omega,q)}, \tag{1.7.8}$$

where the constant $c > 0$ does not depend on u and q.

Similarly, the closure $A_C = A_{C,s,p}$ of the mapping

$$u \mapsto (lu|_{\Omega}, \{D_\nu^{k-1} l_j u|_{\partial\Omega} : j = 1, \ldots, N,\ k = 1, \ldots, \text{æ} - s_j\}, cu|_{\partial\Omega}),$$

$$u \in (C_0^\infty(\overline{\Omega}))^N,$$

acts continuously in the pair of spaces

$$A_C : \widetilde{H}^{T+s,p,(\tau)}(\Omega, q) \to K^{s,p,(\sigma^c)}(\Omega, q), \tag{1.7.9}$$

where

$$K^{s,p,(\sigma^c)}(\Omega, q) := \prod_{k=1}^{N} \widetilde{H}^{s-s_r,p,(\text{æ}-s_r)}(\Omega, q) \times \prod_{h=1}^{m} B^{s-\sigma_h^c-1/p,p}(\partial\Omega, q).$$

In addition

$$\|A_C u\|_{K^{s,p,(\sigma^c)}(\Omega,q)} \leq c\|u\|_{\widetilde{H}^{T+s,p,(\tau)}(\Omega,q)}, \tag{1.7.10}$$

where the constant $c > 0$ does not depend on u and q. Thus, the following theorem is true.

Theorem 1.7.1. *Let problem (1.7.1)–(1.7.2) be elliptic with a parameter, and let $s \in \mathbf{R}$, $p \in]1, \infty[$. Then there exists a number $q_0 > 0$ such that for $q \geq q_0$ the operator A_B (A_C) realizes an isomorphism between the spaces (1.7.7) ((1.7.9)). There exists a constant $c > 0$, independent of u and q ($q \geq q_0$), such that the estimates*

$$\begin{aligned} c^{-1}\|u\|_{\widetilde{H}^{T+s,p,(\tau)}(\Omega,q)} &\leq \|A_B u\|_{K^{s,p,(\sigma)}(\Omega,q)} \leq c\|u\|_{\widetilde{H}^{T+s,p,(\tau)}(\Omega,q)} \\ c^{-1}\|u\|_{\widetilde{H}^{T+s,p,(\tau)}(\Omega,q)} &\leq \|A_C u\|_{K^{s,p,(\sigma^c)}(\Omega,q)} \leq c\|u\|_{\widetilde{H}^{T+s,p,(\tau)}(\Omega,q)} \end{aligned} \tag{1.7.11}$$

hold.

The mapping

$$E(x, D_x, q) : U \mapsto \Phi, \quad q \geq q_0$$

is an isomorphism acting in the pair of spaces

$$B^{T+s,p}(\partial\Omega, q) \to B^{s-S,\sigma,\sigma^c,p}(\partial\Omega, q), \tag{1.7.12}$$

where

$$B^{T+s,p}(\partial\Omega, q) := \prod_{j=1}^{N} \prod_{k=1}^{\tau_j} B^{t_j+s-k+1-1/p,p}(\partial\Omega, q),$$

$$B^{s-S,\sigma,\sigma^c,p}(\partial\Omega, q) := \prod_{r:\text{æ}-s_r\geq 1} \prod_{j=1}^{\text{æ}-s_r} B^{s-s_r-j+1-1/p,p}(\partial\Omega, q)\times$$

$$\prod_{h=1}^{m} B^{s-\sigma_h-1/p,p}(\partial\Omega, q) \times \prod_{h=1}^{m} B^{s-\sigma_h^c-1/p,p}(\partial\Omega, q).$$

In addition, there exists a constant $c > 0$ independet of u and q ($q \geq q_0$) such that the estimates

$$c^{-1}\|U\|_{B^{T+s,p}(\partial\Omega,q)} \leq \|E(x, D_x, q)U\|_{B^{s-S,\sigma,\sigma^c,p}(\partial\Omega,q)} \leq c\|U\|_{B^{T+s,p}(\partial\Omega,q)} \tag{1.7.13}$$

hold.

1.7.2. Recall that if the element

$$u = (u_1, \ldots, u_N) \in \widetilde{H}^{T+s,p,(\tau)}(\Omega, q),$$

with

$$u_j = (u_{j0}, \ldots, u_{j\tau_j}) \in \widetilde{H}^{t_j+s,p,(\tau_j)}(\Omega, q),$$

is a solution of problem (1.7.1)–(1.7.2),

$$cu|_{\partial\Omega} = \psi, \quad e(x, D, q)u|_{\partial\Omega} = \eta,$$

and if

$$U = \{u_{j1}, \ldots, u_{j\tau_j} : j = 1, \ldots, N\}$$

is a vector of the Cauchy data, then the relation

$$E(x, D_x, q)U = \begin{pmatrix} \eta \\ \varphi \\ \psi \end{pmatrix} = \Phi \in B^{s-S,\sigma,\sigma^c,p}(\partial\Omega, q) \tag{1.7.14}$$

is true.

Inversely, if the element $U \in B^{T+s,p}(\partial\Omega, q)$ is a solution of problem (1.7.14), and if

$$(u_0, U) = (u_1, \ldots, u_N), \quad u_j = (u_{j0}, \ldots, u_{j\tau_j}) \in \widetilde{H}^{t_j+s,p,(\tau_j)}(\Omega, q),$$

then the element (u_0, U) is both a solution of problem (1.7.1)–(1.7.2) with $(f, \varphi) \in K^{s,p,(\sigma)}(\Omega, q)$, and a solution of problem (1.7.3)–(1.7.4) with $(f, \psi) \in K^{s,p,(\sigma_c)}(\Omega, q)$.

Thus, the vector Φ uniquely defines the Cauchy data vector

$$U(E(x, D, q))^{-1}\Phi \in B^{T+s,p}(\partial\Omega, q).$$

For the solvability of the Cauchy problem

$$lu = f \quad \text{in } \Omega, \qquad D_\nu^{k-1}u_j\big|_{\partial\Omega} = u_{jk}, \quad j = 1, \ldots, N, \; k = 1, \ldots, \tau_j,$$

in $\widetilde{H}^{T+s,p,(\tau)}(\Omega, q)$ it is necessary and sufficient that there exists a certain relation between the vectors φ and ψ.

In fact, if $bU = \varphi$, then for $q \geq q_0$ the problem

$$lu = f \quad \text{in } \Omega, \qquad bu\big|_{\partial\Omega} = \varphi$$

has the unique solution

$$u = A_B^{-1}(f, \varphi) \in \widetilde{H}^{T+s,p,(\tau)}(\Omega, q)$$

satisfying the equation

$$\psi = cu\big|_{\partial\Omega} = cA_B^{-1}(f, \varphi).$$

For the fixed function f the mapping $\Lambda : \varphi \mapsto \psi$ is a mutually continuous one-to-one correspondence. The inverse mapping is given by the formula

$$\varphi = bA_c^{-1}(f, \psi).$$

The operator $\Lambda : \varphi \mapsto \psi$ is an isomorphism from

$$B^{s-\sigma-1/p,p}(\partial\Omega, q) := \prod_{h=1}^{m} B^{s-\sigma_h-1/p,p}(\partial\Omega, q)$$

onto

$$B^{s-\sigma^c-1/p,p}(\partial\Omega, q) := \prod_{h=1}^{m} B^{s-\sigma_h^c-1/p,p}(\partial\Omega, q).$$

In addition there holds the estimate

$$\|\Lambda\varphi\|_{B^{s-\sigma^c-1/p,p}(\partial\Omega,q)} \le c\|\varphi\|_{B^{s-\sigma-1/p,p}(\partial\Omega,q)},$$

where $c > 0$ does not depend both on φ and q. The operator Λ may be given by matrix $(\Lambda_{jk})_{j,k=1,\dots,m}$. Then $\Lambda_{jk} : \varphi_k \mapsto \psi_j$.

1.7.3.
For any element

$$u^\pm \in \widetilde{H}^{T+s,p,(\tau)}(\mathbf{R}^n_\pm, q)\big(s \in \mathbf{R},\ p \in]1,\infty[\big),$$

we denote by $U^\pm$ the vector of the Cauchy data of the element $u^\pm$ for $x_n = 0$:

$$U^\pm = \Big\{u_{kj} : k = 1,\dots,N,\ j = 1,\dots,\tau_j,\ u^\pm_{kj} = D_n u^\pm_k\big|_{x_n=0}\Big\} \in$$

$$\prod_{k=1}^{N}\prod_{j=1}^{\tau_k} B^{t_k+s-j+1-1/p,p}(\mathbf{R}'_{n-1}, q) =: B^{T+s,p}(\mathbf{R}'_{n-1}, q).$$

Let $\mathcal{U}_{s,p}(\mathbf{R}'_{n-1}, q)$ denote the subspace of the space $B^{T+s,p}(\mathbf{R}'_{n-1}, q)$ consisting of the elements U such that $e(x, D_x, q)U = 0$.

In $\mathbf{R}^n_\pm$ we consider the Cauchy problem

$$l(x, D_x)u^\pm = 0, \qquad x \in \mathbf{R}^n_\pm, \tag{1.7.15}$$

$$D_n^{j-1}u^\pm_k\big|_{x_n=0} = u_{kj} \in B^{t_k+s-j+1-1/p,p}(\mathbf{R}'_{n-1}, q), \tag{1.7.16}$$
$$k = 1,\dots,N,\ j = 1,\dots,\tau_j.$$

Let us establish the solvability conditions of this problem in the space $\widetilde{H}^{T+s,p,(\tau)}(\mathbf{R}^n_\pm, q)$, $s \in \mathbf{R}$, $1 < p < \infty$.

Let

$$u^\pm \in \widetilde{H}^{T+s,p,(\tau)}(\mathbf{R}^n_\pm, q)$$

be a solution of the Cauchy problem (1.7.15)–(1.7.16). According to the vector $U^\pm$ of the Cauchy data we find that

$$e(x, D, q)u^\pm\big|_{x_n=0} = 0,$$

$$b(x, D, q)u^\pm\big|_{x_n=0} = \varphi^\pm \in \prod_{h=1}^{m} B^{æ+s-\sigma_h-1/p,p}(\mathbf{R}'_{n-1}, q),$$

$$c^\pm(x, D, q)u^\pm\big|_{x_n=0} = \psi^\pm \in \prod_{h=1}^{m} B^{æ+s-\sigma^{c\pm}_h-1/p,p}(\mathbf{R}'_{n-1}, q),$$

or, in short,

$$E^{\pm}(x,D,q)U^{\pm}=\Phi^{\pm}=\begin{pmatrix}0\\ \varphi^{\pm}\\ \psi^{\pm}\end{pmatrix}. \tag{1.7.17}$$

By virtue of Theorem 1.7.1, for $q\geq q_0$ the operator

$$E^{\pm}(x,D,q): U^{\pm}\mapsto\Phi^{\pm}$$

is a mutually continuous one-to-one mapping from the subspace

$$\mathcal{U}^{\pm}_{s,p}(\mathbf{R}'_{n-1},q)=\left\{U^{\pm}\in\prod_{k=1}^{N}\prod_{j=1}^{\tau_j}B^{t_k+s-j+1-1/p,p}(\mathbf{R}'_{n-1},q): e(x,D,q)U^{\pm}=0\right\} \tag{1.7.18}$$

of the space

$$\prod_{k=1}^{N}\prod_{j=1}^{\tau_j}B^{t_k+s-j+1-1/p,p}(\mathbf{R}'_{n-1},q)$$

onto the space

$$\prod_{h=1}^{m}B^{æ+s-\sigma_h-1/p,p}(\mathbf{R}'_{n-1},q)\times\prod_{h=1}^{m}B^{æ+s-\sigma_h^c-1/p,p}(\mathbf{R}'_{n-1},q).$$

Furthermore, $U^{\pm}\in\mathcal{U}^{\pm}_{s,p}$ if and only if the relations (1.7.17) hold. It is clear that $\mathcal{U}^{\pm}_{s,p}(\mathbf{R}'_{n-1},q)$ is a closed subspace of the space $\mathcal{U}_{s,p}(\mathbf{R}'_{n-1},q)$.

It is easy to see that

$$\mathcal{U}^{+}_{s,p}(\mathbf{R}'_{n-1},q)\bigcap\mathcal{U}^{-}_{s,p}(\mathbf{R}'_{n-1},q)=\{0\} \tag{1.7.19}$$

In fact, if

$$U\in\mathcal{U}^{+}_{s,p}(\mathbf{R}'_{n-1},q)\bigcap\mathcal{U}^{-}_{s,p}(\mathbf{R}'_{n-1},q),$$

then there exists an element $u^{\pm}\in\widetilde{H}^{T+s,p,(\tau)}(\mathbf{R}^n_{\pm},q)$ with the same Cauchy data U. Then the element

$$u=\begin{cases}u^{+}, & x_n>0\\ u^{-}, & x_n<0\end{cases}$$

belongs to the space $H^{T+s,p}(\mathbf{R}^n,q)$, and $lu=0$. Therefore, $u=0$ (see the end of Section 1.5), and, hence, $U=0$. This proves relation (1.7.19).

It follows from (1.7.19) that

$$\mathcal{U}^{+}_{s,p}(\mathbf{R}'_{n-1},q)\dot{+}\mathcal{U}^{-}_{s,p}(\mathbf{R}'_{n-1},q)\subset\mathcal{U}_{s,p}(\mathbf{R}'_{n-1},q).$$

Theorem 1.7.2. *Let problem (1.7.1)–(1.7.2) be an elliptic with a parameter. Then there exists a number $q_0 > 0$ such that the equality*

$$\mathcal{U}_{s,p}(\mathbf{R}'_{n-1}, q) = \mathcal{U}^+_{s,p}(\mathbf{R}'_{n-1}, q) \dot{+} \mathcal{U}^-_{s,p}(\mathbf{R}'_{n-1}, q), \quad s \in \mathbf{R},\ 1 < p < \infty, \tag{1.7.20}$$

is valid for $q \geq q_0$. The norms of the progection operators

$$P^{\pm} : U \mapsto U^{\pm} : \mathcal{U}_{s,p}(\mathbf{R}'_{n-1}, q) \to \mathcal{U}^{\pm}_{s,p}(\mathbf{R}'_{n-1}, q)$$

are bounded by a constant independent of q $(q \geq q_0)$.

Proof. Let $U \in \mathcal{U}_{s,p}(\mathbf{R}'_{n-1}, q)$. Under the element U we obtain the elements

$$c^{\pm}(x.D.q)U = \psi^{\pm} \in \prod_{h=1}^{m} B^{æ+s-\sigma_h^{\pm}-1/p,p}(\mathbf{R}'_{n-1}, q).$$

Since, by virtue of Lemma 1.7.1, the problem

$$lu = 0,\ x \in \mathbf{R}^n_{\pm}, \qquad c^{\pm}(x, D, q)u\big|_{x_n=0} = \psi^{\pm} \tag{1.7.21}$$

is elliptic with a parameter, there exist a constant $q_0 > 0$ such that for $q \geq q_0$ problem (1.7.21) has one and only one solution $u^{\pm} \in \widetilde{H}^{T+s,p,(\tau)}(\mathbf{R}^n_{\pm}, q)$. Under the element $u^{\pm}$ we construct the Cauchy data vector $U^{\pm}$ satisfying the equation (1.7.17). It is clear that $U^{\pm} \in \mathcal{U}^{\pm}_{s,p}(\mathbf{R}'_{n-1}, q)$, and, hence,

$$U^+ + U^- \in \mathcal{U}^+_{s,p}(\mathbf{R}'_{n-1}, q) \dot{+} \mathcal{U}^-_{s,p}(\mathbf{R}'_{n-1}, q).$$

Let us show that $U = U^+ + U^-$. It is clear that the element U satisfies the system

$$E_C(x, D_x, q)U := \begin{pmatrix} e(x, D, q) \\ c^+(x, D, q) \\ c^-(x, D, q) \end{pmatrix} U = \begin{pmatrix} 0 \\ \psi^+ \\ \psi^- \end{pmatrix} = \Psi, \tag{1.7.22}$$

where E_C is $|\tau| \times |\tau|$-matrix of Douglis-Nirenberg structure, $|\tau| = \tau_1 + \cdots + \tau_N$, and the operator

$$E_C : U \mapsto \Psi$$

is a mutually continuous one-to-one mapping from $\mathcal{U}_{s,p}(\mathbf{R}'_{n-1}, q)$ onto the space

$$B^{s,\sigma^+,\sigma^-}(\mathbf{R}'_{n-1}, q) := \prod_{h=1}^{m} B^{s-\sigma_h^+-1/p,p}(\mathbf{R}'_{n-1}, q) \times \prod_{h=1}^{m} B^{s-\sigma_h^--1/p,p}(\mathbf{R}'_{n-1}, q).$$

System (1.7.22) is parameter elliptic. Then, for $q \geq q_0 > 0$, we obtain:

$$\begin{aligned} U &= (E_C(x, D_x, q))^{-1} \begin{pmatrix} 0 \\ \psi^+ \\ \psi^- \end{pmatrix} \\ &= (E_C(x, D_x, q))^{-1} \begin{pmatrix} 0 \\ \psi^+ \\ 0 \end{pmatrix} + (E_C(x, D_x, q))^{-1} \begin{pmatrix} 0 \\ 0 \\ \psi^- \end{pmatrix}, \end{aligned} \tag{1.7.23}$$

where $E_C^{-1} \begin{pmatrix} 0 \\ \psi^+ \\ 0 \end{pmatrix}$ and $E_C^{-1} \begin{pmatrix} 0 \\ 0 \\ \psi^- \end{pmatrix}$ are the Cauchy data vectors of the parameter elliptic problem (1.7.21). This problem is uniquely solvable for $q \geq q_0$, therefore $E_C^{-1} \begin{pmatrix} 0 \\ \psi^+ \\ 0 \end{pmatrix} = U^+$ and $E_C^{-1} \begin{pmatrix} 0 \\ 0 \\ \psi^- \end{pmatrix} = U^-$. Then the equality (1.7.23) implies that $U = U^+ + U^-$. The equality (1.7.20) is established.

Relations

$$\|P^\pm U\|_{\mathcal{U}^\pm_{s,p}} = \|U^\pm\|_{\mathcal{U}^\pm_{s,p}} \leq c\|\psi^\pm\|_{B^{m+s-\sigma^\pm-1/p,p}} \leq c_1\|\Psi\|_{B^{s,\sigma^+,\sigma^-}} \leq c_2\|U\|_{\mathcal{U}_{s,p}},$$

with the constant independent of U and q ($q \geq q_0$), give us the boundedness of the norms of the projection operators. This completes the proof of the theorem. □

Definition 1.7.1. *The operators P^+ and P^- are called Calderón projections.*

1.7.4. Proof of Lemma 1.7.1. In $\mathbf{R}^n_+$ we consider an elliptic with a parameter problem for one $2m$th-order equation:

$$\begin{aligned} L_0(D, q)u = 0, \quad & x \in \mathbf{R}^n_+, \\ B_{j0}(D, q)u\big|_{x_n=0} = \varphi_j, \quad & j = 1, \dots, m, \end{aligned} \tag{1.7.24}$$

where

$$L_0(D, q) = \sum_{|\alpha|+k=2m} a_{\alpha k} q_1^k D^\alpha, \quad B_{j0}(D, q) = \sum_{|\beta|+k=m_j} b_{jk} q_1^k D^\beta.$$

Let

$$L_0(\zeta) = L_0(\xi', \zeta, q_1) = L_0^+(\xi', \zeta, q_1) L_0^-(\xi', \zeta, q_1).$$

Here the ζ-roots of the polynomial $L_0^+(\zeta) = L_0^+(\xi', \zeta, q_1)$ lie above the real axis.

At the beginning, let $\operatorname{ord} B_j = m_j < 2m$, $j = 1, \ldots, m$. We set:

$$C_j(\xi', \xi_n, q_1) = \xi_n^{j-1} L_0^+(\xi', \xi_n, q_1), \quad j = 1, \ldots, m.$$

The ellipticity with a parameter of problem (1.7.24) means that the polynomials $B_j(\zeta) = B_j(\xi', \zeta, q)$ are linearly independent with respect to the modulo $L_0^-(\zeta)$, $|\xi'| + |q| > 0$. Then the polynomials $C_j(\zeta) = C_j(\xi', \zeta, q)$ are also linearly independent with respect to the modulo $L_0^-(\zeta)$, i.e., the problem

$$\begin{gathered} L_0(D, q)u = 0, \quad x \in \mathbf{R}_+^n, \\ C_j(D, q)u\big|_{x_n=0} = \psi_j, \quad j = 1, \ldots, m, \end{gathered} \tag{1.7.25}$$

is also parameter elliptic. Furthermore, if

$$B_j(\xi', \xi_n, q_1) = \sum_{k=1}^{m_j+1} B_{jk}(\xi', q_1)\xi_n^{k-1}, \quad j = 1, \ldots, m,$$

$$C_j(\xi', \xi_n, q_1) = \sum_{k=1}^{m+j-k} C_{jk}(\xi', q_1)\xi_n^{k-1}, \quad j = 1, \ldots, m,$$

then the system

$$\sum_{k=1}^{m_j+1} B_{jk}(D', q_1)\eta_k = \varphi_j, \qquad \sum_{k=1}^{m+j} C_{jk}(D', q_1)\eta_k = \psi_j, \tag{1.7.26}$$

$$j = 1, \ldots, m, \ \eta_k = D_n^{k-1} u\big|_{x_n=0},$$

is parameter elliptic on the hyperplane $\mathbf{R}'_{n-1} = \{x \in \mathbf{R}^n : x_n = 0\}$.

If

$$r = \max\{2m, m_1 + 1, \ldots, m_m + 1\} > 2m,$$

then it is necessary to join the conditions

$$D_n^{j-1} L_0(D', D_n, q_1)u\big|_{x_n=0} = 0, \quad j = 1, \ldots, r - 2m.$$

to the system (1.2.26). We obtain that the problem

$$\begin{gathered} \sum_{k=1}^{r} L_{0jk}(D', q_1)\eta_k = 0, \quad j = 1, \ldots, r - 2m, \\ \sum_{k=1}^{m_j+1} B_{jk}(D', q_1)\eta_k = \varphi_j, \quad j = 1, \ldots, m, \end{gathered} \tag{1.7.27}$$

$$\sum_{k=1}^{m+j} C_{jk}(D', q_1)\eta_k = \psi_j, \quad j = 1, \ldots, m,$$

is a parameter-elliptic on $\mathbf{R}'_{n-1}$.

Let us introduce the following notation:

$$\left(L_{0jk}(D', q_1)\right)_{\substack{j=1,\ldots,N \\ k=1,\ldots,r-2m}} := \left(e(D', q)\right),$$

$$\left(B_{jk}(D', q_1)\right)_{\substack{j=1,\ldots,N \\ k=1,\ldots,r}} := \left(b(D', q)\right),$$

$$\left(C_{jk}(D', q_1)\right)_{\substack{j=1,\ldots,N \\ k=1,\ldots,r}} := \left(c(D', q_1)\right),$$

where $C_{jk} \equiv 0$ for $k > m + j$.

Then one can rewrite problem (1.7.27) in the form

$$\begin{pmatrix} e(D', q_1) \\ b(D', q_1) \\ c(D', q_1) \end{pmatrix} \eta = \begin{pmatrix} 0 \\ \varphi \\ \psi \end{pmatrix}. \tag{1.7.28}$$

This system is a parameter-elliptic on hyperplane $\mathbf{R}'_{n-1}$.

Now consider the case of the parameter elliptic problem

$$l_0(D', D_n, q_1)u = 0, \quad x \in \mathbf{R}^n_+, \qquad b_0(D', D_n, q_1)u\big|_{x_n=0} = \varphi \tag{1.7.29}$$

for the diagonal system

$$l_0(D', D_n, q_1)u :=$$

$$\begin{pmatrix} l_{0\,11}(D', D_n, q_1), & 0, & \ldots & 0 \\ 0 & l_{0\,22}(D', D_n, q_1) & \ldots & 0 \\ \vdots & & \ddots & \vdots \\ 0 & 0 & \ldots & l_{0\,NN}(D', D_n, q_1) \end{pmatrix}.$$

In this case we have:

$$L(\xi', \xi_n, q_1) = \det\, l_0(\xi', \xi_n, q_1) = \prod_{r=1}^{N} l_{0rr}(\xi', \xi_n, q_1),$$

$$L^{\pm}(\xi', \xi_n, q_1) = \prod_{r=1}^{N} l^{\pm}_{0rr}(\xi', \xi_n, q_1).$$

We take as $C(D, q_1)$ the matrix

$$C(D', D_n, q) = \begin{pmatrix} C_{11}(D', D_n, q_1), & 0 & & \ldots & 0 \\ & \ddots & & 0 & \vdots \\ 0 & 0 & C_{mm}(D', D_n, q_1) & \ldots & 0 \end{pmatrix}.$$

It is clear that the problem

$$l_0(D,q_1)u = 0, \quad x \in \mathbf{R}^n_+, \qquad c(D,q_1)u\big|_{x_n=0} = \psi$$

is elliptic with a parameter. Setting that $c(D', D_n, q_1)u = c(D', q_1)\eta$, we obtain that the system

$$\begin{pmatrix} e(D',q_1) \\ b(D',q_1) \\ c(D',q_1) \end{pmatrix} \eta = \begin{pmatrix} 0 \\ \varphi \\ \psi \end{pmatrix}$$

is elliptic a parameter-elliptic on $\mathbf{R}'_{n-1}$.

Now let us consider the general case. Let the problem

$$l_0(D', D_n, q_1)u = 0, \quad x \in \mathbf{R}^n_+, \qquad b_0(D', D_n, q_1)u\big|_{x_n=0} = \varphi$$

be elliptic with a parameter.

By elementary transformations one can reduce the matrix $l_0(\xi', \xi_n, q_1)$ to diagonal form. In other words (sf. [Vol]) there exist the polynomial matrixes $P(\xi', \xi_n, q_1)$ and $Q(\xi', \xi_n, q_1)$ such that their determinants det P and det Q are nonzero, and the matrix

$$P(\xi', \xi_n, q_1)\, l_0(\xi', \xi_n, q_1)\, Q(\xi', \xi_n, q_1) = \Lambda(\xi', \xi_n, q_1)$$

are diagonal.

Therefore the problem

$$l_0(D,q_1)u = 0, \quad x \in \mathbf{R}^n_+, \qquad b(D,q_1)u\big|_{x_n=0} = \varphi$$

is equivalent to the problem

$$\Lambda(D,q_1)u = 0, \quad x \in \mathbf{R}^n_+, \qquad b(D,q_1)u\big|_{x_n=0} = \varphi,$$

and the problem

$$\Lambda(D,q_1)u = 0, \quad x \in \mathbf{R}^n_+, \qquad c(D,q_1)u\big|_{x_n=0} = \psi$$

is equivalent to the problem

$$l_0(D,q_1)u = 0, \quad x \in \mathbf{R}^n_+, \qquad c(D,q_1)u\big|_{x_n=0} = \psi.$$

Thus, the last problem is also parameter elliptic and Lemma 1.7.1 is proved. □

1.8. Parameter-Elliptic in the Kozhevnikov Sense Mixed-Order Operators on a Closed Manifold in Complete Scale of Sobolev Type Spaces

1.8.1. The paper is devoted to the studies of the mixed-order (Douglis-Nirenberg) elliptic differential operators (see e.g. [WRL, Sect. 9.3] with spectral parameter on closed compact manifolds without boundary. This section is closely related to the works [Agr2], [Agr3], [Ber], [Boi], [DMV], [Gr1], [Kozh1], [Kozh2], [KoY], [R1], and [See], where spectral problems for mixed-order elliptic systems on compact manifolds were studied. We establish here the theorem on complete collection of isomorphisms. More precisely, we introduce a pair of L_p-Sobolev type spaces depending on a parameter $s \in \mathbf{R}$ and show that the parameter-dependent mixed-order differential operator is an isomorphism between these spaces for every s. In the paper by R. Denk, R. Mennicken and L. Volevich [DMV] the isomorphism between a pair of L_2-Sobolev type spaces has been established but only for one value of the parameter $s = 0$.

We establish the isomorphism under some parameter-ellipticity condition for the mixed-order system which is a generalization of the well-known parameter-ellipticity condition for elliptic systems of a single order (see, for instance, R. Seeley [See]). The parameter-ellipticity condition for mixed-order systems on manifolds without boundary was first introduced by A. Kozhevnikov [Kozh1], but in an implicit and rather complicated form. Later it was reformulated in an equivalent more simple form in [Kozh1], [KoY], [Kozh2]. The condition has been called the Kozhevnikov condition and further elaborated by R. Denk, R. Mennicken and L. Volevich where other equivalent forms of the condition have been found [DMV, Sect. 3, Theorem 3.14].

For simplicity we consider here only the case of differential operators. The results are valid also for pseudo-differential operators.

Let Ω be an n-dimensional closed C^∞ compact manifold with a C^∞ positive density dx. Consider a matrix differential operator on Ω. All the necessary definitions concerning differential operator can be found for instance in the monograph by J.T. Wloka, B. Rowley, B. Lawruk [WRL, Ch. 7–9]. Let $T = (t_1, \ldots, t_N)$, $S = (s_1, \ldots, s_N)$ be two multi-indices such that the differential operator A is a mixed-order (or Douglis-Nirenberg) system of order (T, S), i.e. A is a matrix differential operator , $A = \{A_{rj}\}_{r,j=1,\ldots,N}$, such that $\operatorname{ord} A_{rj} \leq t_j + s_r$. This means that its entries $A_{rj} := A_{rj}(x, D)$ are linear differential operators of order $\leq t_j + s_r$. More precisely, let

$$A_{rj}(x, D) := \begin{cases} \sum_{|\alpha| \leq s_r + t_j} a_{rj,\alpha}(x) D^\alpha & (r, j : s_r + t_j \geq 0) \\ 0 & (r, j : s_r + t_j < 0) \end{cases}$$

be differential operators with infinitely smooth coefficients $a_{rj,\alpha}(x)$, $x \in \Omega$. Here we use the usual multi-index notation: $D^\alpha := D_1^{\alpha_1}...D_n^{\alpha_n}$, $|\alpha| = \alpha_1 + ... + \alpha_n$, $D_j := i^{-1}\partial/\partial_j$, where $i := \sqrt{-1}$. Let $\mu_k := t_k + s_k$. Below, we always assume that

$$t_1 \geq ... \geq t_N \geq 0 = s_1 \geq ... \geq s_N$$

(this can be realized by an appropriate simultaneous renumbering of rows and columns of the matrix A). It follows that the orders of the diagonal operators A_{jj} $(j = 1, 2, \ldots, N)$ are non-negative and do not increase, i.e.

$$\mu_1 \geq \mu_2 \geq ... \geq \mu_\rho > \mu_{\rho+1} = ... = \mu_N = 0.$$

Let $a^0(x,\xi)$ be the principal symbol of the operator A. More precisely, $a^0(x,\xi)$ is a matrix with entries $a^0_{jk}(x,\xi)$ which are principal symbols of differential operator A_{jk} as operators of order $t_k + s_j$.

We consider the following spectral problem

$$(A(x,D) - \lambda I)\, u(x) = f(x) \quad x \in \Omega. \tag{1.8.1}$$

To formulate the parameter-ellipticity condition for the mixed-order operator we define the integers $d, k_1, ..., k_d$ as follows:

$$\mu_1 = ... = \mu_{k_1} > \mu_{k_1+1} = ... = \mu_{k_2} > ...$$

$$> \mu_{k_{d-1}+1} = ... = \mu_{k_d} > \mu_{k_d+1} = ... = \mu_N = 0.$$

In particular, we have $k_d = \rho$.

Let L be some ray L on the complex plane emerging from the origin, i.e. $L := \{\lambda \in \mathbf{C} : \lambda = re^{i\varphi},\ r \in [0,\infty),\ \varphi = \text{const}\}$. Suppose that a parameter-ellipticity condition holds along the ray L. This means that for any fixed number $k_r \in \{k_1, k_2, ..., k_d\}$, the determinant of the matrix

$$\begin{pmatrix} a^0_{11} & \cdots & a^0_{1,k_{r-1}} & a^0_{1,k_{r-1}+1} & \cdots & a^0_{1,k_r} \\ \vdots & \ddots & \vdots & \vdots & \vdots & \vdots \\ a^0_{k_{r-1},1} & \cdots & a^0_{k_{r-1},k_{r-1}} & a^0_{k_{r-1},k_{r-1}+1} & \cdots & a^0_{k_{r-1},k_r} \\ a^0_{k_{r-1}+1,1} & \cdots & a^0_{k_{r-1}+1,k_{r-1}} & a^0_{k_{r-1}+1,k_{r-1}+1} - \lambda & \cdots & a^0_{k_{r-1}+1,k_r} \\ \vdots & \cdots & \vdots & \vdots & \ddots & \vdots \\ a^0_{k_r,1} & \cdots & a^0_{k_r,k_{r-1}} & a^0_{k_r,k_{r-1}+1} & \cdots & a^0_{k_r,k_r} - \lambda \end{pmatrix}$$

is nonzero for all (x,ξ) from the cotangent bundle $T^*(\Omega) \setminus \{0\}$ and for all $\lambda \in L$.

Let $p, p' \in (1, \infty)$, $p^{-1} + p'^{-1} = 1$, $s \in \mathbf{R}$. We denote by $H^{s,p}(\mathbf{R}^n)$ the space with the norm

$$\|u\|_{H^{s,p}} := \left\| F^{-1} \left(1 + |\xi|^2\right)^{s/2} Fu \right\|_{L_p(\mathbf{R}^n)}. \tag{1.8.2}$$

Here $F = F_{x \to \xi}$ is the standard Fourier transform and F^{-1} its inverse. Let $q > 0$ and $q^{\mu_\rho} := |\lambda|$. In studying elliptic problems with a parameter λ it is convenient to use equivalent norms which depend on λ.

$$\|u\|_{H^{s,p}} := \left\| F^{-1} \left(q + |\xi|\right)^{\mu_\rho} \left(1 + |\xi|\right)^{s-\mu_\rho} Fu \right\|_{L_p(\mathbf{R}^n)}. \tag{1.8.3}$$

For any $q > 0$ the norms (1.8.2) and (1.8.3) are equivalent. It is convenient to consider only those norms equivalent to (1.8.2) for which the corresponding estimates can be written with constants that do not depend on q. In this sense , the norm (1.8.3) is equivalent to each of the following norms:

$$\|u\|_{H^{s,p}} := \left\| F^{-1} \left(q^2 + |\xi|^2\right)^{\mu_\rho/2} \left(1 + |\xi|^2\right)^{(s-\mu_\rho)/2} Fu \right\|_{L_p(\mathbf{R}^n)},$$

$$\|u\|_{H^{s,p}} := \left\| F^{-1} \left(q^{\mu_\rho} + |\xi|^{\mu_\rho}\right) \left(1 + |\xi|\right)^{s-\mu_\rho} Fu \right\|_{L_p(\mathbf{R}^n)}.$$

Using partition of unity we can define the norm $\|\cdot\|_{H^{s,p}} = \|\cdot\|_{H^{s,p}(\Omega)}$ =and the space

$$H^{s,p}(\Omega, \lambda, \rho)$$

with this norm. Further on we will use the norm depending on the parameter λ.

Theorem 1.8.1. *There exists a number* $|\lambda_0| > 0$ *such that for* $|\lambda| \geq |\lambda_0|$ *the equation*

$$(A(x, D) - \lambda I)\, u(x) = f(x) \in H^{s-S,p}(\Omega, \lambda, \rho), \quad x \in \Omega, \tag{1.8.4}$$

$$H^{s-S,p}(\Omega, \lambda, \rho) := \prod_{j=1}^{N} H^{s-s_j,p}(\Omega, \lambda, \rho),$$

has a unique solution

$$u(x) \in H^{T+s,p}(\Omega, \lambda, \rho) := \prod_{j=1}^{N} H^{t_j+s,p}(\Omega, \lambda, \rho).$$

Moreover, the inequality holds

$$\|u\|_{H^{T+s,p}} \leq C_s \|f\|_{H^{s-S,p}} \quad (s \in \mathbf{R})$$

where a positive constant C_s does not depend on u, f, λ. The function $s \mapsto C_s$ is bounded on any compact set.

Remark 1.8.1. R. Denk, R. Mennicken and L. Volevich [DMV, Sect. 4] introduced the following norms depending on parameter λ :

$$\|u\|^2 = \sum_{j=1}^{N} \left\| \left(|D|^{2\mu_j} + |\lambda|^2 \right)^{t_j/(2\mu_j)} u_j \right\|_{L_2}^2 ,$$

$$\|f\|^2 = \sum_{j=1}^{\rho} \left\| \left(|D|^{2\mu_j} + |\lambda|^2 \right)^{-s_j/(2\mu_j)} f_j \right\|_{L_2}^2 + \sum_{j=\rho+1}^{N} \|f_j\|_{L_2}^2 .$$

It has been proved the solvability of the problem (1.8.1) and the estimate $\|u\| \leq C \|f\|$ where $C > 0$ is a constant independent of λ. These norms are equivalent for any λ to the norms

$$\|u\|_{H^{T+s,p}(\Omega,\lambda,\rho)} \text{ and } \|f\|_{H^{s-S,p}(\Omega,\lambda,\rho)} \text{ for } s = 0, p = 2.$$

1.8.2. Mixed-order parameter-elliptic operator in $\mathbf{R}^n$

We consider an analog of the problem (1.8.1) in $\mathbf{R}^n$

$$(A(D) - \lambda I)\, u(x) = f(x) \quad x \in \mathbf{R}^n,$$

where $A(D)$ is a mixed-order differential operator with constant coefficients. We suppose that the parameter-ellipticity condition is satisfied for the operator $A(D) - \lambda I$ along some ray L.

Let

$$G(\xi, \lambda) := (A(\xi) - \lambda I)^{-1} = \{G_{jk}(\xi, \lambda)\}$$

It has been proved by R. Denk, R. Mennicken and L. Volevich [DMV, Prop.3.10] or by A. Kozhevnikov [Kozh1, Sect. 2, Prop. 12] for the case $\mu_N > 0$, that

$$|G_{jk}(\xi, \lambda)| \leq C (1 + |\xi|)^{s_j + t_k} \left(|\xi|^{\mu_j} + |\lambda| \right)^{-1} \left(|\xi|^{\mu_k} + |\lambda| \right)^{-1} \tag{1.8.5}$$

for $j \neq k$, otherwise

$$|G_{jj}(\xi, \lambda)| \leq C \left(|\xi|^{\mu_j} + |\lambda| \right)^{-1}$$

where a positive constant C does not depend on ξ and λ.

We consider the mapping

$$u \mapsto (A(D) - \lambda I)\, u.$$

It acts continuously from

$$H^{T+s,p}(\mathbf{R}^n,\lambda,\rho) := \prod_{j=1}^{N} H^{t_j+s,p}(\mathbf{R}^n,\lambda,\rho)$$

to

$$H^{s-S,p}(\mathbf{R}^n,\lambda,\rho) := \prod_{j=1}^{N} H^{s-s_j,p}(\mathbf{R}^n,\lambda,\rho).$$

Lemma 1.8.1. *Let $p \in (1,\infty)$, $s \in \mathbf{R}$. Suppose that the model problem is parameter-elliptic on the ray L. Let $f \in H^{s-S,p}(\mathbf{R}^n,\lambda,\rho)$. Then there exists $|\lambda_0| > 0$ such that the model problem has a unique solution $u \in H^{T+s,p}(\mathbf{R}^n,\lambda,\rho)$ for $|\lambda| \geq |\lambda_0|$. Moreover, the following estimate holds*

$$\|u\|_{H^{T+s,p}} \leq C\,\|f\|_{H^{s-S,p}}$$

where the positive constant C does not depend on λ and s.

Proof. Applying the Fourier transform to the model problem we obtain

$$\widetilde{u}_k(\xi) = \sum_{j=1}^{N} G_{kj}(\xi,\lambda)\widetilde{f}_j(\xi) \;\; (k = 1,...,N)$$

Then

$$\begin{aligned}
\|u\|_{H^{T+s,p}} &= \sum_{k=1}^{N} \left\| F^{-1}(|\xi|+q)^{\mu_\rho}(1+|\xi|)^{s+t_k-\mu_\rho} Fu_k \right\|_{L_p} \\
&= \sum_{k=1}^{N}\sum_{j=1}^{N} \left\| F^{-1}(|\xi|+q)^{\mu_\rho}(1+|\xi|)^{s+t_k-\mu_\rho} G_{kj}(\xi,\lambda)\widetilde{f}_j(\xi) \right\|_{L_p} \\
&=: \sum_{k=1}^{N}\sum_{j=1}^{N} J_{kj}
\end{aligned}$$

By (1.8.5), we obtain for $k \neq j$

$$\begin{aligned}
J_{kj} &\leq C \Big\| F^{-1}(|\xi|+q)^{\mu_\rho}(1+|\xi|)^{s+t_k-\mu_\rho}(1+|\xi|)^{s_k+t_j} \\
&\qquad \times \left(|\xi|^{\mu_k}+|\lambda|\right)^{-1}\left(|\xi|^{\mu_j}+|\lambda|\right)^{-1} Ff_j(\xi) \Big\|_{L_p} \\
&= \left\| F^{-1} \frac{(|\xi|+q)^{\mu_\rho}(1+|\xi|)^{s+t_k-\mu_\rho}(1+|\xi|)^{s_k+t_j}}{(|\xi|^{\mu_k}+q^{\mu_\rho})(|\xi|^{\mu_j}+q^{\mu_\rho})(|\xi|+q)^{\mu_\rho}(1+|\xi|)^{s-t_j-\mu_\rho}} \right.
\end{aligned}$$

$$\times FF^{-1}\left(|\xi|+q\right)^{\mu_p}\left(1+|\xi|\right)^{s-s_j-\mu_p}\widetilde{f}_j\left(\xi\right)\Big\|_{L_p}$$

$$= \left\|F^{-1}\frac{\left(1+|\xi|\right)^{\mu_j}\left(1+|\xi|\right)^{\mu_k}}{\left(|\xi|^{\mu_k}+q^{\mu_p}\right)\left(|\xi|^{\mu_j}+q^{\mu_p}\right)}\right.$$

$$\times FF^{-1}\left(|\xi|+q\right)^{\mu_p}\left(1+|\xi|\right)^{s-s_j-\mu_p}\widetilde{f}_j\left(\xi\right)\Big\|_{L_p}$$

$$\leq C\left\|f_j\right\|_{H^{s-s_j,p}},$$

since for large enough q the expression

$$\frac{\left(1+|\xi|\right)^{\mu_j}\left(1+|\xi|\right)^{\mu_k}}{\left(|\xi|^{\mu_k}+q^{\mu_p}\right)\left(|\xi|^{\mu_j}+q^{\mu_p}\right)}$$

is a multiplicator in the space $L_p\left(\mathbf{R}^n\right)$ (see e.g. [R1, Ch. I, Sect.1.3]).

Similarly we have

$$J_{kk} \leq C\left\|F^{-1}\left(|\xi|+q\right)^{\mu_p}\left(1+|\xi|\right)^{s+t_k-\mu_p}G_{kk}\left(\xi,\lambda\right)\widetilde{f}_k\left(\xi\right)\right\|_{L_p}$$

$$\leq C\left\|F^{-1}\left(|\xi|+q\right)^{\mu_p}\left(1+|\xi|\right)^{s+t_k-\mu_p}\left(|\xi|^{\mu_k}+q^{\mu_p}\right)^{-1}\widetilde{f}_k\left(\xi\right)\right\|_{L_p}$$

$$= \left\|F^{-1}\frac{\left(|\xi|+q\right)^{\mu_p}\left(1+|\xi|\right)^{t_k+s-\mu_p}}{\left(|\xi|^{\mu_k}+q^{\mu_p}\right)\left(|\xi|+q\right)^{\mu_p}\left(1+|\xi|\right)^{s-s_k-\mu_p}}\right.$$

$$\times FF^{-1}\left(|\xi|+q\right)^{\mu_p}\left(1+|\xi|\right)^{s-s_k-\mu_p}\widetilde{f}_k\left(\xi\right)\Big\|_{L_p} =$$

$$= \left\|F^{-1}\frac{\left(1+|\xi|\right)^{t_k+s_k}}{\left(|\xi|^{\mu_k}+q^{\mu_p}\right)}\right.$$

$$\times FF^{-1}\left(|\xi|+q\right)^{\mu_p}\left(1+|\xi|\right)^{s-s_k-\mu_p}\widetilde{f}_k\left(\xi\right)\Big\|_{L_p}$$

$$\leq C\left\|f_k\right\|_{H^{s-s_k,p}},$$

since for large enough q the expression

$$\frac{\left(1+|\xi|\right)^{\mu_k}}{\left(|\xi|^{\mu_k}+q^{\mu_p}\right)}$$

is a multiplicator in the space $L_p(\mathbf{R}^n)$ (see e.g. [R1, Ch. I, Sect. 1.3]). Lemma is proved. □

Lemma 1.8.2. *For any*

$$u \in H^{s,p}(\mathbf{R}^n, \lambda, \rho)$$

the following inequality holds:

$$\|u\|_{H^{s-1,p}} \le \frac{C}{q} \|u\|_{H^{s,p}}$$

where $q \to \infty$ for $|\lambda| \to \infty$ and a positive constant C does not depend on u, λ, s.

Proof. By the definition

$$\|u\|_{H^{s,p}} := \left\| F^{-1} (|\xi| + q)^{\mu_\rho} (1 + |\xi|)^{s-\mu_\rho} (Fu)(\xi) \right\|_{L_p},$$

where $q^{\mu_\rho} \ge |\lambda_0|$, $|\lambda_0|$ is large enough. Then since $\dfrac{q}{(|\xi| + q)}$ is a multiplicator in $L_p(\mathbf{R}^n)$,

$$\begin{aligned}
\|u\|_{H^{s-1,p}} &= \left\| F^{-1} (|\xi| + q)^{\mu_\rho} (1 + |\xi|)^{s-1-\mu_\rho} (Fu)(\xi) \right\|_{L_p} \\
&= \frac{1}{q} \left\| F^{-1} \frac{q}{|\xi| + q} F F^{-1} (|\xi| + q)^{\mu_\rho + 1} (1 + |\xi|)^{s-1-\mu_\rho} (Fu)(\xi) \right\|_{L_p} \\
&\le \frac{C}{q} \left\| F^{-1} (|\xi| + q)^{\mu_\rho + 1} (1 + |\xi|)^{s-1-\mu_\rho} (Fu)(\xi) \right\|_{L_p} \\
&= \frac{C}{q} \left\| F^{-1} \frac{(|\xi| + q)^{\mu_\rho + 1}}{q^{\mu_\rho + 1} + |\xi|^{\mu_\rho + 1}} F F^{-1} \left(q^{\mu_\rho + 1} + |\xi|^{\mu_\rho + 1}\right) \right. \\
&\qquad \left. \times (1 + |\xi|)^{s-1-\mu_\rho} (Fu)(\xi) \right\|_{L_p} \\
&\le \frac{C_1}{q} \left\| F^{-1} \frac{q^{\mu_\rho + 1} + |\xi|^{\mu_\rho + 1}}{1 + |\xi|} (1 + |\xi|)^{s-\mu_\rho} (Fu)(\xi) \right\|_{L_p}.
\end{aligned}$$

Let $q^{\mu_\rho + 1} = q_1^{\mu_\rho}$; $q_1 > q$. Then since

$$\frac{q_1^{\mu_\rho} + |\xi|^{\mu_\rho + 1}}{(1 + |\xi|)(q_1^{\mu_\rho} + |\xi|^{\mu_\rho})}$$

is a multiplicator in $L_p(\mathbf{R}^n)$,

$$\begin{aligned}
\|u\|_{H^{s-1,p}} &= \frac{C_1}{q}\left\|F^{-1}\frac{q_1^{\mu_\rho}+|\xi|^{\mu_\rho+1}}{1+|\xi|}(1+|\xi|)^{s-\mu_\rho}(Fu)(\xi)\right\|_{L_p} \\
&= \frac{C_1}{q}\left\|F^{-1}\frac{q_1^{\mu_\rho}+|\xi|^{\mu_\rho+1}}{(1+|\xi|)(q_1^{\mu_\rho}+|\xi|^{\mu_\rho})}FF^{-1}(q_1^{\mu_\rho}+|\xi|^{\mu_\rho})\right. \\
&\qquad \left.\times(1+|\xi|)^{s-\mu_\rho}(Fu)(\xi)\right\|_{L_p} \\
&\le \frac{C_2}{q}\left\|F^{-1}(q_1^{\mu_\rho}+|\xi|^{\mu_\rho})(1+|\xi|)^{s-\mu_\rho}(Fu)(\xi)\right\|_{L_p} \\
&\le \frac{C_3}{q}\|u\|_{H^{s,p}}.
\end{aligned}$$

Lemma 2 is proved. □

Lemma 1.8.3. *For any*

$$u \in H^{s,p}(\mathbf{R}^n, \lambda, \rho)$$

the following inequality holds:

$$\left\|\left(D^\alpha + q^{|\alpha|}\right)u\right\|_{H^{s-|\alpha|,p}} \le C\|u\|_{H^{s,p}}$$

where a positive constant C does not depend on u, λ, s.

Proof.

$$\begin{aligned}
&\left\|\left(D^\alpha + q^{|\alpha|}\right)u\right\|_{H^{s-|\alpha|,p}} \\
&= \left\|F^{-1}(|\xi|+q)^{\mu_\rho}(1+|\xi|)^{s-|\alpha|-\mu_\rho}\left(\xi^\alpha+q^{|\alpha|}\right)\tilde{u}(\xi)\right\|_{L_p} \\
&= \left\|F^{-1}\frac{(\xi^\alpha+q^{|\alpha|})}{(|\xi|+q)^{|\alpha|}}FF^{-1}(|\xi|+q)^{\mu_\rho+|\alpha|}(1+|\xi|)^{s-|\alpha|-\mu_\rho}\tilde{u}(\xi)\right\|_{L_p} \\
&\le C\left\|F^{-1}\left(|\xi|^{\mu_\rho+|\alpha|}+q^{\mu_\rho+|\alpha|}\right)(1+|\xi|)^{s-|\alpha|-\mu_\rho}\tilde{u}(\xi)\right\|_{L_p}.
\end{aligned}$$

Let $q_1^{\mu_\rho} := q^{\mu_\rho+|\alpha|}$, $q_1 > q$. Then since the expression

$$\frac{(|\xi|^{\mu_\rho+|\alpha|}+q_1^{\mu_\rho})}{(1+|\xi|)^{|\alpha|}(|\xi|+q_1)^{\mu_\rho}}$$

is a multiplicator in $L_p(\mathbf{R}^n)$, we have

$$\begin{aligned}
&\left\|\left(D^\alpha + q^{|\alpha|}\right)u\right\|_{H^{s-|\alpha|,p}} \\
&\leq C\left\|F^{-1}\frac{\left(|\xi|^{\mu_p+|\alpha|}+q_1^{\mu_p}\right)}{(1+|\xi|)^{|\alpha|}\left(|\xi|+q_1\right)^{\mu_p}}F^{-1}F\left(|\xi|+q_1\right)^{\mu_p}\right. \\
&\qquad\left.\times(1+|\xi|)^{s-\mu_p}\widetilde{u}(\xi)\right\|_{L_p} \\
&\leq C_1\left\|F\left(|\xi|+q_1\right)^{\mu_p}(1+|\xi|)^{s-\mu_p}\widetilde{u}(\xi)\right\|_{L_p} = \\
&\leq C_1\|u\|_{H^{s,p}}.
\end{aligned}$$

Lemma 3 is proved. □

Similarly (and even simpler) we can prove the following lemma.

Lemma 1.8.4. *For any*

$$u \in H^{s,p}(\mathbf{R}^n, \lambda, \rho)$$

the following inequality holds:

$$\|D^\alpha u\|_{H^{s-|\alpha|,p}} \leq C\|u\|_{H^{s,p}}$$

where a positive constant C *does not depend on* u, λ, s.

Lemma 1.8.5. *If* $\widetilde{\chi} := \widetilde{\chi}(\xi) \in C_0^\infty(\mathbf{R}^n)$, *then for any* $t > 0$ *the function* $\widetilde{\chi}(\xi)(1+|\xi|)^t$ *is a multiplicator in* $L_p(\mathbf{R}^n)$. *Therefore the operator*

$$\chi = F^{-1}\widetilde{\chi}(\xi)F$$

acts continuously from $H^{s,p}(\mathbf{R}^n, \lambda, \rho)$*to* $H^{s+t,p}(\mathbf{R}^n, \lambda, \rho)$, *i.e.,* χ *is an infinitely smoothing operator.*

Proof. The calculations

$$\begin{aligned}
\|\chi u\|_{H^{s+t,p}} &= \left\|F^{-1}\widetilde{\chi}(\xi)(Fu)(\xi)\right\|_{H^{s+t,p}} \\
&= \left\|F^{-1}\widetilde{\chi}(\xi)(|\xi|+q)^{\mu_p}(1+|\xi|)^{s+t-\mu_p}(Fu)(\xi)\right\|_{L_p}
\end{aligned}$$

$$= \left\| F^{-1}\tilde{\chi}(\xi)(1+|\xi|)^t FF^{-1}(|\xi|+q)^{\mu_p}(1+|\xi|)^{s-\mu_p}(Fu)(\xi)\right\|_{L_p}$$

$$\leq C\left\| F^{-1}(|\xi|+q)^{\mu_p}(1+|\xi|)^{s-\mu_p}(Fu)(\xi)\right\|_{L_p}$$

$$= C\|u\|_{H^{s,p}}.$$

proves Lemma 1.8.5. □

Let $\tilde{\chi}(\xi) \in C_0^\infty(\mathbf{R}^n)$, $0 \leq \tilde{\chi}(\xi) \leq 1$,

$$\tilde{\chi}(\xi) = \begin{cases} 1 & \text{for } |\xi| \leq 1, \\ 0 & \text{for } |\xi| \geq 2, \end{cases}$$

and

$$\sup_{\alpha\in K, \xi\in\mathbf{R}^n} |\partial^\alpha \tilde{\chi}(\xi)| = a.$$

Here $K := \{\alpha = (\alpha_1, \ldots, \alpha_n) : \alpha_j = 0, 1;\ j = 1, \ldots, n\}$ denotes the set of n-dimensional multi-indices whose components are equal 0 and 1.

Then

$$\chi_M = F^{-1}\tilde{\chi}(\xi/M)F$$

is a infinitely smoothing operator for any $M > 0$.

Consider the operator

$$1 - \chi_M = F^{-1}(1 - \tilde{\chi}(\xi/M))F$$

Since $M \leq |\xi| \leq 2M$ for $\xi \in \operatorname{supp}\tilde{\chi}(\xi/M)$,

$$\sup_{\alpha\in K,\ \xi\in\mathbf{R}^n} |\xi^\alpha \partial^\alpha(1 - \tilde{\chi}(\xi/M))| \leq 2^n a.$$

Therefore, the following lemma holds:

Lemma 1.8.6. *For any*

$$u \in H^{s,p}(\mathbf{R}^n, \lambda, \rho) \quad (s \in \mathbf{R})$$

the following inequality holds:

$$\|(1 - \chi_M)u\|_{H^{s,p}} \leq b\|u\|_{H^{s,p}}$$

where $b = æ_p 2^n a$, $æ_p$ *is a positive constant which depends only on* p. *Moreover, for any* $\varepsilon > 0$ *and* $\sigma_0 > 0$ *there exists a number* $M(\varepsilon, \sigma_0) > 0$ *such that*

$$\|(1 - \chi_M)u\|_{H^{s-\sigma,p}} \leq \varepsilon\|u\|_{H^{s,p}} \qquad (s \in \mathbf{R},\ \sigma \geq \sigma_0).$$

The proof see Ya. Roitberg [R1, p. 35].

Corollary 1.8.1. *Let $s_1 < s < s_2$. Then for any $\varepsilon > 0$ there exists a constant $C(\varepsilon) > 0$ such that*

$$\|u\|_{H^{s,p}} \leq \varepsilon \|u\|_{H^{s_2,p}} + C(\varepsilon) \|u\|_{H^{s_1,p}} .$$

The proof see [R1, p. 36].

Consider the mapping

$$u \mapsto \chi u$$

in the space $H^{s,p}(\mathbf{R}^n, \lambda, \rho)$, where $\chi \in C^\infty(\mathbf{R}^n)$ and all the derivatives of χ are bounded.

We have

$$\begin{aligned} \|\chi u\|_{H^{s,p}} &= \left\| F^{-1} \left(|\xi|^2 + q^2\right)^{\mu_\rho/2} \left(1 + |\xi|^2\right)^{(s-\mu_\rho)/2} F(\chi u)(\xi) \right\|_{L_p} \\ &= \left\| \left(-\Delta + q^2\right)^{\mu_\rho/2} (1 - \Delta)^{(s-\mu_\rho)/2} (\chi u) \right\|_{L_p} . \end{aligned}$$

Therefore, we have the estimate

$$\|\chi u\|_{H^{s,p}} \leq \|\chi\|_{L_\infty} \|u\|_{H^{s,p}} + C_\chi \|u\|_{H^{s-1,p}} \quad (s \in \mathbf{R}). \tag{1.8.6}$$

Here

$$\|\chi\|_{L_\infty} := \sup_{x \in \mathbf{R}^n} |\chi(x)| ,$$

$C_\chi > 0$ does not depend on χ, s, λ.

1.8.3. Mixed-order operator on compact manifold

To prove Theorem we need the following two lemmas.

Lemma 1.8.7. *Let $x_0 \in \Omega$, $p \in (1, \infty)$, $s \in \mathbf{R}$. For any $\varepsilon > 0$ there exists a neighborhood $U = U(x_0) \subset \Omega$ of $x_0 \in \Omega$ and a number $q_0 > 0$ such that for every function $\varphi \in C_0^\infty(U)$, the expression*

$$\varphi(x)(A(x, D) - \lambda I)$$

can be represented in the form

$$\varphi(x)(A(x, D) - \lambda I) = \varphi(x)\left[(A_0(x_0, D) - \lambda I) + Q(x, D) + A'(x, D)\right].$$

Here $A_0(x, D)$ is the principal part of $A(x, D)$,

$$Q(x, D) := (Q_{jk}(x, D))_{j,k=1,\dots,N}, \quad x \in \mathbf{R}^n,$$

ord $Q_{jk} = s_j + t_k$ for $s_j + t_k \geq 0$, and $Q_{jk} = 0$ for $s_j + t_k < 0$; $Q(x, D)$ is an operator with a small norm, i.e.

$$\|Q(x, D) u\|_{H^{s-s,p}(\mathbf{R}^n)} \leq \varepsilon \|u\|_{H^{T+s,p}(\mathbf{R}^n)} \quad (\forall u \in H^{T+s,p}(\mathbf{R}^n), \ |\lambda| \geq q_0).$$

$$A'(x, D) := \left(A'_{jk}(x, D) \right)_{j,k=1,\ldots,N}, \quad x \in \mathbf{R}^n,$$

ord $A'_{jk} < s_j + t_k$ for $s_j + t_k \geq 1$, and $A'_{jk} = 0$ for $s_j + t_k \leq 0$. The expressions Q and A' do not depend on p, s.

Lemma 1.8.7 is a matrix analog of Y. Roitberg [R1, Ch.9, Lemma 9.3.1] and its proof, in view of Lemmas 1.8.1–1.8.6, is just the same.

The following Lemma 1.8.8 is similar to [R1, Ch.9, Lemma 9.3.2].

Lemma 1.8.8. *Assume that $x_0 \in \Omega$, $p \in (1, \infty)$, and $s \in \mathbf{R}$. There exists a neighborhood $U = U(x_0) \subset \Omega$ of the point x_0 and a number $q_0 > 0$ such that, for every function $\varphi \in C_0^\infty(U)$, the expression*

$$\varphi(x)(A(x, D) - \lambda I)$$

admits a representation

$$\varphi(x)(A(x, D) - \lambda I) = \varphi(x)\left(\tilde{A}(x, D) - \lambda I\right)$$

where $\tilde{A}(x, D)$ $(x \in \mathbf{R}^n)$ is a differential expression of order (T, S) and the mapping $u \mapsto \left(\tilde{A}(x, D) - \lambda I\right) u$ establishes, for $q \geq q_0$, an isomorphism

$$H^{T+s}(\mathbf{R}^n, \lambda, \rho) \mapsto H^{s-S}(\mathbf{R}^n, \lambda, \rho). \tag{1.8.7}$$

Furthermore, there exists a constant $C > 0$ independent of u and q $(q \geq q_0)$ such that

$$C^{-1} \|u\|_{H^{T+s,p}} \leq \left\|\left(\tilde{A}(x, D) - \lambda I\right) u\right\|_{H^{s-S,p}} \leq C|\lambda| \|u\|_{H^{T+s,p}}$$

The expression $\tilde{A}(x, D)$ does not depend on s or p.

To prove Theorem 1.8.3 we can use, in view of Lemmas 1.8.7 and 1.8.8, the standard localization technique, just as it has been done in the proof of Theorem 9.1.1 from the monograph by Y. Roitberg [R1, Ch. 9, Sect. 9.3].

Remark 1.8.2. It follows immediately from Theorem a statement on global and local increase in smoothness of generalized solutions. If $u \in H^{T+s,p}(\Omega)$ is a solution of the problem (1.8.1) and $f \in H^{s-S+k,p}(\Omega)$ $(k \geq 0)$, then

$u \in H^{T+s+k,p}(\Omega)$. If $f \in H^{s-S+k,p}_{\text{loc}}(\Omega_0)$, then $u \in H^{T+s+k,p}_{\text{loc}}(\Omega_0)$ $(\Omega_0 \subset \Omega$ is an open subset).

Remark 1.8.3. Let $f(x)$ has a power singularity near closed smooth i-dimensional manifold $\Gamma \subset \Omega$ $0 \leq i \leq n-1$. Then a regularization f of $f(x)$ belongs to the space $H^{s-S,p}(\Omega)$ with some $s < 0$, depending on the degree of the power singularity of $f(x)$ in the neighborhood of Γ ([R1, Sect. 8.3]) and $u \in H^{T+s,p}(\Omega)$. The statement on the local increase in smoothness enable us to investigate the solution outside Γ and to calculate the degree of its singularity.

Remark 1.8.4. Let $\rho(x) \in C^\infty(\Omega)$ be a function which equals dist(x,Γ) near Γ. The equation $(\rho(x))^k (A(x,D) - \lambda I) u(x) = f(x)$ $\quad x \in \Omega$ $(k > 0$ is an arbitrary fixed number) is strongly degenerate near Γ. Changing this equation by the equivalent one

$$(A(x,D) - \lambda I) u(x) = f(x)(\rho(x))^{-k}$$

we reduce the problem to the problem considered in **2)** (cf. [R1, Sect. 8.5]).

CHAPTER 2

ELLIPTIC BOUNDARY VALUE PROBLEMS FOR GENERAL SYSTEMS OF EQUATIONS WITH ADDITIONAL UNKNOWN FUNCTIONS DEFINED AT THE BOUNDARY OF A DOMAIN

The theory of general elliptic boundary value problems for systems of equations is explained in Chapter 1 (see also [R1, Ch.X], and bibliography there).

In problems of ellastisity theory and hydrodynamics, for example in the works of Aslanjan, Vassiliev and Lidskii [AVL], Garlet [Gar], Nazarov and Pileckas [NaP], there arise boundary value problems for general elliptic systems whose boundary conditions contain both the functions $u_1, \ldots, u_N$ contained in the system and the additional unknown functions $u'_{N+1}, \ldots, u'_{N+k}$ defined at the boundary of the domain. The number of the boundary conditions increases respectively.

The present chapter is devoted to the investigation of the solvability of these problems in complete scales of Banach spaces. These results belong to I. Ya. Roitberg. To explain them let us follow to the works [RI1]–[RI5].

2.1. Statement of a Problem

Let $G \subset \mathbf{R}^n$ be a bounded domain with boundary $\partial G \in C^\infty$. Consider the following boundary value problem:

$$l(x, D)u = f(x) \quad (x \in G); \tag{2.1.1}$$

$$b(x, D)u\big|_{\partial G} + b'(x', D')u' = \varphi(x') \quad (x' \in \partial G). \tag{2.1.2}$$

The system

$$lu := (l_{rj}(x, D))_{r,j=1,\ldots,N}$$

and the boundary differential expressions

$$b(x, D) := (b_{hj}(x, D))_{\substack{h=1,\ldots,m+k'' \\ j=1,\ldots,N}}, \quad b'(x', D') := (b'_{hj}(x', D'))_{\substack{h=1,\ldots,m+k'' \\ j=N+1,\ldots,N+k'}}$$

are described by the following relations:
$\operatorname{ord} l_{rj} \le s_r + t_j$ for $s_r + t_j \ge 0$, $l_{rj} = 0$ for $s_r + t_j < 0$, $r, j = 1, \dots, N$;
$\operatorname{ord} b_{hj} \le \sigma_h + t_j$ for $\sigma_h + t_j \ge 0$, $b_{hj} = 0$ for $\sigma_h + t_j < 0$, $j = 1, \dots, N$, $h = 1, \dots, m + k''$;
$\operatorname{ord} b'_{hj} \le \sigma_h + t_j$ for $\sigma_h + t_j \ge 0$, $b'_{hj} = 0$ for $\sigma_h + t_j < 0$, $j = N+1, \dots, N+k'$, $h = 1, \dots, m + k''$.

Here $t_1, \dots, t_{N+k'}, s_1, \dots, s_N, \sigma_1, \dots, \sigma_{m+k''}$ are given integers such that

$$s_1 + \dots + s_N + t_1 + \dots + t_N = 2m$$

and

$$t_1 \ge \dots \ge t_N \ge 0 = s_1 \ge \dots \ge s_N, \ \sigma_1 \ge \dots \ge \sigma_{m+k''}.$$

In addition, $u(x) = (u_1(x), \dots, u_N(x))$, $x \in \overline{G}$, and, for $x' \in \partial G$,

$$u'(x') = \left(u'_{N+1}(x'), \dots, u'_{N+k'}(x')\right)', \ u'_j(x') = \left(u'_{j1}(x'), \dots, u_{j,\sigma_1+t_j+1}(x')\right)'.$$

Moreover,

$$l_{rj}(x, D) = \begin{cases} \sum_{|\mu| \le s_r + t_j} a_\mu^{rj}(x) D^\mu, & \text{for } s_r + t_j \ge 0 \\ 0, & \text{for } s_r + t_j < 0 \end{cases} \qquad (r, j = 1, \dots, N),$$

and

$$b_{hj}(x, D) u_j(x) = \sum_{k=1}^{\sigma_h + t_j + 1} \Lambda_{hjk}(x', D') u_{jk}(x') \tag{2.1.3}$$

$$(j = 1, \dots, N, \quad h = 1, \dots, m + k''),$$

where $u_{jk} = D_\nu^{k-1} u_j \big|_{\partial G}$, $D_\nu = i\partial/\partial\nu$, ν is a normal to ∂G,

$$b'_{hj}(x', D') u'_j(x') = \sum_{1 \le k \le \sigma_h + t_j + 1} \Lambda_{hjk}(x', D') u'_{jk}(x') \tag{2.1.4}$$

$$(j = N+1, \dots, N+k', \quad h = 1, \dots, m + k''),$$

where $\Lambda_{hjk}(x', D')$ are tangential operators of order $\sigma_h + t_j - k + 1$ for h, j such that $\sigma_h + t_j \ge 0$.

Here and in what follows we assume that the coefficients of all differential expressions and the boundary ∂G are infinitely smooth.

In a natural way we introduce the notion of ellipticity of the problem (2.1.1), (2.1.2).

2.2. Definition of Ellipticity of Problem (2.1.1)–(2.1.2)

Let $L_0(x,\xi) := \det(l_0(x,\xi))$, where $l_0(x,\xi) = (l_{rj0}(x,\xi))$ is the principal symbol of the matrix $l(x,D)$:

$$l^0_{rj}(x,\xi) = \begin{cases} \sum_{|\mu|=s_r+t_j} a^{rj}_\mu(x)D^\mu\,, & \text{for } s_r+t_j \geq 0, \\ 0\,, & \text{for } s_r+t_j < 0, \end{cases} \qquad r,j=1,\dots,N,$$

The expression $l(x,D)$ is called *elliptic* in $\overline{G}$, if

$$L_0(x,\xi) \neq 0 \quad (\forall x \in \overline{G},\ \forall \xi \in \mathbf{R}^N \setminus \{0\}). \tag{2.2.1}$$

Let $x \in \partial G$, let $\tau \neq 0$ be an arbitrary real vector, tangential to ∂G at the point x, and let ν be a unit vector of interior normal to ∂G at this point. If system (2.1.1) is elliptic then the polynomial

$$L_0(\zeta) = L_0(x,\tau+\zeta\nu)$$

of order $r = |S|+|T| = s_1+\cdots+s_N+t_1+\cdots+t_N$ does not have real roots. A system (2.1.1) elliptic in $\overline{G}$ is called *properly elliptic* (see, for example, [R1]), if for every point $x \in \partial G$ and any $\tau \neq 0$ the polynomial $L_0(\zeta)$ has the even order ($r = 2m$), and accurately m of its roots have positive imaginary parts.

For $n > 2$, any elliptic system is properly elliptic.

If the system (2.1.1) is properly elliptic then $L_0(\zeta) = L_+(\zeta)L_-(\zeta)$, where L_+ (L_-) is the mth-order polynomial whose all roots lie in the upper (lower) half-plane. Therefore if $l_0(\zeta) = l_0(x,\tau+\zeta\nu)$ then for every point $x \in \partial G$ and any vector $\tau \neq 0$ tangential to ∂G at the point $x \in \partial G$ the space $\mathcal{M}_+ = \mathcal{M}_+(x,\tau)$ of the stable (i.e., decreasing as $t \longrightarrow \infty$) solutions of the equations $l_0(D_t) = 0$ ($D_t = i\partial/\partial t$) is an m-dimensional space .

On ∂G we consider $m+k''$ boundary conditions (2.1.2). Let $b^0_{hj}(x,D)$ and $b'^0_{hj}(x',D')$ be the principal parts of the expressions $b_{hj}(x,D)$ and $b'_{hj}(x',D')$ respectively, let $b_0(x,D)$ and $b'_0(x',D')$ be the principal parts of the matrixes $b_{hj}(x,D)$ and $b'_{hj}(x',D')$ respectively, and let

$$b_0(\zeta) = b_0(x,\tau+\zeta\nu), \quad b'_0(\tau) = b'_0(x',\tau).$$

Definition 2.2.1. *The ptoblem (2.1.1), (2.1.2) is called elliptic, if system (2.1.1) is properly elliptic in $\overline{G}$ and satisfies Lopatinsky condition:*
(L1). *For every point $x \in \partial G$, for any vector $\tau \neq 0$ tangential to ∂G at the point $x \in \partial G$ and for every $h = (h_1,\dots,h_{m+k''}) \in \mathbf{C}^{m+k''}$ the problem*

$$l_0(D_t)V = 0 \quad (t > 0), \qquad b_0(D_t)V\big|_{t=0} + b'_0(\tau)V' = h \tag{2.2.2}$$

has one and only one solution $(V, V'), V \in \mathcal{M}_+(x, \tau)$.

Since in the Lopatinsky condition the point x is an arbitrary fixed point of the boundary ∂G, vector ν is the interior normal to ∂G at this point, and only the principal parts l_0, b_0 and b_0' whose coefficients are frozen at the point x appear here, it is naturally to consider a problem of the form of (2.1.1)–(2.1.2) in the halfspace $\mathbf{R}^n_+ = \{x = (x', x_n) \in \mathbf{R}^n,\ x_n > 0\}$ in the case where all coefficients are the complex constants, and all expressions are uniform with respect to the derivatives

$$l_0(D)u := (l^0_{ij}(D))_{i,j=1,\dots,N} u(x) = 0, \quad x \in \mathbf{R}^n_+, \tag{2.2.3}$$

$$\begin{gathered} b_0(D)u(x)\big|_{x_n=0} + b_0'(D')u'(x') := \\ (b^0_{hj}(D))_{\substack{h=1,\dots,m+k'' \\ j=1,\dots,N}} u(x)\big|_{x_n=0} + (b'^0_{hj}(D'))_{\substack{h=1,\dots,m+k'' \\ j=N+1,\dots,N+k'}} u'(x') = \varphi(x') \\ (x' \in \mathbf{R}^{n-1} = \{x \in \mathbf{R}^n : x_n = 0\}). \end{gathered} \tag{2.2.4}$$

Using the partial Fourier transform $F' = F'_{x' \mapsto \xi'}$ and setting

$$F'(u(x', x_n)) = \widehat{u}(\xi', x_n), \qquad F'(u'(x'))(\xi') = \widehat{u}'(\xi')$$

we obtain from (2.2.3) and (2.2.4) that

$$l_0(\xi', D_n)\widehat{u}(\xi', x_n) = 0, \qquad\qquad x_n > 0, \tag{2.2.5}$$

$$b_0(\xi', D_n)\widehat{u}(\xi', x_n)\big|_{x_n=0} + b_0(\xi')\widehat{u}'(\xi') = \widehat{\varphi}(\xi'). \tag{2.2.6}$$

Denote by $\mathcal{M}_+ = \mathcal{M}_+(\xi')$ the space of stable solutions of equation (2.2.5), $\xi' \neq 0$. It was above mentioned that the proper ellipticity yields that the space $\mathcal{M}_+$ is m-dimensional. Let the set

$$(e_i(\xi', x_n))_{i=1,\dots,m}$$

be a basis in $\mathcal{M}_+$. Then every stable solution of equation (2.2.5) has a form

$$\widehat{u}(\xi', x_n) = \sum_{i=1}^{m} c_i e_i(\xi', x_n),$$

and we rewrite relation (2.2.6) in the following form:

$$\sum_{i=1}^{m} c_i b_0(\xi', D_n) e_i(\xi', x_n)\big|_{x_n=0} + b_0(\xi')\widehat{u}'(\xi_n') = \widehat{\varphi}(\xi'). \tag{2.2.7}$$

System (2.2.7) is a linear system of $m + k''$ equations with respect to the variables

$$(c_1, \ldots, c_m, \widehat{u}'_{N+1}(\xi'), \ldots, \widehat{u}'_{N+k'}(\xi')), \qquad \widehat{u}'_j = (\widehat{u}'_{j1}, \ldots, \widehat{u}'_{j,\sigma_1+t_j+1}).$$

Thus, the Lopatinskii condition is equivalent to the following condition: system (2.2.7) is square, and its determinant does not equal to zero.

Therefore

$$m + k'' = m + (\sigma_1 + t_{N+1} + 1 + \sigma_1 + t_{N+2} + 1 + \cdots + \sigma_1 + t_{N+k'} + 1),$$

$$k'' = k'(\sigma_1 + 1) + t_{N+1} + \cdots + t_{N+k'}. \tag{2.2.8}$$

That is why in what follows we suppose that in the problem (2.1.1)–(2.1.2) the number k' is connected with k'' by relation (2.2.8).

Hence, one can formulate the Lopatinskii condition in the following forms, equivalent to each other:

Lop1. *For any point $x \in \partial G$, for any vector $\tau \neq 0$ tangential to ∂G at the point x, and for any element $h = (h_1, \ldots, h_{m+k''}) \in \mathbf{C}^{m+k''}$ problem (2.2.2) has a solution (V, V') such that $V \in \mathcal{M}_+(x, \tau)$.*

Lop2. *For any point $x \in \partial G$, for any vector $\tau \neq 0$ tangential to ∂G at the point x, and for any element $h \in \mathbf{C}^{m+k''}$ problem (2.2.2) has not more than one solution (V, V'), $V \in \mathcal{M}_+(x, \tau)$.*

In Subsection 2.5.4 below we give another form of the Lopatinskii condition.

Throughout what follows we assume that the problem (2.1.1)–(2.1.2) is elliptic.

To study this problem we need, beforhand, to study a *model problem* in the half-space $\mathbf{R}^n_+$.

Let us change the operator $D_j = F'^{-1}\xi_j F'$ in (2.2.3) and (2.2.4) by the operator

$$\widehat{D}_j = F'^{-1}\frac{\xi_j}{|\xi'|}(1 + |\xi'|)F', \tag{2.2.9}$$

$$j = 1, \ldots, n-1, \ \xi' = (\xi_1, \ldots, \xi_{n-1}), \ |\xi'| = \left(\xi_1^2 + \cdots + \xi_{n-1}^2\right)^{1/2}.$$

It is clear that (see [R1, §4.2]) the operator $\widehat{D}_j$, $(j = 1, \ldots, n-1)$, is a tangential operator of first order, and the order of the operator $\widehat{D}_j - D_j$ is equal to zero.

For every kth-order linear differential expression

$$C(D) = C(D_1, \ldots, D_{n-1}, D_n) = C(D', D_n)$$

with constant coefficients we set:

$$\widehat{C}(D) = C(\widehat{D}_1, \ldots, \widehat{D}_{n-1}, D_n).$$

It easy to see that the order of the operator $\widehat{C}(D)$ is also k, meanwhile the order of the operator $\widehat{C}(D) - C(D)$ is equal to $k-1$ (see [R1, §4.2]).

The model problem is following:

$$\widehat{l}_0(D)u(x) = f(x), \qquad x \in \mathbf{R}_n^+, \tag{2.2.10}$$

$$\widehat{b}_0(D)u(x)\big|_{x_n=0} + \widehat{b}_0'(D')u'(x') = \varphi(x'), \qquad x' \in \mathbf{R}^{n-1}. \tag{2.2.11}$$

To study both the elliptic problem (2.1.1)–(2.1.2) and the model problem (2.2.10)–(2.2.11) it is necessary to introduce relevant functional spaces.

2.3. Functional Spaces

Let $\mathbf{\Omega} \subset \mathbf{R}^n$ be a domain with infinitely smooth boundary $\partial\mathbf{\Omega}$. We shall consider only the cases where $\mathbf{\Omega} = G$ is a bounded domain with the boundary $\partial\mathbf{\Omega} = \partial G$ or $\mathbf{\Omega} = \mathbf{R}^n_\pm$ with the boundary $\partial\mathbf{\Omega} = \mathbf{R}^{n-1}$.

In what follows we need the spaces $H^{s,p}(\mathbf{\Omega})$, $B^{s,p}(\partial\mathbf{\Omega})$ and $\widetilde{H}^{s,p,(r)}(\mathbf{\Omega})$, $s \in \mathbf{R}$, $1 < p < \infty$, and r is a nonnegative integer. These spaces were defined in Subsection 1.1.2 (see also [R1, Ch. I], [R1, Ch. II]). For convenience we recall here some of the definitions. In addition, here and below we assume that $s, t \in \mathbf{R}$, p, $p' \in (1, \infty)$, and $1/p + 1/p' = 1$.

2.3.1. Space $H^{s,t,p}(\mathbf{R}^n)$.

Let $H^{s,t,p} = H^{s,t,p}(\mathbf{R}^n)$ denote the Banach space of distributions f with the norm

$$\|f\|_{s,t,p} = \|f, \mathbf{R}^n\|_{s,t,p} = \left\|F^{-1}(1+|\xi|^2)^{s/2}(1+|\xi'|^2)^{t/2}Ff\right\|_{L_p}. \tag{2.3.1}$$

The space $H^{s,t,p}$ is dual to the space $H^{-s,-t,p'}$ with respect to the extension $(\cdot,\cdot)_{\mathbf{R}^n}$ of the scalar product in $L_2(\mathbf{R}^n)$ ([R1, §1.4]).
Roughly speaking, the elements of the space $H^{s,t,p}$ have both smoothness s with respect to all the variables and additional smoothness t with respect to the tangential variables. Denote by $H^{s,p} = H^{s,p}(\mathbf{R}^n)$ the space $H^{s,t,p}$ for $t = 0$, i. e. we set:

$$H^{s,p} = H^{s,p}(\mathbf{R}^n) := H^{s,0,p}, \qquad \|f\|_{s,p} = \|s, \mathbf{R}^n\|_{s,p} := \|f\|_{s,0,p}. \tag{2.3.2}$$

For s, $t \in \mathbf{R}$, and $1 < p < \infty$ we denote by $H^{s,t,p}_\pm = H^{s,t,p}_\pm(\mathbf{R}^n)$ the subspace of $H^{s,t,p}$ that consists of the elements whose supports lie in $\overline{\mathbf{R}^n_\pm}$, i.e.,

$$H^{s,t,p}_\pm = \left\{f \in H^{s,t,p}(\mathbf{R}^n) : \operatorname{supp} f \subset \overline{\mathbf{R}^n_\pm}\right\} \subset H^{s,t,p}. \tag{2.3.3}$$

2.3.2. Space $H^{s,t,p}(\mathbf{R}^n_+)$. Let $H^{s,t,p}(\mathbf{R}^n_+)$ $(s,t \in \mathbf{R},\ s > 0,\ p \in]1,\infty[)$ denote the space of restrictions of the elements of $H^{s,t,p}(\mathbf{R}^n)$ to $\mathbf{R}^n_+$ with the norm of factor space, i.e.,

$$\|f, \mathbf{R}^n_+\|_{s,t,p} = \inf_{\substack{g \in H^{s,t,p} \\ g|_{\mathbf{R}^n_+} = f}} \|g, \mathbf{R}^n\|_{s,t,p}, \tag{2.3.4}$$

$$H^{s,t,p}(\mathbf{R}^n_+) = H^{s,t,p}/H^{s,t,p}_-, \qquad s,t \in \mathbf{R}, \quad s \geq 0. \tag{2.3.5}$$

By $H^{-s,-t,p'}(\mathbf{R}^n_+)$ we denote the space dual to $H^{s,t,p}(\mathbf{R}^n_+)$ with respect to the extension $(\cdot,\cdot)_{\mathbf{R}^n_+}$ of the scalar product in $L_2(\mathbf{R}^n_+)$. Equality (2.3.5) yields that the space $H^{s,t,p}(\mathbf{R}^n_+)$ for $s \leq 0$ is isometrically equivalent to the subspace $H^{s,t,p}_+(\mathbf{R}^n)$ of the space $H^{s,t,p}(\mathbf{R}^n)$ (see [R1, §1.7]), i.e.,

$$H^{s,t,p}(\mathbf{R}^n_+) = (H^{-s,-t,p'}(\mathbf{R}^n_+))^* = \left(H^{-s,-t,p'}(\mathbf{R}^n_+)/H^{-s,-t,p'}_-\right)^* \simeq H^{s,t,p}_+(\mathbf{R}^n). \tag{2.3.6}$$

2.3.3. Space $H^{s,p}(\Omega)$. By $H^{s,p}(\Omega)$ $(s \geq 0)$ we denote the space of Bessel potentials, i.e., the space of restrictions of the elements of $H^{s,p}(\mathbf{R}^n)$ onto Ω with the factor topology

$$H^{s,p}(\Omega) = H^{s,p}(\mathbf{R}^n)/H^{s,p}_{C\Omega}(\mathbf{R}^n), \quad s \geq 0, \tag{2.3.7}$$

were

$$H^{s,p}_{C\Omega}(\mathbf{R}^n) = \{f \in H^{s,p}(\mathbf{R}^n) : \operatorname{supp} f \subset C\Omega = \mathbf{R}^n \setminus \Omega\}.$$

Let $H^{-s,p}(\Omega)$ $(s > 0)$ denote the space dual to $H^{s,p'}(\Omega)$ with respect to the extension $(\cdot,\cdot)_\Omega$ of the scalar product in $L_2(\Omega)$, i.e.,

$$H^{-s,p}(\Omega) = (H^{s,p'}(\Omega))^* = \left(H^{s,p'}(\mathbf{R}^n)/H^{s,p'}_{C\Omega}(\mathbf{R}^n)\right)^* \simeq H^{-s,p}_{\overline{\Omega}}(\mathbf{R}^n), \tag{2.3.8}$$

where

$$H^{-s,p}_{\overline{\Omega}}(\mathbf{R}^n) = \{f \in H^{-s,p}(\mathbf{R}^n) : \operatorname{supp} f \subset \overline{\Omega}\}$$

denotes a subspace of $H^{-s,p}(\mathbf{R}^n)$.
Let $\|\cdot\|_{s,p} = \|\cdot,\Omega\|_{s,p}$ denote the norm in $H^{s,p}(\Omega)$, $s \in \mathbf{R}$.

2.3.4. Space $B^{s,p}(\partial\Omega)$. By $B^{s,p}(\partial\Omega)$ $s \in \mathbf{R}^n, p \in]1,\infty[$ we denote the Besov spaces with the norm $\langle\langle\cdot,\partial\Omega\rangle\rangle_{s,p}$. The spaces $B^{s,p}(\partial\Omega)$ and $B^{-s,p'}(\partial\Omega)$ are dual to each other with respect to the extension $\langle\cdot,\cdot\rangle_{\partial\Omega}$ of the scalar product in $L_2(\partial\Omega)$. For $s > 0$, the space $B^{s,p}(\partial\Omega)$ coincides with the space of the traces of elements of $H^{s+1/p,p}(\Omega)$ on $\partial\Omega$ with the factor topology.

2.3.5. Space $\widetilde{H}^{s,p,(r)}(\Omega)$. Let r be a fixed natural number, $s, p \in \mathbf{R}$, $1 < p < \infty$, and $s \neq k + 1/p$ for $k = 0, \dots, r-1$.

By $\widetilde{H}^{s,p,(r)}(\Omega)$ we denote the completion of $C_0^\infty(\overline{\Omega})$ in the norm

$$|||u, \Omega|||_{s,p,(r)} := \left(\|u, \Omega\|_{s,p}^p + \sum_{j=1}^r \langle\langle D_\nu^{j-1} u, \partial\Omega \rangle\rangle_{s-j+1-1/p,p}^p \right)^{1/p}. \qquad (2.3.9)$$

If $r = 0$ then we set

$$\widetilde{H}^{s,p,(0)}(\Omega) := H^{s,p}(\Omega), \quad |||u, \Omega|||_{s,p,(0)} = \|u, \Omega\|_{s,p}.$$

Finally, for excluded values of s, i.e., for $s = k + 1/p$, we define the space $\widetilde{H}^{s,p,(r)}(\Omega)$ and the norm (2.3.9) by the method of complex interpolation (see, for example, [R1, §1.14]).

The closure $S = S_{s,p}$ of the mapping

$$u \mapsto \left(u|_{\overline{\Omega}}, u|_{\partial\Omega}, \dots, D_\nu^{r-1} u|_{\partial\Omega} \right), \qquad u \in C_0^\infty(\overline{\Omega}),$$

is an isometry between $\widetilde{H}^{s,p,(r)}(\Omega)$ and the subspace $\mathcal{H}_0^{s,p,(r)}$ of the space

$$\mathcal{H}^{s,p,(r)} := H^{s,p}(\Omega) \times \prod_{j=1}^r B^{s-j+1-1/p,p}(\partial\Omega). \qquad (2.3.10)$$

In addition (see [R1]),

$$\mathcal{H}_0^{s,p,(r)} = \Big\{ (u_0, u_1, \dots, u_r) \in \mathcal{H}^{s,p,(r)} : u_0 \in H^{s,p}(\Omega),\ u_j \in B^{s-j+1-1/p,p}(\partial\Omega);$$

$$\text{if} \quad s-j+1-1/p > 0 \quad \text{then} \quad u_j = D_\nu^{j-1} u_0|_{\partial\Omega};$$

$$\text{if } s-j+1-1/p < 0 \text{ then } u_j \text{ does not depend on } u_0 \Big\}.$$

In particular, if $s < 1/p$ then $\mathcal{H}_0^{s,p,(r)} = \mathcal{H}^{s,p,(r)}$.

Since the space $\widetilde{H}^{s,p,(r)}(\Omega)$ is isometrically equivalent to $\mathcal{H}_0^{s,p,(r)}(\Omega)$, we identify these spaces with each other. For any element $u \in \widetilde{H}^{s,p,(r)}(\Omega)$, we write

$$u = (u_0, u_1, \dots, u_r) = Su \in \mathcal{H}_0^{s,p,(r)}. \qquad (2.3.11)$$

Thus, the space $\widetilde{H}^{s,p,(r)}(\Omega)$ consists of vectors $u = (u_0, u_1, \dots, u_r)$, such that $u_0 \in H^{s,p}(\Omega)$, $u_j \in B^{s-j+1-1/p,p}(\partial\Omega)$; $u_j = D_\nu^{j-1} u_0|_{\partial\Omega}$ for $s-j+1-1/p > 0$, and u_j does not depend on u_0 for $s-j+1-1/p < 0$.

Let $s \neq k + 1/p$, $k = 0, \ldots, r-1$.

It is clear that for $s > r - 1 + 1/p > 0$, norm (2.3.9) is equivalent to the norm $\|u, \Omega\|_{s,p}$, and the space $\widetilde{H}^{s,p,(r)}(\Omega)$ coincides with $H^{s,p}(\Omega)$.

On the other hand , if $1/p < s < r - 1$ then norm (2.3.9) is equivalent to the norm

$$\left(\|u, \Omega\|^p_{s,p} + \sum_{j=k+1}^{r} \langle\langle D^{j-1}_\nu u, \partial\Omega \rangle\rangle^p_{s-j+1-1/p,p} \right)^{1/p},$$

where $k = [s + 1 - 1/p]$ is the integer part of $s + 1 - 1/p$. Therefore,

$$\widetilde{H}^{s,p,(r)}(\Omega) \simeq H^{s,p}(\Omega) \times \prod_{j=k+1}^{r} B^{s-j+1-1/p,p}(\partial\Omega).$$

Finally, if $s < 1/p$ then

$$\widetilde{H}^{s,p,(r)}(\Omega) \simeq H^{s,p}(\Omega) \times \prod_{j=1}^{r} B^{s-j+1-1/p,p}(\partial\Omega).$$

2.3.6. Let $M(x, D)$ $(x \in \overline{\Omega})$ be a qth-order differential expression with infinitely smooth coefficients; all derivatives of the coefficients are bounded. It turns out (see [R1]) that if $q \leq r$ then the closure M of the mapping

$$u \mapsto Mu|_{\overline{\Omega}} \quad (u \in C^\infty_0(\overline{\Omega})),$$

acts continuously from the whole space $\widetilde{H}^{s,p,(r)}(\Omega)$ into $H^{s-q,p}(\Omega)$, $s \in \mathbf{R}$. If $q \leq r - 1$ then for any $s \in \mathbf{R}$ the closure of the mapping $u \mapsto Mu|_{\partial\Omega}$ acts continuously in the pair of spaces

$$\widetilde{H}^{s,p,(r)}(\Omega) \to B^{s-q-1/p,p}(\partial\Omega).$$

This yields that for every $s \in \mathbf{R}$ the closure of the mapping $u \mapsto Mu$ acts continuously from the whole space $\widetilde{H}^{s,p,(r)}(\Omega)$ into $\widetilde{H}^{s-q,p,(r-q)}(\Omega)$.

Let us give another formulas those give us a possibility to calculate the expressions $Mu|_{\overline{\Omega}}$ and $Mu|_{\partial\Omega}$ for the element

$$u = (u_0, \ldots, u_r) \in \widetilde{H}^{s,p,(r)}(\Omega).$$

Let

$$M(x, D) = \sum_{j=0}^{q} M_j(x, D') D^j_\nu \qquad (x \in \overline{\Omega}), \tag{2.3.12}$$

where $M_j(x, D')$ are tangential differential expressions of orders $\leq q-1$. By integration by parts we easily obtain

$$(Mu, v) = (u, M^+v) - i\sum_{j=1}^{q}\sum_{k=1}^{j}\langle D_\nu^{k-1}u|_{\partial\Omega}, D_\nu^{j-k}M_j^+v\rangle, \qquad (2.3.13)$$

$$u, v \in C_0^\infty(\overline{\Omega}).$$

Here M^+ and M_j^+ are the expressions formally adjoint to the expressions M and M_j, respectively. It follows from equality (2.3.13) that

$$(Mu)_+ = Mu_+ - i\sum_{j=1}^{q}\sum_{k=1}^{j} M_j D_\nu^{j-k}(D_\nu^{k-1}u|_{\partial\Omega} \times \delta(\partial\Omega)), \qquad (2.3.14)$$

where $(Mu)_+$ and u_+ are the extensions by zero of the functions Mu and u to $\mathbf{R}^n$, respectively, and $\delta(\partial\Omega)$ is the Dirac measure, i.e.,

$$\left(\left(D_\nu^{k-1}u|_{\partial\Omega} \times \delta(\partial\Omega)\right), v\right)_{\mathbf{R}^n} = \langle D_\nu^{k-1}u|_{\partial\Omega}, v\rangle_{\partial\Omega}.$$

In the set of the expressions of form (2.3.12) we introduce the operator J:

$$JM(x, D) = \begin{cases} 0, & \text{if } q = 0 \\ \sum_{j=1}^{q} M_j(x, D')D_\nu^{j-1}, & \text{if } q \geq 1. \end{cases}$$

Then one can rewrite (2.3.14) in the form of

$$(Mu)_+ = Mu_+ - i\sum_{k=1}^{q}(J^kM)(D_\nu^{k-1}u|_{\partial\Omega} \times \delta(\partial\Omega)). \qquad (2.3.15)$$

If $u = (u_0, u_1, \ldots, u_r) \in \widetilde{H}^{s,p,(r)}(\Omega)$ $(s \in \mathbf{R})$, then, by passing to the limit, it is easy to see that $Mu|_{\overline{\Omega}}$ belongs to $H^{s-q,p}(\Omega)$ if and only if

$$(Mu)_+ = Mu_{0+} - i\sum_{k=1}^{q}(J^kM)(u_k \times \delta(\partial\Omega)). \qquad (2.3.16)$$

Here note that if $s - q \leq 0$ then $H^{s-q,p}(\Omega) \simeq H_{\overline{\Omega}}^{s-q,p}(\mathbf{R}^n)$ (see (2.3.8)), and $Mu|_{\overline{\Omega}} = (Mu)_+$. Formula (2.3.16) gives us a rule of the calculation of $Mu|_{\overline{\Omega}}$ for the element $u = (u_0, u_1, \ldots, u_r) \in \widetilde{H}^{s,p,(r)}(\Omega)$.

Similarly, by passing to the limit, it is easy to verify that if $q \leq r-1$ then for any

$$u = (u_0, \ldots, u_r) \in \widetilde{H}^{s,p,(r)}(\Omega)$$

the element $Mu|_{\partial\Omega}$ belongs to $B^{s-q-1,p,p}(\partial\Omega)$ if and only if

$$Mu|_{\partial\Omega} = \sum_{j=0}^{q} M_j(x, D')u_{j+1}, \qquad s \in \mathbf{R}. \tag{2.3.17}$$

2.4. Theorem on Complete Collection of Isomorphisms

2.4.1. Consider problem (2.1.1)–(2.1.2).
Let

$$æ = \max\{0, \sigma_1 + 1, \dots, \sigma_{m+k''} + 1\}, \qquad \tau_j = t_j + æ, \tag{2.4.1}$$

$$j = 1, \dots, N, \qquad \tau = (\tau_1, \dots, \tau_N), \qquad |\tau| = \tau_1 + \dots + \tau_N.$$

Since we set that $\sigma_1 \geq \sigma_2 \geq \dots \geq \sigma_{m+k''}$, we have that $æ = \max\{0, \sigma_1 + 1\}$. Denote:

$$\widetilde{H}^{T+s,p,(\tau)} := \prod_{j=1}^{N} \widetilde{H}^{t_j+s,p,(\tau_j)}(G), \tag{2.4.2}$$

$$\mathcal{B}^{T'+s,p} := \prod_{j=N+1}^{N+k''} \prod_{k=1}^{\sigma_1+t_j+1} B^{t_j+s-k+1-1/p,p}(\partial G). \tag{2.4.3}$$

It is easy to see (p. 2.3) that for any $s \in \mathbf{R}$ and $p \in]1, \infty[$ the closure $A = A_{s,p}$ of the mapping

$$U = (u, u') \mapsto (lu, bu + b'u'),$$

$$U = (u, u') \in (C^\infty(\overline{G}))^N \times \prod_{j=N+1}^{N+k'} (C^\infty(\partial G))^{\sigma_1+t_j+1},$$

acts continuously in the pair of spaces

$$\mathcal{H}^{s,p} \to K^{s,p}, \tag{2.4.4}$$

$$\mathcal{H}^{s,p} := \widetilde{H}^{T+s,p,(\tau)} \times \mathcal{B}^{T'+s,p},$$

$$K^{s,p} := \prod_{j=1}^{N} \widetilde{H}^{s-s_j,p,(æ-s_j)}(G) \times \prod_{h=1}^{m+k''} B^{s-\sigma_h-1/p,p}(\partial G).$$

An element $U \in \mathcal{H}^{s,p}$ such that

$$AU = F = (f, \varphi) = (f_1, \dots, f_N, \varphi_1, \dots, \varphi_{m+k''}) \in K^{s,p}, \tag{2.4.5}$$

$$f_j = (f_{j0}, \ldots, f_{j,\text{æ}-s_j}) \in \widetilde{H}^{s-s_j,p,(\text{æ}-s_j)}(G), \quad \varphi_h \in B^{s-\sigma_h-1/p,p}(\partial G),$$

is called a *generalized solution* of problem (2.1.1)–(2.1.2).

The following theorem holds.

Theorem 2.4.1. *Let problem (2.1.1)–(2.1.2) be elliptic. Then for any $s \in$ **R** and $p \in]1, \infty[$ the operator A is Noetherian. In addition, the kernel $\mathfrak{N}$ and the cokernel $\mathfrak{N}^*$ are finite dimensional, do not depend on s and p and consist of infinitely smooth elements, i.e.,*

$$\mathfrak{N} = \{U = (u, u') \in (C^\infty(\overline{G}))^N \times \prod_{J=N+1}^{N+k'} (C^\infty(\partial G))^{\sigma_1+t_j+1} \, : \, A(u, u') = 0\}, \tag{2.4.6}$$

$$\mathfrak{N}^* \subset \{(C^\infty(\overline{G}))^N \times \prod_{j=1}^{N} (C^\infty(\partial G))^{\text{æ}-s_j} \times (C^\infty(\partial G))^{m+k''}\}. \tag{2.4.7}$$

The equation (2.4.5) is solvable in $\mathcal{H}^{s,p}$ if and only if the relation

$$[F, V] := \sum_{j=1}^{N} (f_{j0}, v_{j0}) + \sum_{j=1}^{N} \sum_{k=1}^{\text{æ}-s_j} \langle f_{jk}, v_{jk} \rangle + \sum_{h=1}^{m+k''} \langle \varphi_j, \psi_j \rangle = 0$$

is valid for any element $V \in \mathfrak{N}^*$.

The proof of the Theorem is given in Section 2.6. In Section 2.5 we prove the analog of this theorem for model problem (2.2.10)–(2.2.11).

Here let us show some corollaries obtained from Theorem 2.4.1.

Theorem 2.4.1 is a theorem on complete collection of isomorphisms for problem (2.1.1)–(2.1.2). If the defect is missing, i.e., $\mathfrak{N} = 0$ and $\mathfrak{N}^* = 0$, then for any $s \in \mathbf{R}$ and $p \in]1, \infty[$ the operator A realizes an isomorphism between spaces (2.4.4). In the general case, the isomorphism is defined dy the restriction A_1 of the operator A onto corresponding subspace. Let us show these considerations.

2.4.2. Let

$$H^{T+s,p} = \prod_{j=1}^{N} H^{t_j+s,p}(G), \quad \mathcal{H}^{s,p} = H^{T+s,p} \times \mathcal{B}^{T'+s,p},$$

$$(L_2(G))^N \times (L_2(\partial G))^M = L_2, \quad M = (\sigma_1 + 1)k' + t_{N+1} + \cdots + t_{N+k'},$$

and let $(\cdot, \cdot)$ denote the scalar product in L_2.

The fact that $\mathfrak{N}$ is finite-dimensional yields that the following lemma is true.

Lemma 2.4.1. *Every element $U \in \mathcal{H}^{s,p}$ can be uniquely represented in the form*

$$U = U' + U'', \qquad U' \in \mathfrak{N}, \quad (U'', \mathfrak{N}) = 0. \tag{2.4.8}$$

The projection operator $P : U \mapsto U''$ is continuous in $\mathcal{H}^{s,p}$.

Proof. Since $\mathfrak{N}$ is finite-dimensional, one can consider that $\mathfrak{N}$ is a Hilbert space with the scalar product $(\cdot,\cdot)$. Let $U \in \mathcal{H}^{s,p}$. Then for $V \in \mathfrak{N}$ we have

$$|(U, V)| \le \|U\|_{\mathcal{H}^{s,p}} \|V\|_{(\mathcal{H}^{s,p})^*} \le c\|U\|_{\mathcal{H}^{s,p}} \|V\|_{L_2}.$$

Therefore, in view of the Riesz theorem on general form of a linear functional in a Hilbert space, there exists an element $U' \in \mathfrak{N}$ such that

$$(U, V) = (U', V) \quad (V \in \mathfrak{N}), \qquad \|U'\|_{L_2} \le c\|U\|_{\mathcal{H}^{s,p}}.$$

Setting $U'' = U - U'$, we obtain representation (2.4.8). Since $\mathfrak{N}$ is finite-dimensional, the estimates

$$\|U'\|_{\mathcal{H}^{s,p}} \le c_1\|U'\|_{L_2} \le c_1 c\|U\|_{\mathcal{H}^{s,p}}$$

are true. Therefore

$$\|PU\|_{\mathcal{H}^{s,p}} = \|U''\|_{\mathcal{H}^{s,p}} = \|U - U'\|_{\mathcal{H}^{s,p}}$$

$$\le \|U\|_{\mathcal{H}^{s,p}} + \|U''\|_{\mathcal{H}^{s,p}} \le (1 + c_1 c)\|U\|_{\mathcal{H}^{s,p}},$$

and the continuity of the operator P is, thus, established. □

For any $U = (u', u') \in \widetilde{\mathcal{H}}^{s,p}$ we set: $U_0 = U_{\overline{G}} = (u|_{\overline{G}}, u') \in \mathcal{H}^{s,p}$.

Lemma 2.4.2. *Every element $U \in \widetilde{\mathcal{H}}^{s,p}$ can be uniquely represented in the form*

$$U = U' + U'', \qquad U' \in \mathfrak{N}, \quad (U''_{\overline{G}}, \mathfrak{N}) = 0. \tag{2.4.9}$$

The projection operator $\widetilde{P} : U \mapsto U''$ is continuous in $\widetilde{\mathcal{H}}^{s,p}$.

Proof. In fact, if $U \in \widetilde{\mathcal{H}}^{s,p}$ then $U_0 = U|_{\overline{G}} \in \mathcal{H}^{s,p}$. Let

$$U' = U_0 - PU_0 \in \mathfrak{N}, \quad U'' = U - U' = \widetilde{P}U.$$

Then

$$\|\widetilde{P}U\|_{\widetilde{\mathcal{H}}^{s,p}} \le \|U\|_{\widetilde{\mathcal{H}}^{s,p}} + \|U'\|_{\widetilde{\mathcal{H}}^{s,p}}.$$

Therefore, since $\mathfrak{N}$ is finite-dimensional, the inequalities

$$\|U'\|_{\widetilde{\mathcal{H}}^{s,p}} \le c\|U'\|_{\mathcal{H}^{s,p}} \le c_1\|U_0\|_{\mathcal{H}^{s,p}} \le c_1\|U\|_{\widetilde{\mathcal{H}}^{s,p}}$$

are true. Thus, the estimate

$$\|PU\|_{\widetilde{\mathcal{H}}^{s,p}} \le (1+c_1)\|U\|_{\widetilde{\mathcal{H}}^{s,p}}$$

holds, and the lemma is proved. □

Let

$$\tilde{P}\mathcal{H}^{s,p} = \{\tilde{P}U : U \in \widetilde{\mathcal{H}}^{s,p}\}$$

be a subspace of the space $\widetilde{\mathcal{H}}^{s,p}$, and let $A_1 = A_{1s,p}$ denote a restriction of the operator $A = A_{s,p}$ onto $\tilde{P}\mathcal{H}^{s,p}$. Theorem 2.4.1 directly implies that for any $s \in \mathbf{R}$ and $p \in]1, \infty[$, the operator $A_1 = A_{1s,p}$ realizes an isomorphism

$$A_1 : \tilde{P}\widetilde{\mathcal{H}}^{s,p} \to Q^+K^{s,p}. \tag{2.4.10}$$

Here $Q^+K^{s,p} = \{F \in K^{s,p} : [F, V] = 0,\ V \in \mathfrak{N}^*\}$ denotes a subspace of $K^{s,p}$.

2.4.3. Global increasing of smoothness of generalized solutions

It is clear that for $s_2 \le s_1$ and $1 < p_2 < p_1$, the operators A_{s_2,p_2} and $A_{1\,s_2,p_2}$ are the extensions in continuity of the operators A_{s_1,p_1} and $A_{1\,s_1,p_1}$, respectively. Therefore, Theorem 2.4.1 and representation (2.4.9) imply the assertion on global increasing of the smoothness of generalized solutions of the problem (2.1.1)–(2.1.2). The following statement is true.

Theorem 2.4.2. *Under the assumptions of Theorem 2.4.1, let $U \in \widetilde{\mathcal{H}}^{s,p}$ be a generalized solution of the problem (2.1.1)–(2.1.2), i.e., $AU = F \in K^{s,p}$. If, in fact, $F \in K^{s_1,p_1}$, $s_1 \ge s$, $p_1 \ge p$, then $U \in \widetilde{\mathcal{H}}^{s_1,p_1}$, and the following estimate*

$$\|U\|_{\widetilde{\mathcal{H}}^{s_1,p_1}} \le c(\|U\|_{\mathcal{H}^{s,p}} + \|F\|_{K^{s_1,p_1}}), \tag{2.4.11}$$

where the constant $c > 0$ does not depend on u and F, is valid.

Proof. Representing U in the form of (2.4.9), we obtain that

$$A_{1\,s,p}U'' = F, \qquad U'' = A^{-1}_{1\,s,p}F = A^{-1}_{1\,s_1,p_1}F \in \tilde{P}\widetilde{\mathcal{H}}^{s_1,p_1}.$$

Since $U' \in \mathfrak{N} \subset C^\infty$, the assertion of the theorem is proved. □

2.4.4. Local increasing of smoothness of generalized solutions

Let $\Gamma_1 \in \partial G$ be an open subset of the boundary ∂G, and let $G_1 \subset G$ be a subdomain of G adherent to Γ_1.

Definition 2.4.1. *We say that a function $U \in \widetilde{\mathcal{H}}^{s,p}$ belongs to $\widetilde{\mathcal{H}}^{s_1,p_1}$ with $s_1 \geq s$ and $p_1 \geq p$ locally in G_1 up to the boundary Γ_1 and write $U \in \widetilde{\mathcal{H}}^{s_1,p_1}_{loc}(G_1, \Gamma_1)$ if the inclusion $\chi U \in \widetilde{\mathcal{H}}^{s_1,p_1}$ is true for every sufficiently smooth function χ in $\overline{G}$ vanishing in a some neighborhood of the set $\overline{G} \setminus (G_1 \bigcup \Gamma_1)$ in $\overline{G}$ (such functions are called admissible). Similarly, an element $F \in K_{s,p}$ belongs to K^{s_1,p_1} locally in G_1 up to the boundary Γ_1 and write $F \in K^{s_1,p_1}_{loc}(G_1, \Gamma_1)$ if $\chi F \in K^{s_1,p_1}$.*

The following assertion is true.

Theorem 2.4.3. *Let $U \in \widetilde{\mathcal{H}}^{s,p}$ be a generalized solution of the problem (2.1.1)-(2.1.2) with $F \in K^{s,p}$. If $F \in K^{s_1,p_1}_{loc}(G_1, \Gamma_1)$ with $s_1 \geq s$ and $p_1 \geq p$ then $U \in \widetilde{\mathcal{H}}^{s_1,p_1}_{loc}(G_1, \Gamma_1)$. Furthermore, for each admissible function χ there exists a constant $c > 0$ such that*

$$\|\chi U\|_{\widetilde{\mathcal{H}}^{s_1,p_1}} \leq c(\|\chi F\|_{K^{s_1,p_1}} + \|U\|_{\widetilde{\mathcal{H}}^{s,p}}). \tag{2.4.12}$$

Proof. Note that the proof of the theorem is quite similar to the proof of Theorem 7.2.1 of [R1].

First, let $s_1 > s$, $p_1 = p$. Since $U \in \widetilde{\mathcal{H}}^{s,p}$, we have that $\chi_1 U \in \widetilde{\mathcal{H}}^{s,p}$ for each admissible function $\chi = \chi_1$. Then

$$A(\chi_1 U) = \chi_1 AU + M_{\chi_1} U = \chi_1 F + M_{\chi_1} U = F_1, \tag{2.4.13}$$

where M_{χ_1} is the operator obtained as a result of the transfer of χ_1. Therefore $M_{\chi_1} U \in K^{s_1+1,p}$, and, thus, $F_1 \in K^{\min\{s_1,s+1\},p}$. If $s_1 > s+1$ then, by virtue of Theorem 2.4.2, we obtain that $\chi_1 U \in \widetilde{\mathcal{H}}^{s+1,p}$. Let χ_2 be an admissible function such that $\chi_2\chi_1 = \chi_1$. Then it follows from (2.4.13) that

$$\begin{aligned} A(\chi_1 U) &= A((\chi_1\chi_2)U) = \chi_1(A(\chi_2 U)) + M_{\chi_1}(\chi_2 U) \\ &= \chi_1\chi_2(AU) + \chi_1 M_{\chi_2} U + M_{\chi_1}(\chi_2 U) \\ &= \chi F + M_{\chi_1}(\chi_2 U) + M_{\chi_2}(\chi_1 U) - M_{\chi_1\chi_2} U = F_2 \in K^{\min\{s_1,s+2\},p}. \end{aligned}$$

Then in view of Theorem 2.4.2, we conclude that $\chi_1 U \in \widetilde{\mathcal{H}}^{\min\{s_1,s+2\},p}$. If $s_1 > s+2$ then we repeate this reasoning (sf. [R1, p. 216]). After finitely many steps we obtain that the inclusion $\chi_1 U \in \widetilde{\mathcal{H}}^{s_1,p}$ is valid for each admissible function χ_1. Thus, the theorem is proved for the case where $s_1 > s$ and $p_1 = p$.

Let $s_1 > s$ and $p_1 > p$. Then, first, using the imbedding theorem we prove that $\chi U \in \widetilde{\mathcal{H}}^{s_1,p} \subset \widetilde{\mathcal{H}}^{s,p_2}$ with $p_2 > p$. Futher we conclude that

$\chi U \in \widetilde{\mathcal{H}}^{s_1,p_1} \subset \widetilde{\mathcal{H}}^{s,p_3}$ with $p_3 > p_2$. After finitely many steps we arrive at the inclusion $\chi U \in \widetilde{\mathcal{H}}^{s_1,p_1}$. This completes the proof of the theorem. □

The statement on local increasing of the smoothness of generalized solutions in a srictly internal subdomain G_0 of the domain G can be proved in exactly the similar way. Now one must take admissible functions $\chi \in C^\infty(\overline{G})$ vanishing in some neighborhood of the set $\overline{G} \setminus G_0$ in $\overline{G}$. Let us formulate this result in the form of a separate theorem.

Theorem 2.4.4. *Let $U \in \widetilde{\mathcal{H}}^{s,p}$ be a generalized solution of the problem (2.1.1)–(2.1.2) with $F \in K^{s,p}$. If $F \in K^{s_1,p_1}_{loc}(G_0)$ with $s_1 \geq s$ and $p_1 \geq p$ then $U \in \widetilde{\mathcal{H}}^{s_1,p_1}_{loc}(G_0)$.*

2.4.5. Theorems 2.4.1-2.4.4 give us a possibility to study problem (2.1.1)–(2.1.2)in the case where the right hand side $F(x)$ has arbitrary power singularities near manifolds of various dimensionalities. It means that instead of the function $F(x) = (f(x), \varphi(x))$ (see (2.4.5)) we consider the regularization F which is an element of the space $K^{s,p}$ with some $s < 0$ depending on the singularities of $F(x)$. Theorems 2.4.1-2.4.4 enable us to prove the existence of the solution and istudy its regularity properties near the manifold of the singularities of the function $F(x)$ (sf. [R1, Ch. 8]).

2.4.6. Theorems 2.4.1–2.4.4 give us a possibility to investigate a class of strongly degenerating elliptic problems of the form of (2.1.1)–(2.1.2). For example, consider the problem

$$(\rho(x))^\alpha lu(x) = f(x), \quad x \in G, \qquad bu + b'u' = \varphi, \tag{2.4.14}$$

where $\rho(x) \in C^\infty(\overline{G})$, $\rho(x) > 0$ is equal to the distance of x to certain manifold $\gamma \subset G$ near γ. This problem is equivalent to the problem

$$lu(x) = (\rho(x))^{-\alpha} f(x), \quad x \in G, \qquad bu + b'u' = \varphi.$$

Thus, the problem 2.4.14 coincides with the problem studied in Subsection 2.4.5 (sf. [R1, §8.5]).

2.4.7. Theorem 2.4.1 and Theorem 2.6.1 enable us to prove the existence of the Green's matrix of the problem (2.1.1)–(2.1.2) and study its regularity properties (sf. [R1, §7.4]).

2.5. Proof of Isomorphisms Theorem for Model Problem

2.5.1. First, consider model problem (2.2.10)–(2.2.11) in the half-space

$\mathbf{R}^n_+$, i.e.,

$$\widehat{l}_0(D)u(x) = f(x), \qquad x \in \mathbf{R}^n_+, \tag{2.5.1}$$

$$\widehat{b}_0(D)u(x)\big|_{x_n=0} + \widehat{b}'_0(D')u'(x') = \varphi(x') \qquad x' \in \mathbf{R}^{n-1}. \tag{2.5.2}$$

It directly follows from the considerations of Section 2.3 that for any $s \in \mathbf{R}$ and $p \in (1, \infty)$, the closure $\widehat{A}_0 = \widehat{A}_{0\,s,p}$ of the mapping

$$U = (u, u') \mapsto (\widehat{l}_0 u, \widehat{b}_0 u + \widehat{b}'_0 u'),$$

where

$$U \in (C_0^\infty(\overline{\mathbf{R}^n_+}))^N \times \prod_{j=N+1}^{N+k'} (C_0^\infty(\overline{\mathbf{R}^{n-1}}))^{\sigma_1+t_j+1},$$

acts continuously in the pair of spaces

$$\mathcal{H}^{s,p}(\mathbf{R}^n_+) \to K^{s,p}(\mathbf{R}^n_+), \tag{2.5.3}$$

$$\mathcal{H}^{s,p}(\mathbf{R}^n_+) := \widetilde{H}^{T+s,p,(\tau)}(\mathbf{R}^n_+) \times \mathcal{B}^{T'+s,p}(\mathbf{R}^{n-1}),$$

$$K^{s,p}(\mathbf{R}^n_+) := \prod_{j=1}^{N} \widetilde{H}^{s-s_j,p,(\text{æ}-s_j)}(\mathbf{R}^n_+) \times \prod_{h=1}^{m+k''} B^{s-\sigma_h-1/p,p}(\mathbf{R}^{n-1}).$$

It turns out that the following theorem on complete collection of isomorphisms for the model problem is true.

Theorem 2.5.1. *Let problem (2.1.1)–(2.1.2) be elliptic. Then for any $s \in \mathbf{R}$ and $p \in]1, \infty[$ the operator $\widehat{A}_0 = \widehat{A}_{0\,s,p}$ realizes an isomorphism between the spaces (2.5.3).*

Note that this theorem is playing an important role in the proof of Theorem 2.4.1.

2.5.2. First, let us consider system (2.5.1) in $\mathbf{R}^n$.

Lemma 2.5.1.*([R1, §10.2]) Let $s, \sigma, p \in \mathbf{R}$, $1 < p < \infty$. For any function $\Phi = (\Phi_1, \ldots, \Phi_N)$, $\Phi_r \in H^{s-s_r,\sigma,p}(\mathbf{R}^n)$, there exists one and only one solution $v = (v_1, \ldots, v_N)$, $v_i \in H^{t_j+s,\sigma,p}(\mathbf{R}^n)$ of the problem*

$$\widehat{l}_0(D)v = \Phi. \tag{2.5.4}$$

Furthermore, the inequality

$$\sum_{j=1}^{N} \|v_j, \mathbf{R}_n\|_{t_j+s,\sigma,p} \le c \sum_{j=1}^{N} \|\Phi_j, \mathbf{R}^n\|_{s-s_j,\sigma,p} \tag{2.5.5}$$

holds with a constant $c > 0$ that does not depend on Φ, v, s and σ. (All these spaces and norms are defined in Subsection 2.3.1.)

Proof. Passing to the Fourier images in (2.5.4), we obtain

$$\widehat{l}_0(\xi)\widetilde{v}(\xi) = \left(\widehat{l}^0_{rj}(\xi)\right)_{r,j=1,\dots,N}\widetilde{v}(\xi) = \widetilde{\Phi}(\xi).$$

In view of the fact that the matrix $\widehat{l}_0(\xi)$ is nonsingular, one can write

$$\widetilde{v}(\xi) = \widehat{l}_0^{-1}(\xi)\widetilde{\Phi}, \qquad \widetilde{v}_j(\xi) = \widehat{L}_0^{-1}(\xi)\sum_{r=1}^{N}\widehat{L}_{rj}(\xi)\widetilde{\Phi}_r(\xi) \quad (j = 1, \dots, N), \tag{2.5.6}$$

where $\widehat{L}_0(\xi) = \det \widehat{l}_0(\xi)$, and $\widehat{L}_{rj}(\xi)$ is the cofactor of the element $\widehat{l}_{rj}(\xi)$. It is clear that $L_0(\xi)$ is a $(2m)$th-order homogeneous function that does not equal to 0 for $\xi \neq 0$, and the expression $L_{rj}(\xi)$ is a homogeneous function of order $2m - s_r - t_j$. Therefore

$$c_0^{-1}(1 + |\xi|^2)^m \leq \widehat{L}_0(\xi) \leq c_0(1 + |\xi|^2)^m,$$
$$\widehat{L}_{rj}(\xi) \leq c(1 + |\xi|^2)^{(2m-s_r-t_j)/2}.$$

It now follows from (2.3.1) that

$$\begin{aligned}
&\|v_j, \mathbf{R}^n\|_{t_j+s,\sigma,p}\\
&= \|F^{-1}(1 + |\xi|^2)^{(t_j+s)/2}(1 + |\xi'|^2)^{\sigma/2}Fv_j\|_{L_p}\\
&= \|F^{-1}(1 + |\xi|^2)^{(t_j+s)/2}(1 + |\xi'|^2)^{\sigma/2}\widehat{L}_0^{-1}(\xi)\sum_{r=1}^{N}\widehat{L}_{rj}(\xi)(F\Phi_r)(\xi)\|_{L_p}\\
&\leq \sum_{r=1}^{N}\|F^{-1}(1 + |\xi|^2)^{(t_j+s_r)/2}\widehat{L}_{rj}(\xi)\widehat{L}_0^{-1}(\xi)FF^{-1}(1 + |\xi|^2)^{(s-s_r)/2}\\
&\qquad \times (1 + |\xi'|^2)^{\sigma/2}(F\Phi_r)(\xi)\|_{L_p}.
\end{aligned} \tag{2.5.7}$$

Since the function

$$(1 + |\xi|^2)^{(t_j+s_r)/2}\widehat{L}_{rj}(\xi)\widehat{L}_0^{-1}(\xi)$$

is a multiplicator in L_p (see, for example, [R1, §1.3]), relations (2.5.7) yield the required estimate (2.5.5). Lemma is proved. □

2.5.3. The following question arises:
Under the conditions of Lemma 2.5.1, when does the inclusion $\operatorname{supp}\Phi \subset \overline{\mathbf{R}}^n_+$ *imply the inclusion* $\operatorname{supp} v \subset \overline{\mathbf{R}}^n_+$?

The following lemma, which is proved in [R1, §10.2], gives us an answer for this question.

Lemma 2.5.2. *Let* $s, \sigma, p \in \mathbf{R}$, $1 < p < \infty$. *For the solvability of problem (2.5.4) in*

$$H_+^{T+s,\sigma,p} = \prod_{j=1}^{N} H_+^{t_j+s,\sigma,p}(\mathbf{R}^n) \tag{2.5.8}$$

with

$$\Phi \in \prod_{r=1}^{N} H_+^{s-s_r,\sigma,p}(\mathbf{R}^n) \bigcap \prod_{r=1}^{N} H_+^{s-s_r,\sigma,2}(\mathbf{R}^n) \tag{2.5.9}$$

it is necessary and sufficient that the equalities

$$\int_{-\infty}^{\infty} \hat{L}^{-1}(\xi',\xi_n)\left(\xi_n + i\sqrt{1+|\xi'|^2}\right)^{s+t_j-m} \sum_{r=1}^{N} L_{rk}(\xi',\xi_n)\widetilde{\Phi}_r(\xi',\xi_n)\xi_n^j\, d\xi_n = 0 \tag{2.5.10}$$

$$(k = 1,\ldots,N; \quad j = 0,\ldots,m-1)$$

hold for almost all $\xi' \neq 0$.

Let us note that if, under the conditions of the lemma, Φ is a sufficiently smooth function, then equalities (2.5.10) hold for all $\xi' \neq 0$.

2.5.4. If

$$\begin{aligned} U &= (u_1,\ldots,u_N) \in \widetilde{H}^{T+s,p,(\tau)}(\mathbf{R}^n_+), \\ &\quad u_j = (u_{j0},\ldots,u_{j,\tau_j}) \in \widetilde{H}^{t_j+s,p,(\tau_j)}(\mathbf{R}^n_+), \\ F &= (f,\varphi) = (f_1,\ldots,f_N,\varphi_1,\ldots,\varphi_{m+k''}) \in K^{s,p}(\mathbf{R}^n_+), \\ &\quad f_j = (f_{j0},\ldots,f_{j,\text{æ}-s_j}), \end{aligned}$$

then, by virtue of conditions (2.3.16) and (2.3.17), equality (2.5.1) holds if and only if the relations

$$\left(\hat{l}_{0r}u\right)_+ := \sum_{j=1}^{N} \hat{l}_{rj0}(D)u_{j0+} - \sum_{j:s_r+t_j\geq 1}\ \sum_{k=1}^{s_r+t_j} J^k \hat{l}_{rj0}(u_{jk}\times\delta(x_n)) = f_{r0+} \tag{2.5.11}$$

$$(r = 1, \ldots, N),$$

$$D_n^{h-1}(\widehat{l}_{0r}u)\big|_{x_n=0} := \sum_{j=1}^{N} \sum_{k=1}^{s_r+t_j+1} \widehat{l}_k^{rj}(D')u_{j,k+h}(x') = f_{rh} \tag{2.5.12}$$

$$(h = 1, \ldots, æ - s_r, \qquad r : æ - s_r \geq 1)$$

are valid. By using Lemma 2.5.2, we rewrite equalities (2.5.11) in the form

$$\sum_{\alpha=1}^{N_1} \sum_{\beta=1}^{t_\alpha} \widehat{C}_{kj\alpha\beta}(\xi')\widehat{u}_{\alpha\beta}(\xi') = \widehat{g}_{kj}(\xi') \tag{2.5.13}$$

$$(k = 1, \ldots, N, j = 0, \ldots, m-1),$$

where $N_1 \leq N$ is the number of subscripts j such that $t_j \geq 1$, and

$$\widehat{g}_{kj}(\xi') =$$

$$\int_{-\infty}^{\infty} \widehat{L}^{-1}(\xi', \xi_n)\left(\xi_n + i\sqrt{1 + |\xi'|^2}\right)^{s+t_k-m} \sum_{r=1}^{N} L_{rk}(\xi', \xi_n)\xi_n^j \times \widetilde{f}_{r0+}\, d\xi_n \tag{2.5.14}$$

$$(k = 1, \ldots, N; \quad j = 0, \ldots, m-1).$$

It turns out ([R1, Lemma 10.2.3]) that for any $\xi' \neq 0$, there are m linearly independent conditions in the set of Nm conditions (2.5.13) (or, (2.5.11)), and the other equations can be represented as linear combinations of the indicated equations. Therefore, in what follows, we assume that at every point $\xi' \neq 0$ system (2.5.13) contains only m linearly independent equations and the others are thrown away. Since each mth-order minor is a continuous function of ξ', if m conditions are linearly independent at a point $\xi' \neq 0$, then they are also linearly independent in a certain neighborhood of this point.

Consider system (2.5.13) toobtainher with the Fourier transforms of equalities (2.5.2) and (2.5.13). We obtain a system S which consists of

$$m + (æ - s_1) + \cdots + (æ - s_N) + m + k'' = 2m + Næ - (s_1 + \cdots + s_N) + k''$$

$$= t_1 + \cdots + t_N + Næ + k'' = \tau_1 + \cdots + \tau_N + k'' = |\tau| + k''$$

linear equations of $|\tau|$ unknowns $\widehat{u}_{jk}(\xi')$ and $k'(\sigma_1 + 1) + t_{N+1} + t_{N+k'}$ unknowns $\widehat{u}'_{jk}(\xi')$.

Since, in view of (2.2.8) we have the relation

$$m + k'' = m + k'(\sigma_1 + 1) + t_{N+1} + t_{N+k'},$$

we arive at the conclusion that the obtained system S is square.

By virtue of the fact that the Lopatinskii condition is equivalent to the unique solvability of the system S, we conclud that its determinant does not equal to zero. Thus, the following assertion is true.

Lemma 2.5.3. *The ellipticity of problem (2.5.1), (2.5.2) is equivalent to the fact that the obtained linear system S is quadratic, and its determinant $\widehat{\Delta}(\xi')$ does not equal to zero.*

Lemma 2.5.3 enable us to express and to estimate the elements $\widehat{u}_{jk}(\xi')$ and $\widehat{u}'_{jk}(\xi')$ in terms of the right hand sides, and to prove Theorem 2.5.1.

2.5.5. First, let us prove Theorem 2.5.1 in the case where $t_1 + s < 1/p$. Now (see Section 2.3) the space

$$\widetilde{H}^{T+s,p,(\tau)}(\mathbf{R}^n_+) = \prod_{j=1}^{N} \widetilde{H}^{t_j+s,p,(\tau_j)}(\mathbf{R}^n_+)$$

coincides with the direct product

$$\prod_{j=1}^{N} \left(H^{t_j+s,p}(\mathbf{R}^n_+) \times B^{t_j+s-k+1-1/p,p}(\mathbf{R}^{n-1}) \right),$$

and the space

$$\widetilde{H}^{s-s_j,p,(æ-s_j)}(\mathbf{R}^n_+)$$

coincides with the direct product

$$H^{s-s_j,p}(\mathbf{R}^n_+) \times \prod_{k=1}^{æ-s_j} B^{s-s_j-k+1-1/p,p}(\mathbf{R}^{n-1}).$$

Moreover (see (2.3.6)),

$$H^{t_j+s,p}(\mathbf{R}^n_+) = H_+^{t_j+s,p}(\mathbf{R}^n), \qquad H^{s-s_j,p}(\mathbf{R}^n_+) = H_+^{s-s_j,p}(\mathbf{R}^n).$$

Therefore, in view of (2.5.3), the operator $\widehat{A}_0 = \widehat{A}_{0\,s,p}$ acts continuously in the pair of spaces

$$\mathcal{H}^{s,p}(\mathbf{R}^n_+) \to K^{s,p}, \tag{2.5.15}$$

where

$$\mathcal{H}^{s,p}(\mathbf{R}^n_+) \simeq \prod_{j=1}^{N} \left(H_+^{t_j+s,p}(\mathbf{R}^n) \times \prod_{k=1}^{\tau_j} B^{t_j+s-k+1-1/p,p}(\mathbf{R}^{n-1}) \right) \times \mathcal{B}^{T'+s,p}(\mathbf{R}^{n-1}),$$

and

$$K^{s,p}(\mathbf{R}^n_+) \simeq \prod_{j=1}^{N} \left(H_+^{s-s_j,p}(\mathbf{R}^n) \times \prod_{k=1}^{æ-s_j} B^{s-s_j-k+1-1/p,p}(\mathbf{R}^{n-1}) \right)$$

$$\times \prod_{h=1}^{m+k''} B^{s-\sigma_h-1/p,p}(\mathbf{R}^{n-1}).$$

Lemma 2.5.4. *Let problem (2.5.1)–(2.5.2) be elliptic, $s, p \in \mathbf{R}$, $s < 1/p$, and let*

$$f_{r0} \in C_0^\infty(\overline{\mathbf{R}^n_+}), \qquad r = 1, \ldots, N,$$

$$f_{rh} \in C_0^\infty(\mathbf{R}^{n-1}), \quad h = 1, \ldots, æ - s_r, \quad r : æ - s_r \geq 1, \tag{2.5.16}$$

$$\varphi_h \in C_0^\infty(\mathbf{R}^{n-1}), \quad h = 1, \ldots, m + k''.$$

Then problem (2.5.1)–(2.5.2) possesses a unique solution

$$U = (u, u') = \Big((u_1, \ldots, u_N), (u'_{n+1}, \ldots, u'_{n+k'})\Big), \tag{2.5.17}$$

$$u_j = (u_{j0}, u_{j1}, \ldots, u_{j,\tau_j}) \in H_+^{t_j+s,\sigma,p}(\mathbf{R}^n) \times \prod_{k=1}^{\tau_j} B^{t_j+s-k+1-1/p+\sigma,p}(\mathbf{R}^{n-1})$$

$$(j = 1, \ldots, N),$$

$$u'_j = (u'_{j1}, \ldots, u'_{j,\sigma_1+t_j+1}) \in \prod_{k=1}^{\sigma_1+t_j+1} B^{t_j+s-k+1-1/p+\sigma,p}(\mathbf{R}^{n-1})$$

$$(j = n+1, \ldots, n+k')$$

for any $\sigma > 0$. The inequality

$$\sum_{j=1}^{N} \Big(\|u_{j0}, \mathbf{R}^n\|_{t_j+s,\sigma,p} + \sum_{1 \leq k \leq \tau_j} \langle\langle u_{jk}, \mathbf{R}^{n-1} \rangle\rangle_{t_j+s-k+1-1/p+\sigma,p} \Big)$$

$$+ \sum_{j=n+1}^{n+k'} \sum_{k=1}^{\sigma_1+t_j+1} \langle\langle u'_{jk}, \mathbf{R}^{n-1} \rangle\rangle_{t_j+s-k+1-1/p+\sigma,p}$$

$$\leq c \Big(\sum_{j=1}^{N} (\|f_{j0}, \mathbf{R}^n\|_{s-s_j,\sigma,p} + \sum_{k=1}^{æ-s_j} \langle\langle f_{jk}, \mathbf{R}^{n-1} \rangle\rangle_{t_j+s-k+1-1/p+\sigma,p})$$

$$+ \sum_{h=1}^{m+k''} \langle\langle \varphi_h, \mathbf{R}^{n-1} \rangle\rangle_{s-\sigma_h-1/p+\sigma,p} \Bigg) \tag{2.5.18}$$

holds with a constant c which does not depend on u, u', $\{f_{j0}\}$, $\{f_{jk}\}$ and $\{\varphi_h\}$.

Proof. Consider the system S. Since $\widehat{\Delta}(\xi') \neq 0$ for any $\xi' \neq 0$, one can express and estimate the functions $\widehat{u}_{jk}(\xi')$ and $\widehat{u}'_{jk}(\xi')$ in terms of the right hand sides.

Note that if $f_{r0+} \in H^{s-s_r,\sigma,p}(\mathbf{R}^n)$ $(r = 1, \ldots, N)$ then it follows from (2.5.4) that

$$g_{kj} \in B^{m-j+\sigma-1/p,p}(\mathbf{R}^{n-1}) \qquad (k = 1, \ldots, N, 0 \leq j \leq m-1),$$

and

$$\sum_{k=1}^{N} \sum_{j=0}^{m-1} \langle\langle g_{kj}, \mathbf{R}^{n-1} \rangle\rangle_{m-j+\sigma-1/p,p} \leq \sum_{r=1}^{N} \| f_{r0}, \mathbf{R}^n \|_{s-s_r,\sigma,p}. \tag{2.5.19}$$

The system S is similar to a Douglis-Nirenberg structure system. It means that if we associate the column of the coefficients of $U_{\alpha\beta}$ with the number $t_\alpha - \beta + 1$, the (k,j)th row of system (2.5.13) with the number $-m + j + s$, the (h,r)th row of the Fourier image of system (2.5.12) with the number $s_h + r - 1$, the hth row of the Fourier image of system (2.5.2) with the number σ_h, then the degree of homogeneity of the coefficient of $U_{\alpha\beta}$ located in the intersection of the indicated rows and columns is equal to the sum of the corresponding numbers.

In other words, it is possible to establish the correspondence between the columns of the matrix of the coefficients of the system S and the numbers $t'_1, \ldots, t'_k$ $(k = \tau + k'')$, and between the rows of the matrix and the numbers $s'_1, \ldots, s'_k$, so that the degree of homogeneity of the element $a_{jk}(\xi')$ located in the intersection of the jth row and kth column is equal to $s'_j + t'_k$.

We obtain that $\Delta(\xi')$ is a homogeneous function of ξ' of order $\sum(s'_j + t'_k) = M$. Since $\Delta(\xi') \neq 0$ $(\xi' \neq 0)$, we have

$$c_1|\xi|^M \leq |\Delta(\xi)| \leq c_2|\xi|^M.$$

Hence,

$$c_1(1 + |\xi'|)^M \leq \widehat{\Delta}(\xi') \leq c_2(1 + |\xi'|)^M. \tag{2.5.20}$$

Let $\left(\widehat{a}_{jk}(\xi')\right)_{j,k=1,\ldots,K}$ be a matrix of the coefficients of the system S, and let $\widehat{A}_{jk}(\xi')$ be the cofactor of the element $\widehat{a}_{jk}(\xi')$. It is clear that $\widehat{A}_{jk}(\xi')$ is a homogeneous function of degree $M - s'_j - t'_k$. Therefore the matrix

$$\left(\widehat{A}_{jk}(\xi') (\widehat{\Delta}(\xi'))^{-1} \right)_{j,k=1,\ldots,K}$$

is the inverse of the matrix $(\widehat{a}_{jk})$. In addition,

$$\left|\widehat{A}_{jk}(\xi')(\widehat{\Delta}(\xi'))^{-1}\right| \leq c(1+|\xi'|)^{-s'_k-t'_j}. \tag{2.5.21}$$

Now write the solution of the system S

$$\sum_{k=1}^{K} \widehat{a}_{jk}(\xi')\widehat{V}_k(\xi') = \widehat{\Phi}_j \qquad (j=1,\ldots,K)$$

in the form

$$\widehat{V}_k(\xi') = \sum_{k=1}^{N} \widehat{A}_{kj}(\xi')(\widehat{\Delta}(\xi'))^{-1}\widehat{\Phi}_k,$$

and, using estimates (2.5.21), we arrive at the required estimate (2.5.18) (sf. [R1, §4.4 and §10.2]). Lemma 2.5.4 is, thus, proved. □

By passing to the limit in Lemma 2.5.4, we arrive at the following statement.

Corollary 2.5.1. *Let $p \in]1,\infty[$, $s \in \mathbf{R}$, $t_1 + s < 1/p$, and let problem (2.5.1)–(2.5.2) be elliptic. Assume that*

$$f_{r0} \in H_+^{s-s_r,\sigma,p}(\mathbf{R}^n), \qquad r = 1,\ldots,N,$$

$$f_{rj} \in B^{s+\sigma-s_r-j+1-1/p,p}(\mathbf{R}^{n-1}), \qquad r : æ - s_r \geq 1, \qquad j = 1,\ldots,æ - s_r,$$

$$\varphi_h \in B^{s+\sigma-\sigma_h-1/p,p}(\mathbf{R}'_{n-1}), \qquad h = 1,\ldots,m+k''.$$

Then problem (2.5.1)–(2.5.2) has one and only one solution (2.5.17) satisfying estimate (2.5.18).

Corollary 2.5.1 with $\sigma = 0$ directly implies the assertion of Theorem 2.5.1 for $t_1 + s < 1/p$.

2.5.6.

Now let $t_1 + s > 1/p$ and let $\sigma > 0$ satisfy the inequality $t_1 + s - \sigma < 1/p$.

Since the norm $|||u, \mathbf{\Omega}|||_{s,p,(r)}$ is defined by (2.3.9) for $s \neq k + 1/p$ ($k = 0,\ldots,r-1$), one can suppose that

$$t_j + s \neq k + \frac{1}{p} \quad (k = 1,\ldots,\tau_j - 1, \quad j = 1,\ldots,N). \tag{2.5.22}$$

For excluded values of s, i.e., for $s = k + 1/p$, the assertion of the theorem is then established by the interpolation theorem (see [R1]).

Corollary 2.5.1 immediately implies the validity of the following statement.

Lemma 2.5.5. *Let* $p \in]1,\infty[$, $s \in \mathbf{R}$, $t_1+s > 1/p$, *and let relations (2.5.22) be valid. Assume that problem (2.5.1)–(2.5.2) is elliptic and* $F \in K^{s,p}(\mathbf{R}^n)$. *Then problem (2.5.1)–(2.5.2) possesses one and only one solution (2.5.17) satisfying estimate (2.5.18), with* $s-\sigma$ *instead of* s. *Here* $\sigma > 0$ *satifies the inequality* $t_1 + s - \sigma < 1/p$.

It follows from Lemma 2.5.5 that in the case under consideration (i.e., $t_1 + s > 1/p$) the unique solution $U = (u, u')$,

$$u = (u_1, \dots, u_N), \quad u_j = (u_{j0}, u_{j1}, \dots, u_{j,\tau_j})$$

$$u_j \in H_+^{t_j+s-\sigma,\sigma,p}(\mathbf{R}^n) \times \prod_{k=1}^{\tau_j} B^{t_j+s-k+1-1/p,p}(\mathbf{R}^{n-1}) \qquad (j = 1, \dots, N),$$

$$u' = (u'_{n+1}, \dots, u'_{n+k'}), \quad u'_j = (u'_{j1}, \dots, u'_{j,\sigma_1+t_j+1})$$

$$u'_j \in \prod_{k=1}^{\sigma_1+t_j+1} B^{t_j+s-k+1-1/p,p}(\mathbf{R}^{n-1}) \qquad (j = n+1, \dots, n+k')$$

of problem (2.5.1)–(2.5.2) has the following property: the element $u_{j0} \in H_+^{t_j+s-\sigma,\sigma,p}(\mathbf{R}^n)$ $(j = 1, \dots, N)$ has the smoothness of order $t_j + s - \sigma$ with respect to all the variables, and the additional smoothness of order σ with respect to tangential variables. This gives us a possibility to prove that

$$u_{j0} \in H^{t_j+s,p}(\mathbf{R}^n_+), \qquad u_{jk} = D_\nu^{k-1} u_{j0}\big|_{x_n=0} \quad (\forall k : t_j + s - k + 1 - 1/p > 0),$$

in other words, (see [R1, §10.2])

$$u_j = (u_{j0}, \dots, u_{j,\tau_j}) \in \widetilde{H}^{t_j+s,p,(\tau_j)}(\mathbf{R}^n_+) \quad (j = 1, \dots, N).$$

Hence, the operator $\widehat{A}_{s,p}$ is a continuous one-to-one mapping acting from the space $\mathcal{H}^{s,p}(\mathbf{R}^n_+)$ onto $K^{s,p}(\mathbf{R}^n_+)$. Therefore, in view of the Banach theorem on inverse operator, the operator $(\widehat{A}_{s,p})^{-1}$ is also continuous. This completes the proof of Theorem 2.5.1 for $s \in \mathbf{R}$ and $p \in (1,\infty)$.

2.6. Proof of Theorem 2.4.1

Let

$$p \in (1,\infty), \qquad s \in \mathbf{R},$$
$$t_j + s \neq k + 1/p \qquad (k = 0, \dots, \tau_j - 1, j = 1, \dots, N). \tag{2.6.1}$$

Assume that the assumption of Theorem 2.4.1 are valid.

The operator $A = A_{s,p}$ acts continuously between the spaces $\widetilde{\mathcal{H}}^{s,p}$ and $K^{s,p}$ (see (2.4.4) and (2.4.5)).

Definition 2.6.1. *The operator $R = R_{s,p}$, continuously acting from $K^{s,p}$ into $\widetilde{\mathcal{H}}^{s,p}$, is called an regularizer of the operator $A = A_{s,p}$ if*

$$RA = I_1 + T_1, \qquad AR = I_2 + T_2, \tag{2.6.2}$$

where I_1 and I_2 are the identity operators in $\widetilde{\mathcal{H}}^{s,p}$ and $K^{s,p}$, respectively, T_1 is a smoothing operator in $\widetilde{\mathcal{H}}^{s,p}$ (i.e., a continuous operator acting from $\widetilde{\mathcal{H}}^{s-1,p}$ into $\widetilde{\mathcal{H}}^{s,p}$), and T_2 is a smoothing operator in $K^{s,p}$ (i.e., a continuous operator acting from $K^{s,p}$ into $K^{s+1,p}$).

The following theorem is true:

Theorem 2.6.1. *Under the assumptions of Theorem 2.4.1, let conditions (2.6.1) are satisfied. Then there exists the regularizer $R = R_{s,p}$ of the operator $A = A_{s,p}$. For $s_1 \le s_2 < æ$ or $æ \le s_1 \le s_2$ (see (2.4.1)) the operator $R_{s_1,p}$ is an extension of the operator $R_{s_2,p}$ by continuity.*

The proof of the theorem is quite similar to the proof of Theorem 10.3.1 from [R1], where the analogous assertion is proved for the case of boudary value problem without additional functions at the boudary.

We split the proof of Theorem 2.6.1 into several lemmas.

Lemma 2.6.1. *Let $p \in]1, +\infty[$ and $x_0 \in G$. For any $\varepsilon > 0$ and any bounded number set $E \subset \mathbf{R}$, one can idicate a neighborhood $U = U(x_0) \subset G$ of the point x_0 such that, for every function $\varphi \in C_0^\infty(U)$, the expression $\varphi(x)l(x, D)$ admits the representation*

$$\varphi(x)\, l(x, D) = \varphi(x)(l_0(x_0, D) + Q(x, D) + l'(x, D)). \tag{2.6.3}$$

Here

$$l_0(x_0, D) = \left(l_{rj}^0(x_0, D)\right)_{r,j=1,\dots,N},$$

$l_{rj}^0(x_0, D)$ is a homogeneous expression of order $s_r + t_j$ with constant (fixed at the point x_0) coefficients, $l_{rj}^0 = 0$ for $s_r + t_j < 0$,

$$Q(x, D) = \left(Q_{rj}(x, D)\right)_{r,j=1,\dots,N} \qquad (x \in \mathbf{R}^n)$$

is a matrix operator of the form like $l(x, D)$ with small norm

$$\|Q\|_{\prod_{j=1}^N H^{t_j+s,p}(\mathbf{R}^n) \to \prod_{j=1}^N H^{s-s_j,p}(\mathbf{R}^n)} \le \varepsilon \qquad (\forall s \in E),$$

$\operatorname{ord} Q_{rj}(x,D) \le s_r + t_j$, $Q_{rj} \equiv 0$ for $s_r + t_j = 0$, and

$$l'(x,D) = \big(l'_{rj}(x,D)\big)_{r,j=1,\dots,N} \qquad (x \in \mathbf{R}^n)$$

is a matrix differential expression of lower order, i.e.,

$$\operatorname{ord} l'_{rj}(x,D) \le s_r + t_j - 1$$

and $l'_{rj} \equiv 0$ whenever $s_r + t_j - 1 \le 0$.

The operators Q and l' do not depend on $s \in E$.

Proof. Let $x_0 \in G$, let $U = U(x_0) \subset G$ be a neighborhood of this point, and let $\varphi \in C_0^\infty(U)$. Assume that

$$l(x,D) = l_0(x,D) + l''(x,D), \quad l_0(x,D) = \big(l^0_{rj}(x,D)\big)_{r,j=1,\dots,N} \tag{2.6.4}$$

$$l^0_{rk}(x,D) = \begin{cases} \sum_{|\mu|=s_r+t_j} a^{rj}_\mu(x) D^\mu\,, & \text{for } s_r + t_j \ge 0 \\ 0, & \text{for } s_r + t_j < 0, \end{cases}$$

in other words, $l_0(x,D)$ is the principal part of $l(x,D)$.

We represent l_0 in the form

$$\begin{aligned} l_0(x,D) &= l_0(x_0,D) + (l_0(x,D) - l_0(x_0,D)) \\ &= l_0(x_0,D) + Q_0(x,D). \end{aligned} \tag{2.6.5}$$

It is clear that the coefficients of expressions of the matrix $Q_0(x,D)$ are small in a sufficiently small neighbourhood of the point x_0.

Let $\psi(x) \in C_0^\infty(U)$ and let $\psi = 1$ in a neighborhood of the support of function φ, $0 \le \psi(x) \le 1$. It is clear that $\psi\varphi = \varphi$. As a result, we obtain

$$\begin{aligned} \varphi(x) l(x,D) u &= \varphi(x) l(x,D)(\psi u) \\ &= \varphi\Big(l_0(x_0,D)u + Q_0(x,D)(\psi u) + l''(x,D)(\psi u)\Big). \end{aligned} \tag{2.6.6}$$

Now one can assume that the expressions $\varphi l(x,D), l_0(x_0,D), Q_0(x,D)(\psi,\cdot)$, $l''(x,D)(\psi,\cdot)$ are defined in the whole space $\mathbf{R}^n$. Let χ_N be a smoothing operator (see [R1, §1.6]), i.e.,

$$\chi_N = F^{-1}\tilde{\chi}(\xi/N)F, \qquad \tilde{\chi}(\xi) \in C_0^\infty(\mathbf{R}^n), \qquad 0 \le \chi(\xi) \le 1,$$

$$\tilde{\chi}(\xi) = \begin{cases} 1 & \text{for } |\xi| \le 1, \\ 0 & \text{for } |\xi| \ge 2, \end{cases} \qquad \sup_{k \in \mathcal{K}, \xi \in \mathbf{R}^n} |\partial^k(\tilde{\chi}(\xi))| = a,$$

$$\mathcal{K} = \Big\{k = (k_1, \dots, k_n) : k_j = 0, 1,\ j = 1, \dots, N\Big\}.$$

Then

$$\begin{aligned} Q_0(x, D)\psi u &= Q_0(x, D)\psi(I - \chi_N)u + Q_0(x, D)\chi_N u \\ &= Q_N(x, D)u + Q_{1N}(x, D)u. \end{aligned} \tag{2.6.7}$$

It turns out that, for sufficiently small neighborhood U an sufficiently large N, the order of the operator Q_{1N} is lower than the order of the operator l_0 by one, and Q_N is an operator with small norm. Now (2.6.3) follows from (2.6.6) and (2.6.7), and the required lemma is proved. □

In exactly the similar way one can prove the following assertion (sf. [R1, §10.3]):

Lemma 2.6.2. *Let $x_0 \partial G$ and $p \in (1, \infty)$. For any $\varepsilon > 0$ and any bounded set $E \subset \mathbf{R}$, there exists a neighborhood $U = U(x_0)$ of the point x_0 in $\mathbf{R}^n$ such that, for every function $\varphi \in C_0^\infty(U)$ and $s \in E$, the expression*

$$\varphi A(x, D) = (\varphi l(x, D), \varphi|_{\partial G}(b + b'))$$

admits the representation

$$\varphi A(x, D) = \varphi\Big(A_0(x_0, D) + Q(x, D) + A'(x, D)\Big), \tag{2.6.8}$$

where $A_0(x_0, D)$ is the principal part of the operator A with the constant (frozen at the point x_0) coefficients, $Q(x, D)$ $(x \in \mathbf{R}^n)$ is an operator with small norm

$$\|Q\|_{\widetilde{H}^{s,p}(\mathbf{R}^n_+) \to K^{s,p}(\mathbf{R}^n_+)} \leq \varepsilon,$$

acting continuously from $\widetilde{H}^{s,p}(\mathbf{R}^n_+)$ into $K^{s,p}(\mathbf{R}^n_+)$, and A' is an operator acting continuously from the space $\widetilde{H}^{s,p}(\mathbf{R}^n_+)$ into $K^{s+1,p}(\mathbf{R}^n_+)$. The operators Q and A' do not depend on s and p.

Now we can replace in (2.6.3) and (2.6.8) the expressions $l_0(x_0, D)$ and $A_0(x_0, D)$ by $\widehat{l}_0(x_0, D)$ and $\widehat{A}_0(x_0, D)$, respectively. As a result, l' and A' are affected only. Therefore, under the assumption of Lemma 2.6.1, there holds the representation

$$\varphi(x) l(x, D) = \varphi(x)\Big(\widehat{l}_0(x_0, D) + Q(x, D) + l'(x, D)\Big), \tag{2.6.9}$$

and under the assumption of Lemma 2.6.2, there holds the representation

$$\varphi(x)A(x,D)=\varphi(x)\Big(\widehat{A}_0(x_0,D)+Q(x,D)+A'(x,D)\Big). \tag{2.6.10}$$

For convenience, we formulate the results established above as a lemma.

Lemma 2.6.3. *Under the conditions of Lemma 2.6.1, the expression $\varphi(x)l(x,D)$ admits the representations (2.6.3) and (2.6.9). Under the conditions of Lemma 2.6.2, the expression $\varphi(x)A(x,D)$ admits the representations (2.6.8) and (2.6.10).*

Now let us complete the proof of Theorem 2.6.1. First, let $x_0 \in G$ be an internal point of G. By Lemma 2.6.3, there exists a neighborhood $U(x_0)$ such that any function $\varphi \in C_0^\infty(U(x_0))$ admits representation (2.6.9) with the operator Q whose norm is arbitrary small.

In view of Lemma 2.5.1, the operator

$$\widehat{l}_0 : H^{T+s,p} := \prod_{j=1}^{N} H^{t_j+s,p}(\mathbf{R}^n) \to H^{s-S,p} := \prod_{j=1}^{N} H^{s-s_j,p}(\mathbf{R}^n) \tag{2.6.11}$$

is an isomorphism. Then, taking into account that

$$\widehat{l}_0+Q=\widehat{l}_0(I_1+\widehat{l}_0^{-1}Q)=(I_2+Q\widehat{l}_0^{-1})\widehat{l}_0,$$

where I_1 and I_2 are the identity operators in $H^{T+s,p}$ and $H^{s-S,p}$, respectively, and setting $\varepsilon \le \left(2\|\widehat{l}_0^{-1}\|\right)^{-1}$, we arrive at the conclusion that $\widehat{l}_0+Q$ is an invertible operator and the operator $R_0=(\widehat{l}_0+Q)^{-1}$ realizes an isomorphism

$$R_0=(\widehat{l}_0+Q)^{-1} : H^{s-S,p} \to H^{T+s,p}. \tag{2.6.12}$$

Then (2.6.9) implies that

$$R_0(\widehat{l}_0+Q+l')=I_1+T_0', \qquad (\widehat{l}_0+Q+l')R_0=I_2+T_0'', \tag{2.6.13}$$

where

$$T_0' = R_0 l' : H^{T+s-1,p} \to H^{T+s,p},$$

$$T_0'' = l' R_0 : H^{s-S,p} \to H^{s+1-S,p}$$

are smoothing operators.

Now let $x_0 \in \partial G$. By Lemma 2.6.3, there exists a neighborhood $U(x_0)$ of the point x_0 in $\mathbf{R}^n$ such that any function $\varphi \in C_0^\infty(U(x_0))$ admits representation (2.6.10) with the operator Q whose norm is arbitrary small. In view of Theorem 2.4.1 the operator $\widehat{A}_0=\widehat{A}_0(x,D)$ realizes an isomorphism

$$\widehat{A}_0 : \widetilde{\mathcal{H}}^{s,p}(\mathbf{R}^n_+) \to K^{s,p}(\mathbf{R}^n_+). \tag{2.6.14}$$

Therefore, since

$$\widehat{A}_0 + Q = \widehat{A}_0(I_1 + \widehat{A}_0^{-1}Q) = (I_2 + Q\widehat{A}_0^{-1})\widehat{A}_0,$$

where I_1 and I_2 are the identity operators in $\widetilde{\mathcal{H}}^{s,p}(\mathbf{R}^n_+)$ and $K^{s,p}(\mathbf{R}^n_+)$, respectively, one can conclude that if $\|Q\| < \left(2\|\widehat{A}_0^{-1}\|\right)^{-1}$ then $\widehat{A}_0 + Q$ is an invertible operator. By choosing $\varepsilon \le \left(2\|\widehat{A}_0^{-1}\|\right)^{-1}$ in Lemma 2.6.3, we obtain that the operator $\widehat{A}_0 + Q$ realizes an isomorphism between spaces (2.6.14). Then the operator

$$R_0 = (\widehat{A}_0 + Q)^{-1} \tag{2.6.15}$$

establishes an isomorphism

$$R_0 : K^{s,p}(\mathbf{R}^n_+) \to \widetilde{\mathcal{H}}^{s,p}(\mathbf{R}^n_+). \tag{2.6.16}$$

Therefore, it follows from (2.6.15) and (2.6.10) that

$$\begin{gathered} R_0(\widehat{A}_0(x,D) + Q(x,D) + A'(x,D)) = I_1 + T_0' \\ (\widehat{A}_0(x,D) + Q(x,D) + A'(x,D))R_0 = I_2 + T_0'', \end{gathered} \tag{2.6.17}$$

where I_1 and I_2 are the identity operators in $\widetilde{\mathcal{H}}^{s,p}(\mathbf{R}^n_+)$ and $K^{s,p}(\mathbf{R}^n_+)$, respectively, and

$$T_0' : \widetilde{\mathcal{H}}^{s-1,p}(\mathbf{R}^n_+) \to \widetilde{\mathcal{H}}^{s,p}(\mathbf{R}^n_+), \quad T_0'' : K^{s,p}(\mathbf{R}^n_+) \to K^{s+1,p}(\mathbf{R}^n_+) \tag{2.6.18}$$

are smoothing operators.

Thus, for any point $x_0 \in G$ there exists a sufficiently small neighborhood $U(x_0) = U \subset G$ such that in this neighborhood the assertions of Lemmas 2.6.1 and 2.6.1 are true, and the operator R_0 defined by (2.6.12) satisfies relations (2.6.13). For any point $x_0 \in \partial G$ there exists a sufficiently small neighborhood $U(x_0) = U \subset G$ in $\mathbf{R}^n$ in which the assertions of Lemmas 2.6.2 and 2.6.3 are true, and the operator R_0 defined by (2.6.15) satisfies relations (2.6.17).

Hence, we have a finite covering of the compact set $\overline{G} = G \bigcup \partial G$ by the neighborhoods. Let us select a finite subcovering. As a result, we obtain the finite set of points $\{x^j : j = 1, \dots, \nu\}$ and their neighborhoods

$$\{U^j : j = 1, \dots, \nu\}, \qquad U^j = U(x^j),$$

whose properties are formulated in Lemma 2.6.3. For $x^j \in G$ we denote by R_j the operator of the form (2.6.12), and if $x^j \in \partial G$, we denote by R_j the operator (2.6.15).

Let $\{\varphi^j : j = 1, \ldots, \nu\}$ be the decomposition of unity subordinate to the covering $\{U^j\}$,

$$\varphi_j \in C_0^\infty(U^j), \qquad \sum_{j=1}^{\nu} \varphi_j(x) = 1. \tag{2.6.19}$$

Let $\psi_j \in C_0^\infty(U^j)$, let $0 \le \psi_j \le 1$, and let $\psi_j(x) = 1$ in some neighborhood of the support of φ_j. It is clear that $\psi_j \varphi_j = \varphi_j \quad (j = 1, \ldots, \nu)$.

We set

$$RF = R_{s,p}F = \sum_{j=1}^{\nu} \psi_j R_j \varphi_j F \qquad (F \in K^{s,p}(G)). \tag{2.6.20}$$

One can easily verify that R is a regularizer of the operator A. Indeed, by commuting A with ψ_j, we obtain

$$ARF = \sum_{j=1}^{\nu} A\psi_j R_j \varphi_j F = \sum_{j=1}^{\nu} (\psi_j A R_j \varphi_j F + A_j'' R_j \varphi_j F), \tag{2.6.21}$$

where the order of the operator A_j'' is lower than the order of A by one. Then it follows from representations (2.6.13) and (2.6.17) that

$$ARF = \sum_{j=1}^{\nu} (\psi_j I_2 \varphi_j F + T_j'' \varphi_j F + A_j'' R_j \varphi_j F) = I_2 F + T_2 F, \tag{2.6.22}$$

where T_2 acts continuously from $K^{s,p}(G)$ into $K^{s+1,p}(G)$. This establishes the second relation in (2.6.2).

Let $u \in \widetilde{\mathcal{H}}^{s,p}$. By commuting A with φ_j, we obtain

$$RAu = \sum_{j=1}^{\nu} \psi_j R_j \varphi_j Au = \sum_{j=1}^{\nu} (\psi_j R_j A \varphi_j u + \psi_j R_j A_j'' u) \tag{2.6.23}$$

If $x^j \in \partial G$ then, from equalities (2.6.17) and (2.6.10), we obtain

$$\begin{aligned} R_j A \varphi_j u &= R_j A \psi_j \varphi_j u \\ &= R_j \Big(\widehat{A}_0(x', D) + Q(x, D) + A_j'(x, D)\Big)\varphi_j u = (I_1 + T_j')\varphi_j u. \end{aligned}$$

The case of $x^j \in G$ is considered in the similar way. Theorem 2.6.1 is, thus, proved.

It follows from Theorem 2.6.1 that the operator $A_{s,p}$ is Noetherian: the kernel $\mathfrak{N}_{s,p}$ and the cokernel $\mathfrak{N}_{s,p}^*$ are finite-dimensional, and the range of the operator $A_{s,p}$ is closed in $K^{s,p}$ (see, for example, [R1, Sec. 1.15]).

To complete the proof of Theorem 2.4.1, it remains to prove that $\mathfrak{N}_{s,p}$ and $\mathfrak{N}^*_{s,p}$ do not depend on s and p and consist of infinitely smooth elements. The following assertion is true.

Lemma 2.6.4. *Under the conditions of Theorem 2.6.1, let relations (2.6.1) be valid,* $u \in \widetilde{\mathcal{H}}^{s,p}$ *and* $Au = F \in K^{s,p}$. *If* $F \in K^{s+t,p}$ $(t > 0)$ *then* $u \in \widetilde{\mathcal{H}}^{s+t,p}$.

Proof. It follows from Theorem 2.6.1 that

$$u = -T_1 u + RF. \tag{2.6.24}$$

The operator $T_1 = T_{1\,s,p}$ is a smoothing operator acting continuously from $\widetilde{\mathcal{H}}^{s-1,p}$ into $\widetilde{\mathcal{H}}^{s,p}$. This implies that $u \in \widetilde{\mathcal{H}}^{\min\{s+t,s+1\},p}$. If $t \geq 1$ then $u \in \widetilde{\mathcal{H}}^{s+1,p}$. Therefore, (2.6.24) implies that $u \in \widetilde{\mathcal{H}}^{\min\{s+t,s+2\},p}$, etc. After finitely many steps we arrive at the inclusion $u \in \widetilde{\mathcal{H}}^{s+t,p}$. □

Corollary 2.6.1. *If* $u \in \widetilde{\mathcal{H}}^{s,p}$ *and* $Au = 0$ *then* $u \in \widetilde{\mathcal{H}}^{s+t,p}$, *for any* $t > 0$. *Hence,* $\mathfrak{N}_{s,p} = \mathfrak{N} \subset (C^\infty(\overline{G}))^N$.

Let us prove inclusion (2.4.6).

The operator $A = A_{s,p}$ is continuous in the pair of spaces (2.4.4). Then the adjoint operator $A^* = A^*_{s,p}$ acts continuously from $(K^{s,p})^*$ into $(\widetilde{\mathcal{H}}^{s,p})^*$.

If $s < 1/p$ then (see Sec. 2.3)

$$\widetilde{\mathcal{H}}^{s,p} = \prod_{j=1}^{N}\left(H^{t_j+s,p}(G) \times \prod_{j=1}^{r_j} B^{t_j+s-k+1-1/p,p}(\partial G)\right)$$

$$\times \prod_{j=N+1}^{N+k''} \prod_{k=1}^{\sigma_1+t_j+1} B^{t_j+s-k+1-1/p,p}(\partial G),$$

$$K^{s,p} := \prod_{j=1}^{N}\left(H^{s-s_j,p}(G) \times \prod_{k=1}^{æ-s_j} B^{s-s_j-k+1-1/p,p}(\partial G)\right)$$

$$\times \prod_{h=1}^{m+k''} B^{s-\sigma_h-1/p,p}(\partial G),$$

and, hence,

$$(\widetilde{\mathcal{H}}^{s,p})^* = \prod_{j=1}^{N}\left(H^{-t_j-s,p'}(G) \times \prod_{j=1}^{r_j} B^{-(t_j+s-k+1-1/p),p'}(\partial G)\right)$$

$$\times \prod_{j=N+1}^{N+k''} \prod_{k=1}^{\sigma_1+t_j+1} B^{-(t_j+s-k+1-1/p),p'}(\partial G), \qquad (2.6.25)$$

$$(K^{s,p})^* := \prod_{j=1}^{N}\Big(H^{-(s-s_j),p'}(G) \times \prod_{k=1}^{æ-s_j} B^{-(s-s_j-k+1-1/p),p'}(\partial G)\Big)$$

$$\times \prod_{h=1}^{m+k''} B^{-(s-\sigma_h-1/p),p'}(\partial G). \qquad (2.6.26)$$

One can obtain the similar representations for the spaces $(\widetilde{\mathcal{H}}^{s,p})^*$ and $(K^{s,p})^*$ in the case where $s > 1/p$. It follows from the facts that the space $\widetilde{H}^{s,p,(r)}(G)$ is isomorphic to the direct product

$$H^{s,p}(G) \times \prod_{j=k+1}^{r} B^{s-j+1-1/p,p}(\partial G) \quad (k = [s+1-1/p]),$$

for $1/p < s < r-1+1/p$ and noninteger $s-1/p$, and the space $\widetilde{H}^{s,p,(r)}(G)$ is isomorphic to the space $H^{s,p}(G)$ for $s > r-1+1/p$ (see the end of Subsection 2.3.4).

It follows from the second relation in (2.6.2) that

$$(R_{s,p})^*(A_{s,p})^* = I_2^* + T_{2\,s,p}^*. \qquad (2.6.27)$$

Here I_2^* is the identity operator in $(K^{s,p})^*$, $R^* = (R_{s,p})^*$ is the operator adjoint to $R = R_{s,p}$ which acts continuously in the pair of spaces

$$(\widetilde{\mathcal{H}}^{s,p})^* \to (K^{s,p})^*,$$

$T_{2\,s,p}^*$ is the operator adjoint to $T_{2\,s,p}$ which acts continuously in the pair of spaces

$$(K^{s+1,p})^* \to (K^{s,p})^*,$$

i.e., it is a smoothing operator.

Repeating the reasoning of the proof of Lemma 2.6.4, we obtain that (2.6.27) implies the assertion on increasing in smoothness for the equation

$$(A_{s,p})^*V = \mathcal{G} \in (\widetilde{\mathcal{H}}^{s,p})^*,$$

i.e., if $\mathcal{G} \in (\widetilde{\mathcal{H}}^{s-t,p})^*$ $(t > 0)$ then $V \in (K^{s-t,p})^*$. In particular, if $A^*V = 0$ then V is an infinitely smooth element.

Thus, Theorem 2.4.1 is completely proved.

2.7. Green's Formula for problem (2.1.1)–(2.1.2)

2.7.1. Let

$$M = \sum_{j=N+1}^{N+k'} (\sigma_1 + t_j + 1),$$

and let

$$U = (u, u') \in \left(C^\infty(\overline{G})\right)^N \times \left(C^\infty(\partial G)\right)^M$$

be an infinitely smooth solution of problem (2.1.1)–(2.1.2) with infinitely smooth f and φ. Then (2.1.1) yields that

$$D_\nu^{h-1} l_r u\big|_{\partial G} = D_\nu^{h-1} f_r\big|_{\partial G} \qquad (h = 1, \ldots, æ - s_r,\ r : æ - s_r \geq 1). \quad (2.7.1)$$

Here

$$l_r u = \sum_{j=1}^{N} l_{rj}(x, D) u_j$$

is an rth row of system (2.1.1), æ is defined by formula (2.4.1). We set

$$\begin{aligned} &D_\nu^{h-1} f_r\big|_{\partial G} = f_{rh}(x') \qquad (h = 1, \ldots, æ - s_r,\ r : æ - s_r \geq 1), \\ &D_\nu^{h-1} u_j\big|_{\partial G} = u_{jh}(x') \quad (j = 1, \ldots, N,\ h = 1, \ldots, \tau_j,\ \tau_j = t_j + æ). \end{aligned} \quad (2.7.2)$$

Then, in certain neighborhood in $\overline{G}$ of the boundary ∂G we represent the expression $D_\nu^{h-1} l_r u$ in the form

$$D_\nu^{h-1} l_r u = \sum_{j=1}^{N} \sum_{k=1}^{s_r+t_j+h} l_{rhjk}(x, D') D_\nu^{k-1} u_j,$$

where $l_{rhjk}(x, D')$ are tangential operators of orders $s_r + t_j + h - k$. Then one can rewrite equalities (2.7.1) in the form

$$\sum_{j=1}^{N} \sum_{k=1}^{s_r+t_j+h} l_{rhjk}(x', D') u_{jk} = f_{rh}(x') \quad (2.7.3)$$

$$(h = 1, \ldots, æ - s_r,\ r : æ - s_r \geq 1).$$

By using equalities (2.1.3) and (2.1.4), we reduce equalities (2.1.2) to the form

$$b_h u + b_h' u' = \sum_{j=1}^{N} \sum_{k=1}^{\sigma_h+t_j+1} \Lambda_{hjk}(x', D') u_{jk}(x')$$

$$+\sum_{j=N+1}^{N+k'}\sum_{k=1}^{\sigma_h+t_j+1}\Lambda_{hjk}(x',D')u'_{jk}(x')=\varphi_h(x').$$

If we set here that $\Lambda_{hjk}=0$ for $\sigma_h+t_j+1<k\le æ+t_j=\tau_j$, then we can rewrite these equalities in the following form:

$$b_h u+b'_h u'=\sum_{j=1}^{N}\sum_{k=1}^{\tau_j}\Lambda_{hjk}(x',D')u_{jk}$$

$$+\sum_{j=N+1}^{N+k'}\sum_{k=1}^{\sigma_h+t_j+1}\Lambda_{hjk}(x',D')u'_{jk}(x')=\varphi_h(x') \tag{2.7.4}$$

$$(h=1,\ldots,m+k'').$$

Here $\Lambda_{hjk}(x',D')$ are tangential operators of orders σ_h+t_j-k+1 $(\forall h,j:\sigma_h+t_j\ge 0,\ k=1,\ldots,\sigma_h+t_j+1)$.
It follows from the considerations in Section 2.5, that one can complement equalities (2.7.3) and (2.7.4) by the equalities

$$c_h u:=\sum_{j=1}^{N}\sum_{k=1}^{\sigma_h^c+t_j+1}\Lambda^c_{hjk}(x',D')u_{jk}(x')=\psi_h(x')(h=1,\ldots,m) \tag{2.7.5}$$

$$(h=1,\ldots,m)$$

so that expressions (2.7.3)–(2.7.5) form a Douglis-Nirenberg elliptic system on ∂G. Here $\Lambda^c_{hjk}(x',D')$ are tangential operators of orders $\sigma^c_h+t_j-k+1$ $(\forall h,j:\sigma^c_h+t_j\ge 0,\ k=1,\ldots,\sigma^c_h+t_j+1)$, all the numbers σ^c_h are negative.

Let us write system (2.7.3)–(2.7.5) in the form

$$\begin{cases} e(x',D')U &= F,\\ \Lambda U+\Lambda' U' &= \varphi,\\ cU &= \psi.\end{cases} \tag{2.7.6}$$

Here

$$U = (U_1,\ldots,U_N)',\qquad U_j = (u_{j1},\ldots,u_{j\tau_j})',$$

$$U' = (U'_{N+1},\ldots,U'_{N+k'})',\qquad U'_j = (u'_{j1},\ldots,u'_{j,\tau_j})',$$

$$F = (F_1,\ldots,F_N)',\qquad F_j = (f_{j1},\ldots,f_{j,æ-s_j})',$$

$$\varphi = (\varphi_1,\ldots,\varphi_{m+k''})',\qquad \psi = (\psi_1,\ldots\psi_m)'.$$

We also assume that, for $\sigma_h + t_j + 1 < k \leq \tau_j$, the expressions Λ_{hjk} and Λ'_{hjk} are equal to zero. Denote:

$$E(x', D') = \begin{pmatrix} e(x', D') & 0 \\ \Lambda(x', D') & \Lambda'(x', D') \\ c(x', D') & 0 \end{pmatrix}. \tag{2.7.7}$$

Then one can rewrite system (2.7.3)–(2.7.5) or system (2.7.6) in the form

$$E(x', D')\mathbf{U} = \left(E_{jk}(x', D')\right)_{j,k=1,\ldots,K}\mathbf{U} = \mathbf{\Phi}. \tag{2.7.8}$$

Here $\mathbf{U}$ is the column $(U, U')'$ and $\mathbf{\Phi}$ is the column $(F, \varphi, \psi)'$.

It was above mentioned that system (2.7.8) is a linear system of

$$N æ + t_1 + \cdots + t_N + k'' = \tau_1 + \cdots, \tau_N + k'' = |\tau| + k'' = K$$

equations in the same number of unknown functions. It is an elliptic system in the Douglis-Nirenberg sence, i.e., there exist numbers $T_1, \ldots, T_K$, $S_1, \ldots, S_K$ such that $\operatorname{ord} E_{jk} \leq T_k + S_j$ for $T_k + S_j \geq 0$ and $E_{jk} = 0$ for $T_k + S_j < 0$, and, finally,

$$\det E_0(x', \xi') \neq 0 \qquad (\forall \xi' \in \mathbf{R}^{n-1} \setminus \{0\}).$$

Here

$$E_0(x', \xi') = \left(E^0_{jk}(x', \xi')\right)_{j,k=1,\ldots,K}$$

is the principal part of the matrix E, $\operatorname{ord} E^0_{jk} = T_k + S_j$ for $T_k + S_j \geq 0$ and $E^0_{jk} = 0$ for $T_k + S_j < 0$.

For any $s \in \mathbf{R}$ and $p \in]1, +\infty[$, the closure $E = E_{s,p}$ of the mapping

$$\mathbf{U} \mapsto E(x', \xi')\mathbf{U} \qquad \left(\mathbf{U} \in \left(C^\infty(\partial G)\right)^K\right)$$

acts continuously in the pair of spaces

$$B^{T+s,p} := \prod_{j=1}^{K} B^{T_j+s,p}(\partial G) \to \prod_{j=1}^{K} B^{s-S_j,p}(\partial G) =: B^{s-S,p}. \tag{2.7.9}$$

Since (2.7.8) is a Douglis-Nirenberg elliptic system on ∂G, the operator $E = E_{s,p}$ is Noetherian, the kernel $\mathfrak{N}(E)$ and the cokernel $\mathfrak{N}^*(E) = \mathfrak{N}(E^*)$ are finite-dimensional, do not depend on s and p and cosist of infinitely smooth elements, i.e.,

$$\mathfrak{N}(E) = \{\mathbf{U} \in \left(C^\infty(\partial G)\right)^K : E\mathbf{U} = 0\}, \tag{2.7.10}$$

$$\mathfrak{N}^*(E) = \{\mathbf{U} \in (C^\infty(\partial G))^K : E^*\mathbf{U} = (E^*_{kj}(x', D'))_{j,k=1,\dots,K}\mathbf{U} = 0\}. \tag{2.7.11}$$

If the defect is missing, i.e.,

$$\mathfrak{N}(E) = 0, \qquad \mathfrak{N}^*(E) = 0,$$

then the operator $E = E_{s,p}$ establishes an isomorphism between spaces (2.7.9). In the general case, the isomorphism between the corresponding subspaces, whose defects are finite and independent of s and p, is realized by the restriction E_1 of the operator E:

$$E_1 : PB^{T+s,p} \to P^+B^{s-S,p}. \tag{2.7.12}$$

Here

$$PB^{T+s,p} = \{\mathbf{U} \in B^{T+s,p} : (\mathbf{U}, V) = 0 \ \ (V \in \mathfrak{N}(E))\}$$

is a subspace of $B^{T+s,p}$, and

$$P^+B^{s-S,p} = \{\Phi \in B^{s-S,p} : (\Phi, V) = 0 \ \ (V \in \mathfrak{N}^*(E))\}$$

is a subspace of $B^{s-S,p}$; $(\cdot,\cdot)$ denotes the extension of the scalar product in $(L_2(\partial G))^K$. The projection operators P and P^+ are constructed in exactly the same way as in [R6, Section 4.1]

2.7.2. Let us deduce Green's formula for problem (2.1.2)–(2.1.2) under the additional assumption

$$\mathfrak{N}(E) = 0, \qquad \mathfrak{N}^*(E) = 0. \tag{2.7.13}$$

It was already mentioned that the operator $E = E_{s,p}$ realizes an isomorphism between spaces (2.7.9). Therefore, if

$$E\mathbf{U} = \Phi \in B^{s-S,p},$$

then

$$\mathbf{U} = E^{-1}\Phi \in B^{T+s,p}. \tag{2.7.14}$$

By integrating by parts (see (1.2.10)), we calculate

$$(lu, v) - (u, l^+v) = \sum_{j:t_j\geq 1}\sum_{s=1}^{t_j}\langle D_\nu^{s-1}u_j, M_j^s v\rangle, \tag{2.7.15}$$

$$M_j^s v = -i \sum_{r:s_r+t_j\geq s}\ \sum_{k=s}^{s_r+t_j} D_\nu^{k-j}(l_k^{rj}(x, D'))^+ v_r.$$

We associate the element $v = (v_1, \dots, v_N)' \in \left(C^\infty(\overline{G})\right)^N$ with the vector

$$Mv = (\zeta_1, \dots, \zeta_K),$$

$$\zeta_j = (\zeta_{j1}, \dots, \zeta_{j,\tau_j}), \quad \zeta_{jk} = \begin{cases} M_j^k v\big|_{\partial G} & \text{for } k = 1, \dots, t_j,\ j = 1, \dots, N, \\ 0 & \text{for } t_j < k \le \tau_j,\ j = 1, \dots, N, \end{cases}$$

$$\zeta_j = 0 \qquad (j = N+1, \dots, K).$$

Then one can rewrite formula (2.7.15) in the form

$$(lu, v) - -(u, l^+ v) = \langle \mathbf{U}, Mv \rangle \qquad \left(u, v \in \left(C^\infty(\overline{G})\right)^N\right), \tag{2.7.16}$$

where $\langle \cdot, \cdot \rangle$ denotes the scalar product in $\left(L_2(\partial G)\right)^K$. Futher,

$$\begin{aligned} \langle \mathbf{U}, Mv \rangle &= \langle E^{-1} E \mathbf{U}, Mv \rangle = \langle E\mathbf{U}, (E^{-1})^* Mv \rangle \\ &= \left\langle \begin{pmatrix} e(x', D') & 0 \\ \Lambda(x', D') & \Lambda'(x', D') \\ c(x', D') & 0 \end{pmatrix} \begin{pmatrix} U \\ U' \end{pmatrix}, (E^{-1})^* Mv \right\rangle. \end{aligned} \tag{2.7.17}$$

As a result, (2.7.16)–(2.7.17) imply that there holds Green's formula (sf. (1.2.17))

$$\begin{aligned} (lu, v) &+ \sum_{r=1}^{N} \sum_{k=1}^{-s_r + \text{æ}} \langle D_\nu^{k-1} l_r u, e'_{kr} v \rangle + \sum_{h=1}^{m+k''} \langle b_h u + b'_h u', c'_h v \rangle \\ &= (u, l^+ v) + \sum_{h=1}^{m} \langle c_h u, b_h^+ v \rangle, \end{aligned} \tag{2.7.18}$$

where u and v are infinitely smooth vector-functions in $\overline{G}$, and u' is an infinitely smooth vector-function on ∂G.

Thus, it is proved the following theorem:

Theorem 2.7.1. *Let problem (2.1.2)–(2.1.2) be elliptic and let conditions (2.7.13) be valid. Then Green's formula (2.7.18) holds.*

2.7.3. Now let $\mathfrak{N}(E) \neq 0$. Every element $\mathbf{U} \in B^{T+s,p}$ is representable in the form

$$\mathbf{U} = \mathbf{U}' + \mathbf{U}'', \qquad \mathbf{U}'' \in \mathfrak{N}(E), \qquad \mathbf{U}' \perp \mathfrak{N}(E). \tag{2.7.19}$$

The operator $P : \mathbf{U} \mapsto \mathbf{U}'$ is continuous in the space $B^{T+s,p}$. The restriction E_1 of the operator E establishes an isomorphism (2.7.12). Then, it follows from (2.7.16) that

$$\begin{aligned}
(lu, v) - (u, l^+v) &= \langle \mathbf{U}' + \mathbf{U}'', Mv\rangle \\
&= \langle \mathbf{U}', Mv\rangle + \langle \mathbf{U}'', Mv\rangle \\
&= \langle E_1^{-1}E_1\mathbf{U}', Mv\rangle + \langle \mathbf{U}'', Mv\rangle \\
&= \langle E_1^{-1}E\mathbf{U}, Mv\rangle + \langle \mathbf{U}'', Mv\rangle \\
&= \langle E\mathbf{U}, (E_1^{-1})^* Mv\rangle + \langle \mathbf{U}'', Mv\rangle \\
&= \left\langle \begin{pmatrix} e(x', D') & 0 \\ \Lambda(x', D') & \Lambda'(x', D') \\ c(x', D') & 0 \end{pmatrix} \begin{pmatrix} U \\ U' \end{pmatrix}, (E^{-1})^* Mv \right\rangle \\
&\quad + \langle \mathbf{U}'', Mv\rangle,
\end{aligned}$$

and we can write the Green's formula in the following form

$$\begin{aligned}
(lu, v) &+ \sum_{r=1}^{N} \sum_{k=1}^{-s_r+æ} \langle D_\nu^{k-1} l_r u, e'_{kr} v\rangle + \sum_{h=1}^{m+k''} \langle b_h u + b'_h u', c'_h v\rangle \\
&= (u, l^+ v) + \sum_{h=1}^{m} \langle c_h u, b_h^+ v\rangle + \langle \mathbf{U}'', Mv\rangle,
\end{aligned} \tag{2.7.20}$$

where u v and u' are infinitely smooth vector-functions in $\overline{G}$ and on ∂G, respectively, and $\mathbf{U}'' = \mathbf{U} - P\mathbf{U} \in \mathfrak{N}(E)$.
Thus, we have proved the theorem:

Theorem 2.7.2. *Let problem (2.1.2)–(2.1.2) be elliptic and let $\mathfrak{N}(E) \neq 0$. Then Green's formula (2.7.20) holds.*

2.7.4. For simplicity, in what follows we assume that conditions (2.7.13). Else, in the general case, all the considerations became more cumbersome (sf. [R2]).

Definition 2.7.1. *The problem*

$$l^+ v = g \quad (in\ \ G), \qquad b_h^+ v\big|_{\partial G} = \psi_h \quad (h = 1, \ldots, m) \tag{2.7.21}$$

is called formally adjoint to problem (2.1.2)–(2.1.2) with respect to Green's formula (2.7.18).

Let

$$\mathfrak{N}^+ = \left\{ v \in (C^\infty(\overline{G}))^N : l^+ v = 0,\ b_h^+ v\big|_{\partial G} = 0\ (h = 1, \dots, m) \right\} \quad (2.7.22)$$

be the kernel of problem (2.7.21).

By repeating the reasoning from the proof of Theorem 1.2.3, we verify the validity of the following theorem:

Theorem 2.7.3. *Under the assumptions of Theorem 2.7.1, let $s \in \mathbf{R}$ and $p \in (1, \infty)$. The element*

$$U = (u, u') \in \widetilde{\mathcal{H}}^{s,p}$$

is a generalized solution of problem (2.1.2)–(2.1.2) with $F \in K^{s,p}$ (see (2.4.5)) if and only if the equality

$$(f_0, v) + \sum_{r=1}^{N} \sum_{k=1}^{-s_r+\text{æ}} \langle f_{rk}, e'_{kr} v \rangle + \sum_{h=1}^{m+k''} \langle \varphi_h, c'_h v \rangle = 0 \quad (\forall v \in \mathfrak{N}^+) \quad (2.7.23)$$

holds. Furthermore, $\mathfrak{N}^+$ is finite-dimensional.

By analogy, repeating the arguments from the proof of Theorem 1.2.4, we verify the validity of the following theorem:

Theorem 2.7.4. *Let the assumptions of Theorem 2.7.1 be valid. Then problem (2.7.21), formally adjoint to problem (2.1.2)–(2.1.2) with respect to Green's formula (2.7.18), is also elliptic.*

The analogs of the assertions of Section 1.3 remain true here with, clearly, natural changes. These results are not used below and, hence, we omit corresponding precise formulations.

Thus, the analogs of all Theorems on isomorphisms, which are known for the case of one equation with normal boundary conditions, remain true for boundary value problems (2.1.2)–(2.1.2).

2.8. Parameter-Elliptic boundary value Problems for General Systems of Equations with Additional Unknown Functions Defined at the Boundary of Domain

Considerations of the present section are quite analogous to those given in Section 1.5 which is devoted to parameter-elliptic problems for general systems of equations whose boundary conditions do not includ additional unknown functions.

2.8.1. In the bounded domain $G \subset \mathbf{R}^n$ with the boundary $\partial G \in C^\infty$ we consider the problem

$$l(x, D, q_1)u := \left(l_{rj}(x, D, q_1)\right)_{r,j=1,\dots,N} u(x) = f(x) \quad (x \in G), \tag{2.8.1}$$

$$b(x, D, q_1)u + b'(x', D', q_1)u' = \varphi(x') \quad (x' \in \partial G), \tag{2.8.2}$$

$$b(x, D, q_1) = \left(b_{hj}(x, D, q_1)\right)_{\substack{h=1,\dots,m+k'' \\ j=1,\dots,N}},$$

$$b'(x', D', q_1) = \left(b'_{hj}(x', D', q_1)\right)_{\substack{h=1,\dots,m+k'' \\ j=N+1,\dots,N+k'}}.$$

Here $q_1 = qe^{i\theta}$, $q \in \mathbf{R}$, $\theta \in [\theta_1, \theta_2]$ (the case where $\theta = \theta_1 = \theta_2$ is not excluded), the system $l = l(x, D, q_1)$ and the boundary expressions

$$b(x, D, q_1) := \left(b_{hj}(x, D, q_1)\right)_{\substack{h=1,\dots,m+k'' \\ j=1,\dots,N}}$$

and

$$b'(x', D', q_1) := \left(b'_{hj}(x', D', q_1)\right)_{\substack{h=1,\dots,m+k'' \\ j=N+1,\dots,N+k'}}$$

are described by the following relations:

$$\operatorname{ord} l_{rj} \le s_r + t_j \text{ for } s_r + t_j \ge 0,\ l_{rj} = 0 \text{ for } s_r + t_j < 0\ (r, j = 1, \dots, N);$$

$$\operatorname{ord} b_{hj} \le \sigma_h + t_j \text{ for } \sigma_h + t_j \ge 0,\ b_{hj} = 0 \text{ for } \sigma_h + t_j < 0$$
$$(j = 1, \dots, N, \quad h = 1, \dots, m + k'');$$

$$\operatorname{ord} b'_{hj} \le \sigma_h + t_j \text{ for } \sigma_h + t_j \ge 0,\ b'_{hj} = 0 \text{ for } \sigma_h + t_j < 0$$
$$(j = N+1, \dots, N+k', \quad h = 1, \dots, m + k'').$$

Here $t_1, \dots, t_{N+k'}, s_1, \dots, s_N, \sigma_1, \dots, \sigma_{m+k''}$ are given integers such that

$$s_1 + \dots + s_N + t_1 + \dots + t_N = 2m,$$

$$t_1 \geq \cdots \geq t_N \geq 0 = s_1 \geq \ldots \geq s_N, \qquad \sigma_1 \geq \ldots \geq \sigma_{m+k''}.$$

Recall that $u(x) = (u_1(x), \ldots, u_N(x))$, $x \in \overline{G}$, and

$$u'(x') = (u'_{N+1}(x'), \ldots, u'_{N+k'}(x')),$$

$$u'_j(x') = (u'_{j1}(x'), \ldots, u_{j,\sigma_h+t_j+1}(x')).$$

Moreover,

$$l_{rj}(x, D, q_1) = \begin{cases} \sum_{|\mu|+k \leq s_r+t_j} a^{rj}_{\mu k}(x) q_1^k D^\mu, & \text{for } r, j : s_r + t_j \geq 0, \\ 0, & \text{for } r, j : s_r + t_j < 0 \end{cases} \tag{2.8.3}$$

$$(\mu = (\mu_1, \ldots, \mu_n),\ |\mu| = \mu_1 + \cdots + \mu_n),$$

$$b_{hj}(x, D, q_1) = \begin{cases} \sum_{|\alpha|+k \leq \sigma_h+t_j} b^{hj}_{\alpha k}(x) q_1^k D^\alpha, & \text{for } h, j : \sigma_h + t_j \geq 0, \\ 0, & \text{for } h, j : \sigma_h + t_j < 0 \end{cases} \tag{2.8.4}$$

$$(j = 1, \ldots, N),$$

$$b'_{hj}(x, D, q_1) = \begin{cases} \sum_{|\alpha|+k \leq \sigma_h+t_j} b'^{hj}_{\alpha k}(x') q_1^k D^\alpha, & \text{for } h, j : \sigma_h + t_j \geq 0, \\ 0, & \text{for } h, j : \sigma_h + t_j < 0 \end{cases} \tag{2.8.5}$$

$$(j = N+1, \ldots, N+k').$$

Here and in what follows we assume that the coefficients of all differential expressions and the boundary ∂G are infinitely smooth in G and ∂G, respectively.

Assume that problem (2.8.1)–(2.8.2) is elliptic with a parameter. This means that the problem

$$\begin{aligned} l(x, D_x, e^{i\theta} D_t)u &= f \qquad \text{(in } G \times \mathbf{R}), \\ b(x, D_x, e^{i\theta} D_t)u + b'(x', D'_x, e^{i\theta} D_t)u' &= \varphi \qquad \text{(in } \partial G \times \mathbf{R}), \end{aligned} \tag{2.8.6}$$

where $\theta \in [\theta_1, \theta_2]$ and $D_t = i\partial/\partial t$, is elliptic in the cylinder $\overline{G} \times \mathbf{R}$.

let us now write Definition 2.2.1 for problem (2.8.6) and obtain the equivalent definition of parameter-ellipticity of problem (2.8.1)–(2.8.2). Let $L(x, \xi, q_1)$ be defined by formula (1.5.5):

$$L(x, \xi, q_1) = \det(l^0_{rj}(x, \xi, q_1)). \tag{2.8.7}$$

System (2.8.1) is called elliptic with a parameter (or, shortly, parameter-elliptic) if for every point $x \in \overline{G}$, and any $\xi \in \mathbf{R}^n$, $q \in \mathbf{R}$ $(|\xi| + |q| > 0)$, and $\theta \in [\theta_1, \theta_2]$, we have

$$L(x, \xi, q_1) \neq 0.$$

The parameter-elliptic system (2.8.1) is called properly elliptic with a paremeter (or, shortly, properly parameter-elliptic) if for every point $x \in \overline{G}$, $q \in \mathbf{R}$ $(|\xi| + |q| > 0)$, $\theta \in [\theta_1, \theta_2]$, and every vector $\tau \in \mathbf{R}^n$ tangential to ∂G at the point x, the polynomial $L(\eta) = L(x, \tau + \eta\nu, q_1)$ has even order $2m = \sum_{j=1}^{N}(s_j + t_j)$, and accurately m of its roots have positive (negative) imaginary parts.

Then $L(\eta) = L_+(\eta)L_-(\eta)$, where L_+ (L_-) is an mth-order polynomial whose all roots lie in the upper (lower) half-plane.

Let $\mathcal{M}^+ = \mathcal{M}^+(x, \tau, q_1)$ be the space of stable solutions (i.e., solutions decreasing as $t \to +\infty$) of the equation $l_0(D_t) = 0$ $(l_0(\zeta) = l_0(x, \tau + \zeta\nu, q_1))$, and let $b_0(x, D, q_1)$ and $b_0'(x, D, q_1)$ be the principal parts of the matrixes b and b', respectively. Assume that

$$b_0(\zeta) = b_0(x, \tau + \zeta\nu, q_1), \quad b_0'(\tau) = b_0(x, \tau, q_1) \quad (|\tau| + |q| > 0,\ \theta \in [\theta_1, \theta_2]).$$

Definition 2.8.1. *The boundary value problem (2.8.1)–(2.8.2) is called parameter-elliptic problem if system (2.8.1) is properly parameter-elliptic, and the Lopatinskii condition is sutisfied, i.e., at every point $x \in \partial G$, for any τ, q $(|\tau| + |q| > 0)$ and $\theta \in [\theta_1, \theta_2]$, and any $h = (h_1, \ldots, h_{m+k''}) \in C^{m+k''}$, the problem*

$$l_0(D_t)V = 0 \quad (t > 0), \qquad b_0(D_t)V\big|_{t=0} + b_0'(\tau)V' = h \tag{2.8.8}$$

has one and only one solution $V \in \mathcal{M}^+$.

Throughout this subsection, we assume that problem (2.8.1)–(2.8.2) is parameter-elliptic. Therefore condition (2.2.8) holds. Under this condition, the Lopatinskii condition is equivalent to either the solvability of problem (2.8.8) for any h, or the unique solvability of this problem (see Section (2.2)).

2.8.2. Functional spaces and norms that depend on the parameter $q \in \mathbf{R}$ are already introduced in Subsection 1.5.2. Let

$$æ = \max\{0, \sigma_1 + 1, \ldots, \sigma_{m+k''} + 1\}, \qquad \tau_j = t_j + æ, \tag{2.8.9}$$

$$j = 1, \ldots, N, \qquad \tau = (\tau_1, \ldots, \tau_N), \qquad |\tau| = \tau_1 + \cdots + \tau_N.$$

Since we assume that $\sigma_1 \geq \sigma_2 \geq \cdots \geq \sigma_{m+k''}$, we obtain: æ $= \max\{0, \sigma_1 + 1\}$. Denote:

$$\widetilde{H}^{T+s,p,(\tau)}(q) := \prod_{j=1}^{N} \widetilde{H}^{t_j+s,p,(\tau_j)}(G,q), \tag{2.8.10}$$

$$\mathcal{B}^{T'+s,p}(q) := \prod_{j=N+1}^{N+k'} \prod_{k=1}^{\sigma_1+t_j+1} B^{t_j+s-k+1-1/p,p}(\partial G, q). \tag{2.8.11}$$

It directly follows from the materials of Subsection 1.5.2 that, for any $s \in \mathbf{R}$ and $p \in]1,\infty[$, the closure $A(q) = A_{s,p}(q)$ of the mapping

$$U = (u, u') \mapsto \Big(l(x, D_x, q_1)u,\ b(x, D_x, q_1)u\big|_{\partial G} + b'(x', D', q_1)u'\Big),$$

$$U = (u, u') \in (C^\infty(\overline{G}))^N \times \prod_{j=N+1}^{N+k'} (C^\infty(\partial G))^{\sigma_1+t_j+1},$$

acts continuously in the pair of spaces

$$\widetilde{\mathcal{H}}^{s,p}(q) \to K^{s,p}(q), \tag{2.8.12}$$

$$\widetilde{\mathcal{H}}^{s,p}(q) := \widetilde{H}^{T+s,p,(\tau)}(q) \times \mathcal{B}^{T'+s,p}(q),$$

$$K^{s,p}(q) := \prod_{j=1}^{N} \widetilde{H}^{s-s_j,p,(æ-s_j)}(G,q) \times \prod_{h=1}^{m+k''} B^{s-\sigma_h-1/p,p}(\partial G, q).$$

In addition, there exists a constant $c > 0$, which does not depend on u, q and $\theta \in [\theta_1, \theta_2]$, such that the following estimate is valid:

$$\|A(q_1)U\|_{K^{s,p}} \leq c\|U\|_{\widetilde{\mathcal{H}}^{s,p}(q)} \qquad (\forall U \in \widetilde{\mathcal{H}}^{s,p}(q)). \tag{2.8.13}$$

For problem (2.8.1)–(2.8.2), the following analog of Theorem 1.5.1 holds (sf., Theorem 2.4.1)

Theorem 2.8.1. *Let $p \in]1,\infty[$, $s \in \mathbf{R}$, and let problem (2.8.1)–(2.8.2) be parameter elliptic. Assume that both the coefficients and the boundary are infinitely smooth. Then there exists a number $q_0 > 0$ such that, for $q \geq q_0$ and $\theta \in [\theta_1, \theta_2]$, the operator $A_{s,p} = A_{s,p}(q_1)$, acting continuously in the pair of spaces (2.8.12), realizes an isomorphism between these spaces. Moreover, there exists a constant $c_s > 0$, which does not depend on u, q $(|q| \geq q_0)$ and $\theta \in [\theta_1, \theta_2]$, such that the following estimate is valid:*

$$c_s^{-1}\|U\|_{\widetilde{\mathcal{H}}^{s,p}(q)} \leq \|A_{s,p}U\|_{K^{s,p}} \leq c_s\|U\|_{\widetilde{\mathcal{H}}^{s,p}(q)} \qquad (\forall U \in \widetilde{\mathcal{H}}^{s,p}(q)),$$

and, in addition, the function $s \mapsto c_s$ $(s \in \mathbf{R})$ *is bounded for every compact set.*

The proof of this theorem is quite analoguous to the proof of Theorem 1.5.1.

2.8.3. One can obtain the Green's formula for parameter-elliptic problem (2.8.1)–(2.8.2). Now system (2.7.6) is elliptic with a parameter on ∂G, and, therefore, it is uniquely solvable problem for $q \geq q_0$ with sufficiently large $q_0 > 0$. Then conditions (2.7.13) hold, and the Greens formula is written in the form (2.7.18). Further, formally adjoint problem (2.7.21) is also parameter-elliptic and, therefore, $\mathfrak{N}^+ = 0$.

CHAPTER 3

THE SOBOLEV PROBLEM

Let $G \subset \mathbf{R}^n$ be a bounded domain, and let $\partial G = \Gamma_0 \cup \Gamma_1 \cup \ldots \Gamma_{\overline{k}} \in C^\infty$ be the boundary of G. Assume that Γ_0 denotes an $(n-1)$-dimensional compact set that is the exterior boundary of the domain G. Denote by Γ_j $(j = 1, \ldots, \overline{k})$ the i_j-dimensional manifold without boundary lying inside of Γ_0, $0 \le i_j \le n-1$. Let $i'_j = n - i_j$ denotes the codimensionality of Γ_j. Assume that $\Gamma_j \in C^\infty$ $(j = 0, \ldots, \overline{k})$, and $\Gamma_j \cap \Gamma_k = \emptyset$ for $j \neq k$.

The Sobolev problem is a boundary-value problem in the domain G where boundary expressions are given at the manifolds $\Gamma_0, \Gamma_1 \ldots, \Gamma_{\overline{k}}$.

The Sobolev problem is studied completely in the classes of sufficiently smooth functions (see [Sob], [St1], [St2] and bibliography there). This problem was investigated in complete scales of Banach spaces in [RSk1], [RSk2]; in those works the authors assumes that the boundary expressions on Γ_k $(k = 1, \ldots, \overline{k})$ form the Dirichlet system, and the orders of boundary expressions on Γ_0 do not exceed the order of the equation.

In the present chapter these restrictions are thrown out. In addition, all the expressions are (generally speaking) pseudo differential along ∂G. The solvability of the Sobolev problem in complete scales of Banach spaces is obtained also for elliptic problems with a parameter (Section 3.2) and parabolic problems. We consider also a number of applications of this theory.

3.1. The Sobolev Problem in the Complete Scale of Banach Spaces

3.1.1. Generalized solution of the Sobolev problem

We consider the Sobolev problem

$$L(x, D)u(x) = f(x) \qquad (x \in G;\ \operatorname{ord} L = 2m) \tag{3.1.1}$$

$$B_{j0}(x, D)u\big|_{\Gamma_0} = \varphi_{j0} \qquad (j = 1, \ldots, m;\ \operatorname{ord} B_{j0} = q_{j0}), \tag{3.1.2}$$

$$B_{jk}(x, D)u\big|_{\Gamma_k} = \varphi_{jk} \quad (k = 1, \ldots, \overline{k};\ j = 1, \ldots, m_k;\ \operatorname{ord} B_{jk} = q_{jk}). \tag{3.1.3}$$

Let

$$\begin{aligned} r &= \max\{2m, q_{10} + 1, \ldots, q_{m0} + 1\}, \\ q_k &= q_{1k} \ge \cdots \ge q_{m_k,k} \quad (k = 1, \ldots, \overline{k}). \end{aligned} \tag{3.1.4}$$

In a neighborhood

$$G_{0\delta} = \{x \in G : \mathrm{dist}\,(x, \Gamma_0) < \delta\},$$

where $\delta > 0$ is sufficiently small, let us introduce special local coordinates

$$x = (x_1, \ldots, x_{n-1}, x_n) = (x', x_n) \in U(\partial G)$$

such that $(x', 0)$ are local coordinates on Γ_0, and x_n is the distance between the point x and Γ_0. Using these coordinates, we can represent the expressions $L(x, D)$ and $D_\nu^{j-1} L(x, D)|_{\Gamma_0}$ in the following form:

$$L(x, D) = \sum_{k=0}^{2m} L_k(x, D') D_\nu^k, \tag{3.1.5}$$

$$D_\nu^{j-1} L(x, D)|_{\Gamma_0} = \sum_{k=0}^{2m+j-1} L_{kj}(x', D') D_\nu^k \quad (j = 1, \ldots, r - 2m + 1), \tag{3.1.6}$$

Here L_k, L_{kj} are tangential differential (or pseudo diffrntial) expressions of orders $2m - k$ and $2m - k + j$, respectively; $D_\nu = i\partial/\partial\nu$, ν is a normal to Γ_0.

In the similar way, in a neighborhood

$$G_{k\delta} = \{x \in G : \mathrm{dist}(x, \Gamma_k) < \delta\} = \Gamma_k \times \{|(y_1, \ldots, y_{i'_k})| < \delta\}$$

of mainfold Γ_k we introduce local coordinates

$$(t, y) = (t_1, \ldots, t_{i_k}, y_1, \ldots, y_{i'_k})$$

so that that $(t, 0) = (t_1, \ldots, t_{i_k}, 0)$ are local coordinates on Γ_k, and $(0, y) = (0, y_1, \ldots, y_{i'_k})$ are local coordinates in the ball $|y| < \delta$. Now one can write expressions (3.1.2) and (3.1.3) in the following form:

$$B_{j0}(x, D) = \sum_{l=1}^{q_{j0}+1} B_{jl}(x, D') D_\nu^{l-1} \quad (j = 1, \ldots, m), \tag{3.1.7}$$

where B_{jl} are tangential differential (or pseudo differential) expressions of orders $q_{j0} - l + 1$, and

$$B_{jk}(x, D) = \sum_{|\beta| \le q_{jk}} T_{kj\beta}(t, D_t) D_y^\beta \tag{3.1.8}$$

$$(D_y^\beta = D_{y_1}^{\beta_1} \cdots D_{y_{i'_k}}^{\beta_{i'_k}}, x = (t, y),\ (t, 0) \in \Gamma_k,\ |y| = |(y_0, \ldots, y_{i'_k})|,$$

$$j = 1, \ldots, m_k; \ k = 1, \ldots, \overline{k}).$$

For simplisity, the coefficients (the symbols) of all the differential expressions are assumed to be infinitely smooth in $\overline{G}$ and on Γ_k, respectively.

Further, in what follows we assume that problem (3.1.1), (3.1.2) is elliptic in $\overline{G}$. This means that the expression L is properely elliptic in $\overline{G}$ and the boundary expressions $\{B_{j0}\}$ satisfy the Lopatinskii condition on Γ_0 (see, for example, [Ber],[R1]). Moreover, we assume that for every $k \in \{1, \ldots, \overline{k}\}$ boundary conditions (3.1.3) form the Douglis-Nirenbergy eliptic system on Γ_k (see Subsection 3.1.4 below).

Let us now introduce the notion of a generalized solution of problem (3.1.1)–(3.1.3).

Integrating by parts, we obtain

$$(Lu, v) = (u, L^+ v) - i \sum_{j=1}^{2m} \langle D_\nu^{j-1} u, M_j(x, D) v \rangle_{\Gamma_0} \ (u, v \in C^\infty(\overline{G})), \quad (3.1.9)$$

where

$$M_j v = \sum_{k=j}^{2m} D_\nu^{k-j} L_k^+(x, D') v, \quad \operatorname{ord} M_j = 2m - j.$$

Here and below, the symbols $(\cdot, \cdot)$ and $\langle \cdot, \cdot \rangle_k$ denote the scalar products (or their extensions) in $L_2(G)$ and $L_2(\Gamma_k)$, respectively. The expressions L^+ and L_k^+ are formally adjoint to the expressions L and L_k, respectively.

We identify an element $u \in C^\infty(\overline{G})$ with the vector

$$\begin{gathered} u = \Big(u_0, u_1, \ldots, u_r, (u_{k\beta} : k = 1, \ldots, \overline{k}, |\beta| \le q_k) \Big); \\ u_0 = u|_{\overline{G}}, \quad u_j = D_\nu^{j-1} u|_{\Gamma_0} \ (j = 1, \ldots, r), \quad u_{k\beta} = D_y^\beta u|_{\Gamma_k}, \end{gathered} \quad (3.1.10)$$

and the element $f \in C^\infty(\overline{G})$ with the vector

$$\begin{gathered} f = (f_0, \ldots, f_{r-2m}), \\ f_0 = f|_{\overline{G}}, \quad f_j = D_\nu^{j-1} f|_{\Gamma_0} \ (1 \le j \le r - 2m). \end{gathered} \quad (3.1.11)$$

Since an element $u \in C^\infty(\overline{G})$ is a solution of equation (3.1.1) if the equalities

$$(Lu, v)_G = (f, v)_G \quad (v \in C^\infty(\overline{G})),$$

$$D_\nu^{j-1} Lu|_{\Gamma_0} = D_\nu^{j-1} f|_{\Gamma_0} \quad (j = 1, \ldots, r - 2m)$$

are valid, formulas (3.1.5)–(3.1.9) imply that the vector $u \in C^\infty(\overline{G})$ (3.1.10) is a solution of problem (3.1.1)–(3.1.3) if there hold the following relations

$$(u_0, L^+ v) - i\sum_{j=1}^{2m} \langle u_j, M_j v\rangle = (f_0, v) \quad (v \in C^\infty(\overline{G})), \tag{3.1.12}$$

$$\sum_{k=0}^{2m+j-1} L_{kj}(x', D')u_{k+1} = f_j \quad (j : 1 \le j \le r - 2m), \tag{3.1.13}$$

$$B_{j0}u|_{\Gamma_0} \equiv \sum_{l=1}^{q_{j0}+1} B_{jl}(x, D')u_l = \varphi_j \quad (j = 1, \dots, m), \tag{3.1.14}$$

$$B_{jk}u|_{\Gamma_k} \equiv \sum_{|\beta| \le q_{jk}} T_{kj\beta}(t, D_t)u_{k\beta} = \varphi_{jk} \quad (k = 1, \dots, \overline{k}; j = 1, \dots m_k). \tag{3.1.15}$$

Now consider the vectors u and f defined by formulas (3.1.10) and (3.1.11), where u_0 and f_0 are distributions in $\overline{G}$, the elements $\{u_j, f_l\}$ and $\{u_{k\beta}\}$ are distributions in Γ_0 and Γ_k respectively. If relations (3.1.12)–(3.1.15) hold then the vector u is called a generalized solution of problem (3.1.1)–(3.1.3). Relations (3.1.12)–(3.1.15) define the mapping

$$A : u \mapsto F = (f, \varphi_0, \varphi_1, \dots, \varphi_{\overline{k}}), \tag{3.1.16}$$

$$f = (f_0, \dots, f_{r-2m}), \quad \varphi_0 = (\varphi_{10}, \dots, \varphi_{m0}), \quad \varphi_k = (\varphi_{1k}, \dots, \varphi_{m_k k}),$$

connected with problem (3.1.1)–(3.1.3). To study this mapping and to precise the definition of a generalized solution, we introduce relevant functional spaces.

3.1.2. Functional spaces *(see Chapter 1)*

Let $p, p' \in (1, \infty)$, $1/p + 1/p' = 1$, $s \in \mathbf{R}$. We denote by $H^{s,p}(R^n)$ the spaces of Bessel potentials (see, for example, [R1], [Gr2]); $\|f, R^n\|_{s,p}$ denotes the norm in $H^{s,p}(R^n)$. The spaces $H^{s,p}(R^n)$ and $H^{-s,p'}(R^n)$ are dual to each other with respect to the extension of the scalar product in $L_2(\mathbf{R}^n)$.

Denote by $H^{s,p}(G)$ $(s \ge 0)$ the space of restrictions of elements of $H^{s,p}(R^n)$ to G with quotient-space topology;

$$\|u, G\|_{s,p} = \inf_{v \in H^{s,p}(\mathbf{R}^n):\, v|_G = u} \|v, \mathbf{R}^n\|_{s,p}$$

denotes the norm in $H^{s,p}(G)$. Therefore,

$$H^{s,p}(G) = H^{s,p}(R^n)/H^{s,p}_{CG}(R^n), \tag{3.1.17}$$

where

$$H^{s,p}_{CG}(R^n) = \{v \in H^{s,p}(R^n) : \operatorname{supp} v \subset \mathbf{R}^n \setminus G\}$$

is a subspace of the space $H^{s,p}(R^n)$. Let $H^{-s,p}(G)$ $(s \geq 0)$ denote the space dual to $H^{s,p'}(G)$ with respect to the extension $(\cdot,\cdot)$ of the scalar product in $L_2(G)$. It then follows from (3.1.17) (see [R1], [Gr2]) that

$$H^{-s,p}(G) \simeq H^{-s,p}_{\overline{G}}(R^n) = \{f \in H^{-s,p}(R^n) : \operatorname{supp} f \subset \overline{G}\}.$$

Since for $s \geq 0$ there are not elements concentrated at Γ_k $(k = 1,\dots,\overline{k})$ in $H^{s,p}(\mathbf{R}^n)$, the relation

$$H^{s,p}(G) \simeq H^{s,p}(G \cup \Gamma_1 \cup \ldots \cup \Gamma_k) \quad (s \in \mathbf{R})$$

is true; $||u, G||_{s,p}$ is the norm in $H^{s,p}(G)$ for $s \in \mathbf{R}$.

We denote by $B^{s,p}(\Gamma_k)$ $(k = 0,\dots,\overline{k})$ the Besov space, and by $\langle\langle \varphi, \Gamma_k \rangle\rangle_{s,p}$ the norm in $B^{s,p}(\Gamma_k)$. The spaces $B^{s,p}(\Gamma_k)$ and $B^{-s,p'}(\Gamma_k)$ are dual to each other with respect to the extension $\langle\cdot,\cdot\rangle_{\Gamma_k}$ of the scalar product in $L_2(\Gamma_k)$ (see [R1],[Gr2]).

Let

$$s \in \mathbf{R}, \quad s \neq j + \frac{i_{k'}}{p} \quad (j = 1,\dots,q_k, \ k = 0,\dots,\overline{k}, \ q_0 = r). \tag{3.1.18}$$

Let $\widetilde{H}^{s,p}$ denote the completion of $C^\infty(\overline{G})$ in the norm

$$|||u|||_{s,p} = \Bigg(||u||^p_{s,p} + \sum_{j=1}^{r} \langle\langle D_\nu^{j-1} u, \Gamma_0 \rangle\rangle^p_{s-j+1-1/p,p} + \sum_{k=1}^{\overline{k}} \sum_{|\alpha| \leq q_k} \langle\langle D_y^\alpha u, \Gamma_k \rangle\rangle^p_{s-|\alpha|-i'_k/p,p} \Bigg)^{1/p}. \tag{3.1.19}$$

The space $\widetilde{H}^{s,p}$ is isometric to the closure $\mathfrak{H}_0^{s,p}$ of the space of vectors $u \in C^\infty(\overline{G})$ (defined by (3.1.10)) in the direct product

$$\mathcal{H}^{s,p} = H^{s,p}(G) \times \prod_{j=1}^{r} B^{s-j+1-1/p,p}(\Gamma_0) \times \prod_{k=1}^{\overline{k}} \prod_{|\alpha| \leq q_k} B^{s-|\alpha|-i'_k/p,p}(\Gamma_k). \tag{3.1.20}$$

Hence, the space $\widetilde{H}^{s,p}$ coincides with a subspace $\mathcal{H}_0^{s,p}$ of the space of vectors

$$u = \Big(u_0, u_1, \dots, u_r, \{u_{\alpha k} : k = 1,\dots,\overline{k}, \ |\alpha| \leq q_k\} \Big) \in \mathcal{H}^{s,p}. \tag{3.1.21}$$

In addition, an element $u \in \mathcal{H}^{s,p}$ belongs to $\mathcal{H}_0^{s,p} \simeq \widetilde{H}^{s,p}$ if and only if

$$\begin{aligned} u_j &= D_\nu^{j-1} u_0|_{\Gamma_0} \ (\forall j : s - j + 1 - 1/p > 0), \\ u_{\alpha k} &= D_y^\alpha u_0|_{\Gamma_k} \ (\forall \alpha, k : s - |\alpha| - i'_k/p > 0) \end{aligned} \tag{3.1.22}$$

(sf. [R1], [R11]). If $s < 1/p$, then $\mathcal{H}_0^{s,p} = \mathcal{H}^{s,p} = \widetilde{H}^{s,p}$.

Next, let $\widetilde{H}^{s,p,(r)}$ (see [R1], [R11]) denote the completion of the space $C^\infty(\overline{G})$ in the norm

$$|||u|||_{s,p,(r)} = \left(||u||_{s,p}^p + \sum_{j=1}^r \langle\langle D_\nu^{j-1} u, \Gamma_0 \rangle\rangle_{s-j+1-1/p,p}^p \right)^{1/p}. \tag{3.1.23}$$

Thus, the space $\widetilde{H}^{s,p,(r)}$ consists of vectors

$$(u_0, \dots, u_r) \in H^{s,p}(G) \times \prod_{j=1}^r B^{s-j+1-1/p,p}(\Gamma_0)$$

such that

$$u_j = D_\nu^{j-1} u_0|_{\Gamma_0} \quad (\forall j : s - j + 1 - 1/p > 0).$$

For the rest of values of s the spaces $\widetilde{H}^{s,p}$, $\widetilde{H}^{s,p,(r)}$ and norms (3.1.18), (3.1.22) are defined by complex interpolation.

Theorem 3.1.1. *(sf. [R1], [RSk1], [RSk2], [R11]) For each $s \in \mathbf{R}$ and $p \in (1, \infty)$ the closure $\mathcal{A} = \mathcal{A}_{s,p}$ of the mapping*

$$u \to \Big(Lu|_{\overline{G}}, (D_\nu^{j-1} Lu|_{\Gamma_0} \,|\, j : 1 \le j \le r - 2m), B_1 u|_{\Gamma_0}, \dots, B_m u|_{\Gamma_0},$$

$$\{B_{jk} u|_{\Gamma_k} : k = 1, \dots, \overline{k}, \, j = 1, \dots, m_k\} \Big) \ (u \in C^\infty(\overline{G}))$$

acts continuously in the pair of spaces

$$\widetilde{H}^{s,p} \to K^{s,p} \tag{3.1.24}$$

$$K^{s,p} := \widetilde{H}^{s-2m,p,(r-2m)} \times \prod_{j=1}^m B^{s-q_{j0}-1/p,p}(\Gamma_0) \times \prod_{k=1}^{\overline{k}} \prod_{j=1}^{m_k} B^{s-q_{jk}-i'_k/p,p}(\Gamma_k)$$

If $s_1 \le s$ and $p_1 \le p$, then the operator $\mathcal{A} = \mathcal{A}_{s_1,p_1}$ is an extension of the operator $\mathcal{A}_{s_2,p_2}$ in continuity.

Definition 3.1.1. *An element* $u \in \widetilde{H}^{s,p}$ *is called a generalized solution of problem (3.1.1)–(3.1.3) if*

$$\mathcal{A}_{s,p}u = F = (f, \varphi_{10}, \ldots, \varphi_{m0}, \{\varphi_{jk}\}) \in K^{s,p}.$$

By passing to the limit, it is easy to verify that the following statement is true.

Theorem 3.1.2. *An element* $u \in \widetilde{H}^{s,p}$ *is a generalized solution of the problem (1)–(3) if and only if relations (12)–(15) hold.*

It should be noted that if element $u \in \widetilde{H}^{s,p}$ given by (3.1.21) is the generalized solution of problem (3.1.1)–(3.1.3), then the element

$$\widetilde{u} = (u_0, \ldots, u_r) \in \widetilde{H}^{s,p,(r)}$$

is the general solution of problem (3.1.1)–(3.1.2) (this means that the element $\widetilde{u}$ satisfies relations (3.1.12)–(3.1.14)). Therefore, solving problem (3.1.1)–(3.1.2) we find the first $r+1$ components of the solution of problem (3.1.1)–(3.1.3). Conditions (3.1.3) (or (3.1.15)) give us the rest of the components of the solution $u \in \widetilde{H}^{s,p}$ of problem (3.1.1)–(3.1.2).

3.1.3. Theorem on complete collection of izomorphisms for problem (3.1.1)–(3.1.2)

In $\overline{G}$ we consider elliptic problem (3.1.1)–(3.1.2). It turns out (see Chapter 1 of the present book, or [R1, Chapter IV]) that the following four statements are true :

i) For any $s \in \mathbf{R}$ and $p \in (1, \infty)$ the closure $A_{s,p}$ of the mapping

$$u \to (Lu, B_1 u|_{\Gamma_0}, \ldots, B_m u|_{\Gamma_0}) \quad (u \in C^\infty(\overline{G}))$$

acts continuously in the pair of spaces

$$\widetilde{H}^{s,p,(r)} \to \widetilde{H}^{s-2m,p,(r-2m)} \times \prod_{j=1}^{m} B^{s-q_{j0}-1/p,p}(\Gamma_0) =: K^{s-2m,p,(r-2m)}.$$

ii) An element $\widetilde{u} = (u_0, u_1, \ldots, u_r) \in \widetilde{H}^{s,p,(r)}$, is called a generalized solution of problem (3.1.1)–(3.1.2) if $A_{s,p}\widetilde{u} = (f, \varphi) \in K^{s-2m,p,(r-2m)}$. Furthermore, $A_{s,p}\widetilde{u} = (f, \varphi)$ if and only if equalities (3.1.12)–(3.1.14) hold. Hence, if element (3.1.21) $u \in \widetilde{H}^{s,p}$ is a generalized solution of problem (3.1.1)–(3.1.3), then the element

$$\widetilde{u} = (u_0, u_1, \ldots, u_r) \in \widetilde{H}^{s,p,(r)}$$

is a generalized solution of problem (3.1.1)–(3.1.2).

iii) The operator $A_{s,p}$ is Noetherian; the kernel $\mathfrak{N}$ and the cokernel $\mathfrak{N}^*$ are finite-dimensional, independent of s and p, and consist of infinitely smooth elements :

$$\mathfrak{N} = \{u \in C^\infty(\overline{G}) : Au = 0\},$$
$$\mathfrak{N}^* = \{V = (v_0, \{v_k : 1 \le k \le r-2m\},$$
$$\{v_j : 1 \le j \le m\})\} \subset C^\infty(\overline{G}) \times (C^\infty(\Gamma_0))^{r-2m} \times (C^\infty(\Gamma_0))^m.$$

The problem

$$A_{s,p}\tilde{u} = (f, \varphi) \in K^{s-2m,p,(r-2m)},$$
$$f = (f_0, f_1, \ldots, f_{r-2m}), \quad \varphi = (\varphi_1, \ldots, \varphi_m)$$

is sovable in $\widetilde{H}^{s,p,(r)}$ if and only if the following relation is valid:

$$[(f,\varphi), V] =: (f_0, v_0) + \sum_{j=1}^{r-2m} \langle f_j, v_j\rangle_{\Gamma_0} + \sum_{k=1}^{m} \langle \varphi_k, v_k\rangle_{\Gamma_0} = 0 \; (\forall V \in \mathfrak{N}^*). \tag{3.1.25}$$

iv) The restriction $A_1 = A_{1\,s,p}$ of the operator $A_{s,p}$ onto the subspace $P\widetilde{H}^{s,p,(r)} = \{(u_0, u_1, \ldots, u_r) \in \widetilde{H}^{s,p,(r)} : (u_0, \mathfrak{N}) = 0\}$ of the space $\widetilde{H}^{s,p,(r)}$ realizes an isomorphism

$$A_1 : P\widetilde{H}^{s,p,(r)} \to Q^+K^{s-2m,p,(r-2m)}. \tag{3.1.26}$$

Here $Q^+K^{s-2m,p,(r-2m)}$ is a subspace of the space $K^{s-2m,p,(r-2m)}$ which consist of all elements (f, φ) satisfying relations (3.1.25).

3.1.4. Let us formulate the assumptions on the boundary expressions on Γ_k $(k = 1, \ldots, \overline{k})$. Assume that

$$u = (u_0, u_1, \ldots, u_r, \{u_{\alpha k}|\ k = 1, \ldots, \overline{k},\ |\alpha| \le q_k\}) \in \widetilde{H}^{s,p}$$

be a generalized solution of problem (3.1.1)–(3.1.3). Then, according to the Theorem 3.1.2, the elements $\{u_{\alpha k}\}$ satisfy equalites (3.1.15). For any $k \in \{1, \ldots, \overline{k}\}$ we set

$$\widetilde{T}_{kj\beta}(t, D_t) = \begin{cases} T_{kj\beta}(t, D_t) & \text{for } |\beta| \le q_{jk} \\ 0 & \text{for } q_{jk} < |\beta| \le q_k \end{cases},$$

where the numbers q_k are defined by formula (3.1.4). Then one can rewrite eqalities (3.1.15) in the form

$$B_{jk}u|_{\Gamma_k} \equiv \sum_{|\beta| \le q_k} \widetilde{T}_{kj\beta}(t, D_t)u_{k\beta} = \varphi_{jk} \tag{3.1.27}$$

$$(j = 1, \ldots, m_k,\ k = 1, \ldots, \overline{k}),$$

or, in short,

$$\widetilde{T}_k(t, D_t)U_k = \varphi_k \quad \Big(\widetilde{T}_k = (\widetilde{T}_{kj\beta})_{\substack{j=1,\ldots,m_k \\ |\beta|\le q_k}},\ \varphi_k = (\varphi_{1k}, \ldots, \varphi_{m_k k})\Big) \quad (3.1.27')$$

For any k the matrix $(\widetilde{T}_{kj\beta}(t, D_t))$ is a matrix of Douglis-Nirenberg structure. This means that if we associate the number $s_j = q_{jk} - q_k$ with the row with index j, the number $t_\beta = q_k - |\beta|$ with the column with index β, then we obtain $\operatorname{ord} \widetilde{T}_{kj\beta} \le s_j + t_\beta$ for $s_j + t_\beta \ge 0$, and $\widetilde{T}_{kj\beta} \equiv 0$ for $s_j + t_\beta < 0$. We now assume that for any $k \in \{1, \ldots, \overline{k}\}$ system (3.1.27) is a Douglis-Nirenberg elliptic system on Γ_k, i.e. the system is square and

$$\det(\widetilde{T}^0_{kj\beta}(t, \sigma)) \ne 0 \quad (\sigma = (\sigma_1, \ldots, \sigma_{i_k}) \ne 0,\ t \in \Gamma_k),$$

where $\widetilde{T}^0_{kj\beta}$ is the principal symbol of the expression $\widetilde{T}_{kj\beta}(t, D_t)$.

For each $s \in \mathbf{R}$ and $p \in (1, \infty)$ the mapping

$$T_k = T_{k,s,p} : U \mapsto T_k U : B^{T+s,p}(\Gamma_k) \to B^{s-S,p}(\Gamma_k), \qquad (3.1.28)$$

$$B^{T+s,p}(\Gamma_k) := \prod_{j=1}^{m_k} B^{t_j+s,p}(\Gamma_k), \quad B^{s-S,p}(\Gamma_k) := \prod_{j=1}^{m_k} B^{s-s_j,p}(\Gamma_k).$$

is a continuous operator.

It is clear that the system

$$\sum_{j=1}^{m_k} \widetilde{T}^+_{k\beta j}(t, D_t)v_{kj} = \psi_{\beta k},$$

or, in short,

$$T^+_k V = T_{k,s,p} V = \psi_k,$$

formally adjoint to system (3.1.27), is an elliptic system in the sense of Douglis–Nirenberg. Therefore (sf. [R11]), for each $s \in \mathbf{R}$, $p' \in (1, \infty)$, and $k = 1, \ldots, \overline{k}$ the mapping

$$T^+_k = T^+_{k,s,p'} : V \mapsto T^+_k V : \prod_{j=1}^{m_k} B^{t'_j+s,p'}(\Gamma_k) \to \prod_{j=1}^{m_k} B^{s-s'_j,p'}(\Gamma_k)$$

$$(s'_j = t_1 - t_j, \quad t'_j = t_1 + s_j)$$

is continuous. In addition,

$$(T_k U, V)_{\Gamma_k} = (u, T^+_k V)_{\Gamma_k} \quad (u \in B^{T+s,p}(\Gamma_k),\ V \in B^{-s+S,p'}(\Gamma_k)),$$

where $(\cdot,\cdot)_{\Gamma_k}$ is the extension of the scalar product in $(L_2(\Gamma_k))^{m_k}$.

It follows from the ellipticity of system (3.1.27) that (see [R1]) the operator T_k is Noetherian. This means that the range of values $\mathcal{R}(T_k)$ is closed in $B^{s-S,p}(\Gamma_k)$, the kernel $\mathfrak{N}_k$ and the cokernel $\mathfrak{N}_k^*$ are finite-dimensional, independent of s and p and consist of infinitely smooth elements:

$$\mathfrak{N}_k \subset (C^\infty(\Gamma_k))^{m_k}, \quad \mathfrak{N}_k^* \subset (C^\infty(\Gamma_k))^{m_k}.$$

The equation $T_k U = \varphi_k \in B^{s-S,p}$ has a solution $U \in B^{T+s,p}(\Gamma_k)$ if and only if the equality

$$(\varphi_k, V)_{\Gamma_k} = 0 \quad (\forall V \in \mathfrak{N}_k^*; k = 1, \ldots, \overline{k}) \tag{3.1.29}$$

is valid. In addition the cokernel $\mathfrak{N}_k^*$ of the operator T_k coincides with the kernel of the operator T_k^*. The restriction T_k^1 of the operator T_k onto the subspace

$$PB^{T+s,p}(\Gamma_k) = \{U \in B^{T+s,p}(\Gamma_k) : (U, W)_{\Gamma_k} = 0 \ (\forall W \in \mathfrak{N}_k)\}$$

of the space $B^{T+s,p}(\Gamma_k)$ realizes an isomorphism

$$T_k^1 : PB^{T+s,p}(\Gamma_k) \to Q^+B^{s-S,p}(\Gamma_k) \quad (s \in \mathbf{R}, p \in (1,\infty)). \tag{3.1.30}$$

Here $Q^+B^{s-S,p}(\Gamma_k) = \{\varphi \in B^{s-S,p}(\Gamma_k) : (\varphi, V)_{\Gamma_k} = 0 \ (\forall V \in \mathfrak{N}_k^*)\}$ is a subspace of $B^{T+s,p}(\Gamma_k)$.

3.1.5. Solvability of Sobolev problem

Theorem 3.1.3. *Let $s \in \mathbf{R}$, $p \in (1,\infty)$. For the solvability of problem (3.1.1)–(3.1.3) in $\widetilde{H}^{s,p}$ it is nesessary that relations (3.1.25) and (3.1.29) are valid; this condition is also sufficient if, in addition, compatibility conditions (3.1.34) hold for $s > 1/p$.*

Proof. If the element $u \in \widetilde{H}^{s,p}$ (see (3.1.21)) is a solution of the equation

$$\mathcal{A}u = F = (f, \varphi_0, \varphi_1, \ldots, \varphi_{\overline{k}}) \in K^{s,p} \tag{3.1.31}$$

$$\Bigg(f = (f_0, \ldots, f_{r-2m}), \ \varphi_0 = (\varphi_{10}, \ldots, \varphi_{m0}), \ \varphi_k = (\varphi_{1k}, \ldots, \varphi_{m_k,k})$$

$$f \in \widetilde{H}^{s,p,(r-2m)}, \ \varphi_0 \in \prod_{j=1}^{m} B^{s-q_{j0}-1/p,p}(\Gamma_0), \ \varphi_k \in \prod_{j=1}^{m_k} B^{s-q_{jk}-i'_k/p,p}(\Gamma_k)\Bigg),$$

then the vector $(u_1, \dots, u_r)$ is a solution of problem (3.1.1)–(3.1.3) in $\overline{G}$. Hence, relations (3.1.25) hold (see Subsection 3.1.3). Moreover, the vector

$$U_k = \{u_{\alpha k} : |\alpha| \leq q_k\} \in B^{T+s,p}(\Gamma)$$

satisfies the equation

$$T_k U_k = \varphi_k \qquad (k = 1, \dots, \overline{k}). \tag{3.1.32}$$

Therefore, relations (3.1.29) are valid. The necessity has been proved.

Let us prove the sufficiency. Let the element $F \in K^{s,p}$ satisfies conditions (3.1.25) and (3.1.29). Assume that, first, $s < 1/p$, and relations (3.1.18) hold. In this case the space $\widetilde{H}^{s,p}$ coincides with the direct product $\mathcal{H}^{s,p}$ defined by (3.1.20). Solving the problem (3.1.1)–(3.1.2), we find the element

$$\widetilde{u} = A_1^{-1}(f, \varphi) = (u_0, \dots, u_r) \in P\widetilde{H}^{s,p,(r)}$$

(see (3.1.26)). Then the folmula

$$\widetilde{u} + w \ (W \in \mathfrak{N})$$

is the general form of a solution of problem (3.1.1)–(3.1.2). The element

$$\widetilde{u} + w = (u_0 + w_0, \dots, u_r + w_r)$$

gives us the first $r + 1$ components of the solution $u \in \widetilde{H}^{s,p}$ required. We find the rest of the components by solving the problem (3.1.32), i.e.,

$$U_k = (T_k)^{-1}\varphi_k \in PB^{T+s,p}(\Gamma_k) \quad (k = 1, \dots, \overline{k}).$$

We obtain the general solution $U_k + W_k$ $(W_k \in \mathfrak{N}_k)$ of problem (3.1.32). The vector

$$\left(\widetilde{u} + w, \left\{U_k + W_k : k = 1, \dots, \overline{k}\right\}\right) \in \widetilde{H}^{s,p} \tag{3.1.33}$$

is the general solution of equation (3.1.31).

If $s > 1/p$, then $u_0 \in H^{s,p}(G)$, and, hence, there exist the traces

$$D_y^\alpha(u_0 + w_0)\big|_{\Gamma_k} \quad (\forall \alpha, k : s - |\alpha| - i'_k/p > 0).$$

In order to the solution obtained above be belonging to the space $\widetilde{H}^{s,p}$, it is nesessary and sufficient that the compatibility conditions

$$u_{\alpha k} + w_{\alpha k} = D_y^\alpha(u_0 + w_0)|_{\Gamma_k} \quad (\forall \alpha, k : s - |\alpha| - i'_k/p > 0). \tag{3.1.34}$$

hold. This completes the proof of the theorem. □

Remark 3.1.1. Assume that $s < 1/p$, the set $(e_{10}, \ldots, e_{t0})$ is a basis in $\mathfrak{N}$, and $(e_{1k}, \ldots, e_{t_k,k})$ is a basis in $\mathfrak{N}_k$ $(k = 1, \ldots, \overline{k})$. Then, it follows from the proof above that the set

$$(e_{10}, \ldots, e_{t0}, e_{11}, \ldots, e_{t_1,1}, \ldots, e_{1\overline{k}}, \ldots, e_{t_{\overline{k}},\overline{k}})$$

forms a basis in the kernel $\mathfrak{N}(\mathcal{A})$ of the operator $\mathcal{A}$. The similar statement is valid also for the cokernel of the operator $\mathcal{A}$. Therefore

$$\operatorname{ind} \mathcal{A} = \operatorname{ind} A + \sum_{k=1}^{\overline{k}} \operatorname{ind} T_k. \tag{3.1.35}$$

3.1.6. Special case of $\overline{k} = 1$ and $s < 2m - i'/p'$

Under the conditions $\overline{k} = 1$, $\Gamma_1 = \Gamma$, $i_1 = i$, $i'_1 = i'$, and $s < 2m - i'/p'$, we consider problem (3.1.1)–(3.1.3). In this case, if $f = (f_0, \ldots, f_r) \in \widetilde{H}^{s-2m,p,(r-2m)}$, then the element $f_0 \in H^{s-2m,p}(G)$ can be concentrated at Γ. It turns out that we can add to f_0 the element f'_0 concentrated at Γ so that the compatbility conditions hold automatically for problem (3.1.1)–(3.1.3) with

$$\widetilde{F} = (f_0 + f'_0, f_1, \ldots, f_r, \phi_1, \ldots, \phi_m, \{\phi_{r1}\}) \in K^{s,p}.$$

Note that the element f_0 coincides with $f_0 + f'_0$ inside of G.

The following theorem is true.

Theorem 3.1.4. *Assume that $1/p < s < 2m - i'/p$, and the element $F \in K^{s,p}$ satisfies relations (3.1.25) and (3.1.29). Then one can add to f_0 the element f'_0 concentrated at Γ, so that the compatbility condition hold automatically for the problem $\mathcal{A}u = \widetilde{F}$, $\widetilde{F} = F + (f'_0, 0, \ldots, 0)$, and this problem is solvable in $\widetilde{H}^{s,p}$.*

Proof. Let us use the folloving statement.

Lemma 3.1.1. *([R1], Ch. 6) The non-zero element $g \in H^{t,p}(G)$ is concentrated at Γ if and only if $t < -i'/p'$ and there exist the elements*

$$w_\mu \in B^{s+|\mu|+i'/p',p}(\Gamma) \quad (|\mu| \le \kappa_1 = [-t - i'/p']^-)$$

(here $[\tau]^-$ is the closest integer to τ which does not exceed τ), such that

$$g = \sum_{|\mu|\le\kappa_1} D_y^\mu (w_\mu \times \delta(\Gamma)),$$

where, by the definition of the Dirac measure $\delta(\Gamma)$,

$$(g,v)=\sum_{|\mu|\le\kappa_1}\langle w_\mu, D_y^\mu v\rangle_\Gamma \quad (\forall v\in C^\infty(\overline{G})).$$

In addition, there exists a constant $c>0$, *such that*

$$c^{-1}||g||_{s,p}\le\sum_{|\mu|\le\kappa_1}\langle\langle w_\mu,\Gamma\rangle\rangle_{s+|\mu|+i'/p,p}\le c||g||_{s,p}.$$

Now let $f_0'\in H^{s-2m,p}(G)$ $(s<2m-i'/p')$ is concentrated at Γ. Then, according to lemma above,

$$f_0'=\sum_{|\mu|\le\kappa}D_y^\mu(w_\mu\times\delta(\Gamma)),\ \kappa=[-s+2m-i'/p']^-,\ w_\mu\in B^{s-2m+|\mu|+i'/p',p}(\Gamma). \tag{3.1.36}$$

Let us assume, first, that the defect of problem (3.1.1)–(3.1.3) is missing. We denote by $R(w_\mu,x)\in\widetilde{H}^{s,p,(r)}$ $(s<2m-i'/p')$ a solution of the problem

$$\begin{aligned} L(x,D)R(w_\mu,x)&=(D_y^\mu(w_\mu\times\delta(\Gamma)),0,\ldots,0)\in\widetilde{H}^{s-2m,p,(r-2m)}\\ B_jR(w_\mu,x)|_{\Gamma_0}&=0\qquad(j=1,\ldots,m).\end{aligned} \tag{3.1.37}$$

In view of the Theorem on local increasing in smoothness ([R1, Sec.7.2]), we hawe that $R(w_\mu,x)\in C^\infty(\overline{G}\setminus\Gamma)$, and, in addition,

$$D_y^\alpha R(w_\mu,x)|_\Gamma\in B^{s-|\alpha|-i'/p,p}(\Gamma)\quad(|\alpha|<s-i'/p). \tag{3.1.38}$$

Thus, by vector

$$w=(w_\mu:|\mu|\le\kappa=[2m-s-i'/p]^-),\ w_\mu\in B^{s-2m+|\mu|+i'/p',p}(\Gamma)$$

we constract the vector

$$R=(R(w_\mu,x):\ |\mu|\le\kappa).$$

Then, by vector R we constract the matrix

$$V_{\alpha\beta}=D_y^\alpha R(w_\mu,x)|_\Gamma\in B^{s-|\alpha|-i'/p,p}(\Gamma)\quad(|\mu|\le\kappa,\alpha:s-|\alpha|-i'/p>0).$$

The operator $P_{\alpha\mu}:w_\mu\mapsto V_{\alpha\mu}$ acts continuously from the space $B^{s-2m+|\mu|+i'/p',p}(\Gamma)$ into $B^{s-|\alpha|-i'/p,p}(\Gamma)$. Therefore, the mapping

$$P:w\mapsto V=\Big(V_\alpha=\sum_{|\mu|\le\kappa}V_{\alpha\mu}:|\alpha|<s-i'/p\Big) \tag{3.1.39}$$

acts continuously in the pair of spaces

$$\prod_{|\mu|\leq\kappa} B^{s-2m+|\mu|+i'/p',p}(\Gamma)\to \prod_{|\alpha|\leq s-i'/p} B^{s-|\alpha|-i'/p,p}(\Gamma). \qquad (3.1.40)$$

The function

$$u(x)=\sum_{|\mu|\leq\kappa} R(w_\mu,x)\big|_G \in H^{s,p}(G) \qquad (3.1.41)$$

belongs to $C^\infty(\overline{G}\setminus\Gamma)$ and satisfies the problem

$$\begin{gathered} L(x,D)u(x)=0 \qquad (x\in G\setminus\Gamma),\\ B_j u|_{\Gamma_0}=0\ (j=1,\dots,m),\quad D_y^\alpha u|_\Gamma = V_\alpha\ (|\alpha|<s-i'/p). \end{gathered} \qquad (3.1.42)$$

Since problem (3.1.42) is elliptic ([St1], [St2]), and each its solution is represented in the form (3.1.41), then the value set of mapping (3.1.39) give us all the values of

$$V\in\prod_{|\alpha|\leq s-i'/p} B^{s-|\alpha|-i'/p,p}(\Gamma),$$

such that problem (3.1.42) is solvable. Furthermore, each solution of problem (3.1.42) is generated by the solution $u\in\widetilde{H}^{s,p,(r)}(G)$ of problem (3.1.37) with the right-hand side concentrated at Γ. Now let $u'=(u_0,\dots,u_r)\in\widetilde{H}^{s,p,(r)}$ be a solution of problem (3.1.1)–(3.1.2). Then, setting that $w=u-u'$, we reduce problem (3.1.1)–(3.1.3) with$\overline{k}=1$ to the problem

$$\begin{gathered} Lw=0,\quad B_j w|_{\Gamma_0}=0\ (j=1,\dots,m);\\ B_{r1}w|\Gamma=\varphi_{r1}-B_{r1}u'=\psi_{r1}\in B^{s-q_1-i'/p,p}(\Gamma)\ (r=1,\dots,m_1). \end{gathered} \qquad (3.1.43)$$

By the isomorphism (3.1.30), this problem is equivalent to the problem of the form (3.1.42). Then, each solution of this problem is generated by the solution $u\in\widetilde{H}^{s,p,(r)}$ of problem (3.1.37) with the right-hand side concentrated at Γ. The theorem is proved in the case under consideration.

If the defect is non-zero, then we have to replace the element

$$g=(D_y^\mu(w_\mu\times\delta(\Gamma),0,\dots,0)\in\widetilde{H}^{s-2m,p,(r-2m))}$$

in (3.1.37) by the element Q^+g such that Q^+gbelongs to the same space and is ortohonal to the cokernel $\mathfrak{N}^*$. Then we choose the solution $R(w_\mu,x)$ of problem (3.1.37) belonging to $P\widetilde{H}^{s,p,(r)}$. In view of the isomorphism

(3.1.26), such solution exists and is unique. This enables us to use the reasonings above to this case. □

Now let us show the Sobolev example ([Sob, p. 114]). In the domain $G = \{x \in \mathbf{R}^3 : 0 < |x| < 1\}$ we consider the problem

$$\Delta^2 u = 0\,(in\, G),\ u|_{|x|=1} = \frac{\partial u}{\partial \nu}|_{|x|=1} = 0,\ u(0) = 1,$$

where Δ denotes the Laplace operator. The function $u = (1 - |x|)^2$ is the unique solution of this problem which belongs to $H^{2,2}(G)$. Moreover, $u \notin H^{3,2}(G)$. It easy to verify that in $G \cup \{0\}$ we have

$$\Delta^2 u = c\delta(0) \in H^{-3/2-\varepsilon,2}(G),$$

where $\delta(0)$ is a Dirac measure concentrated at the point 0, and $\varepsilon > 0$ can be choosen arbitrary small. Therefore, $u \in H^{5/2-\varepsilon,2}(G)$. To obtain the smoother solution it is necessary to require that the compatibility condition $u(0) = 0$ holds. Then $u \equiv 0$.

3.1.7.
All the statements obtained remain true for the parameter-elliptic Sobolev problem (sf. [R1, Ch. IX], [AgV]). In this case, instead of the Noetherity property we have the unique solvability for large values of the parameter.

This case will be considered in detail in Section 3.2 below.

3.2. The Sobolev Problem with a Parameter in Complete Scale of Banach Spaces

In the present section we study the Sobolev problem with a parameter in complete scales of Banach spaces. In this case, instead of the Noetherian property we have the unique solvability for large values of the parameter. The solvability of the parabolic Sobolev problems follows directly from the results obtained (sf. [AgV], [RR2], [EZh]).

3.2.1. Let $G \subset R^n$ be a bounded domain with the boundary ∂G, where $\partial G = \Gamma_0 \cup \Gamma_1 \cup \dots \Gamma_{\overline{k}} \in C^\infty$, Γ_0 is an $(n-1)$-dimensional compact that is the external boundary of G, and Γ_j $(j = 1, \dots, \overline{k})$ is a i_j-dimensional manifold without boundary lying inside Γ_0, $(0 \le i_j \le n-1)$. Assume that $i'_j = n - i_j$ denotes the codimensionality of Γ_j. In addition, let $\Gamma_j \in C^\infty$ $(j = 0, \dots, \overline{k})$, and $\Gamma_j \cap \Gamma_k = \emptyset$ $(j \ne k)$.

Now let us consider the Sobolev problem with a parameter

$$L(x,D,q_1)u(x) \equiv \sum_{|\mu|+d\leq 2m} a_{d\mu}(x)q_1^d D^\mu u(x) = f(x) \quad (x\in G), \tag{3.2.1}$$

$$B_{j0}(x,D,q_1)u\big|_{\Gamma_0} \equiv \sum_{|\mu|+d\leq \widetilde{q}_{j0}} b_{jd\mu}(x)q_1^d D^\mu u\big|_{\Gamma_0} = \varphi_{j0} \quad (j=1,\dots,m), \tag{3.2.2}$$

$$B_{jk}(x,D,q_1)u\big|_{\Gamma_k} \equiv \sum_{|\mu|+d\leq \widetilde{q}_{jk}} b_{jd\mu k}(x)q_1^d D^\mu u\big|_{\Gamma_k} = \varphi_{jk} \tag{3.2.3}$$

$$(k=1,\dots,\overline{k};\quad j=1,\dots,m_k),$$

where $q_1 = qe^{i\theta}$, $q\in\mathbf{R}$, $\theta\in[\theta_1,\theta_2]$ (the case where $\theta_1=\theta=\theta_2$ is not excluded), and

$$D^\mu = D_1^{\mu_1},\cdots,D_n^{\mu_n},\quad D_j = i\frac{\partial}{\partial x_j},\quad |\mu| = \mu_1+\cdots+\mu_n.$$

Let

$$r = \max\{2m, \widetilde{q}_{10}+1,\dots,\widetilde{q}_{m0}+1\},\quad \widetilde{q}_k = \widetilde{q}_{1k}\geq\cdots\geq\widetilde{q}_{m_k,k} \tag{3.2.4}$$

$$(k=1,\dots,\overline{k}).$$

Assume that in a neighborhood

$$G_{0\delta} = \{x\in G: \operatorname{dist}(x,\Gamma_0)<\delta\}$$

with sufficiency small $\delta>0$ the folloving representations are true:

$$L(x,D,q_1) = \sum_{k=0}^{2m} L_k(x,D',q_1)D_\nu^k, \tag{3.2.5}$$

$$D_\nu^{j-1}L(x,D,q_1)|_{\Gamma_0} = \sum_{k=0}^{2m+j-1} L_{kj}(x',D',q_1)D_\nu^k \quad (j=1,\dots,r-2m+1). \tag{3.2.6}$$

$$B_{j0}(x,D,q_1) = \sum_{l=1}^{\widetilde{q}_{j0}+1} B_{jl}(x,D',q_1)D_\nu^{l-1} \quad (j=1,\dots,m) \tag{3.2.7}$$

Here L_k, L_{kj} and B_{jl} are tangential differential (or pseudo differential) expressions of corresponding orders $2m-k$, $2m-k+j$ and $\widetilde{q}_{j0}-l+1$ with respect to (D,q_1), and $D_\nu = i\partial/\partial\nu$, ν is a normal to Γ_0.

Assume, furter, that in a neighborhood

$$G_{k\delta} = \{x\in G: \operatorname{dist}(x,\Gamma_k)<\delta\} = \Gamma_k\times\{|(y_1,\dots,y_{i'_k})|<\delta\}$$

of mainfold Γ_k one can write

$$B_{jk}(x,D,q_1) = \sum_{|\beta|\le\widetilde{q}_{jk}} T_{kj\beta}(t,D_t,q_1)D_y^\beta, \tag{3.2.8}$$

where $D_y^\beta = D_{y_1}^{\beta_1}\cdots D_{y_{i'_k}}^{\beta_{i'_k}}$, and

$$T_{kj\beta}(t,D_t,q_1) \equiv \sum_{|\alpha|+d\le\widetilde{q}_{jk}-|\beta|} b_{jd\alpha k}(x)q_1^d D_t^\alpha$$

$$D_t^\alpha = D_{t_1}^{\alpha_1}\cdots D_{t_{i_k}}^{\alpha_{i_k}},\quad x=(t,y),\quad (t,0)\in\Gamma_k,\quad |y| = |(y_0,\dots,y_{i'_k})|,$$

$$j=1,\dots,m_k,\quad k=1,\dots,\overline{k}.$$

For simplisity, the coefficients (the symbols) of all the expressions are assumed to be infinitely smooth in $\overline{G}$. Moreover, in what follows we assume that problem (3.2.1)–(3.1.2) is elliptic with a parameter in $\overline{G}\times\mathbf{R}$; this means that the expression L is properly elliptic with a parameter in $\overline{G}\times\mathbf{R}$, and the expressions $\{B_{j0}\}$ satisfy the Lopatinskii conditions with a parameter on $\Gamma_0\times\mathbf{R}$ (see, for example, [R1, Ch.X]). At last, we assume that for each $k\in\{1,\dots,\overline{k}\}$ conditions (3.2.3) form a Douglis–Nirenberg eliptic system with a parameter on $\Gamma_k\times\mathbf{R}$ (see Subsection 3.2.5 below).

3.2.2. Let us now introduce the notion of a generalized solution of problem (3.1.1)–(3.1.3). By integration by parts we obtain

$$(Lu,v) = (u,L^+v) - i\sum_{j=1}^{2m}\langle D_\nu^{j-1}u, M_j(x,D,q_1)v\rangle_{\Gamma_0}\ (u,v\in C^\infty(\overline{G})), \tag{3.2.9}$$

$$M_jv = \sum_{k=j}^{2m} D_\nu^{k-j}L_k^+(x,D',q)v,\quad \operatorname{ord}M_j = 2m-j.$$

Recall that the symbols $(\cdot,\cdot)$ and $\langle\cdot,\cdot\rangle_k$ denote the scalar products (or their extensions) in $L_2(G)$ and in $L_2(\Gamma_k)$, correspondingly. The expressions L^+ and L_k^+ are formally adjoint to L and L_k, respectively.
We identify the element $u\in C^\infty(\overline{G})$ with the vector

$$u = \Big(u_0,u_1,\dots,u_r,\big(u_{k\beta}: k=1,\dots,\overline{k}, |\beta|\le\widetilde{q}_k\big)\Big)$$
$$\Big(u_0 = u|_{\overline{G}},\quad u_j = D_\nu^{j-1}u|_{\Gamma_0}\ (j=1,\dots,r),\quad u_{k\beta} = D_y^\beta u|_{\Gamma_k}\Big), \tag{3.2.10}$$

and the element $f\in C^\infty(\overline{G})$ with the vector

$$f = (f_0,\dots,f_{r-2m})\ \Big(f_0 = f|_{\overline{G}},\quad f_j = D_\nu^{j-1}f|_{\Gamma_0}\ (1\le j\le r-2m)\Big). \tag{3.2.11}$$

It follows from (3.2.5)–(3.2.9) that vector (3.2.10) $u \in C^\infty(\overline{G})$ is a solution of problem (3.2.1)–(3.2.3) if there hold the following relations

$$(u_0, L^+v) - i\sum_{j=1}^{2m}\langle u_j, M_jv\rangle = (f_0, v) \quad (v \in C^\infty(\overline{G})), \tag{3.2.12}$$

$$\sum_{k=0}^{2m+j-1} L_{kj}(x', D', q_1)u_{k+1} = f_j \quad (j : 1 \leq j \leq r - 2m). \tag{3.2.13}$$

$$B_{j0}u|_{\Gamma_0} \equiv \sum_{l=1}^{\widetilde{q}_{j0}+1} B_{jl}(x, D', q_1)u_l = \varphi_j \quad (j = 1, \ldots, m), \tag{3.2.14}$$

$$B_{jk}u|_{\Gamma_k} \equiv \sum_{|\beta| \leq \widetilde{q}_{jk}} T_{kj\beta}(t, D_t, q_1)u_{k\beta} = \varphi_{jk} \quad (k = 1, \ldots, \overline{k}; j = 1, \ldots m_k). \tag{3.2.15}$$

Now let u and f be vectors defined by formulas (3.2.10) and (3.2.11), where u_0 and f_0 are generalized functions in $\overline{G}$, and $\{u_j, f_l\}$ and $\{u_{k\beta}\}$ are generalized functions in Γ_0 and Γ_k, respectively. If relations (3.2.12)–(3.2.15) hold, then the vector u is called a generalized solution of problem (3.2.1)–(3.2.3). Relations (3.2.12)–(3.2.15) generate the mapping

$$\mathcal{A}(q_1) : u \mapsto F = (f, \varphi_0, \varphi_1, \ldots, \varphi_{\overline{k}}), \tag{3.2.16}$$

$$f = (f_0, \ldots, f_{r-2m}), \quad \varphi_0 = (\varphi_{10}, \ldots, \varphi_{m0}), \quad \varphi_k = (\varphi_{1k}, \ldots, \varphi_{m_kk})$$

To study this mapping and to precise the definition of generalized solution, we need some functional spaces.

3.2.3. Let $p, p' \in (1, \infty)$, $1/p + 1/p' = 1$, $s \in \mathbf{R}$. We denote by $H^{s,p}(\mathbf{R}^n)$ the spaces of Bessel potentials (see, for example, [R1], [Gr2]) with the norm

$$||f, R^n||_{s,p} = ||F^{-1}(1 + |\xi|^2)^{s/2}Ff||_{L_p(\mathbf{R}^n)} \tag{3.2.17}$$

Here $(Ff)(\xi)$ is the Fourier transformation of element f, F^{-1} is the inverse one.

The spaces $H^{s,p}(\mathbf{R}^n)$ and $H^{-s,p'}(\mathbf{R}^n)$ are dual to each other with respect to the extension of the scalar product in $L_2(\mathbf{R}^n)$. Let $q \in \mathbf{R}$. In the space $H^{s,p}(\mathbf{R}^n)$ we introduce a norm depending on the parameter $q \in \mathbf{R}$. We set:

$$||f, R^n, q||_{s,p} = ||F^{-1}(1 + |\xi|^2 + q^2)^{s/2}Ff||_{L_p(\mathbf{R}^n)} \tag{3.2.18}$$

The space $H^{s,p}(\mathbf{R}^n)$ with norm (3.2.18) is denoted by $H^{s,p}(\mathbf{R}^n, q)$. For each $q \in \mathbf{R}$ norm (3.2.18) is equialent to norm (3.2.17). As a rule, it is convenient to consider only the norms equivalent to (3.2.18) for which the corresponding bilateral estimates can be written with constants independent of q. If $|q| \geq q^0 > 0$, then (3.2.18) is equivalent to the norm

$$||F^{-1}(|\xi|^2 + q^2)^{s/2} F f||_{L_p(\mathbf{R}^n)}.$$

Denote by $H^{s,p}(G,q)$ ($s \geq 0$) the space of restrictions of elements of $H^{s,p}(\mathbf{R}^n, q)$ onto G with quotient space topology

$$H^{s,p}(G,q) = H^{s,p}(\mathbf{R}^n, q)/H^{s,p}_{CG}(\mathbf{R}^n, q), \tag{3.2.19}$$

where

$$H^{s,p}_{CG}(\mathbf{R}^n, q) = \{v \in H^{s,p}(\mathbf{R}^n, q) : \operatorname{supp} v \subset \mathbf{R}^n \setminus G\}$$

is a subspace of the space $H^{s,p}(\mathbf{R}^n, q)$. Let $H^{-s,p}(G,q)$ ($s \geq 0$) denote the space dual to $H^{s,p'}(G,q)$ with respect to the extension $(\cdot,\cdot)$ of the scalar product in $L_2(G)$. It follow from (3.2.19) (see [R1]) that

$$H^{-s,p}(G,q) \simeq H^{-s,p}_{\overline{G}}(\mathbf{R}^n, q) = \{f \in H^{-s,p}(\mathbf{R}^n, q) : \operatorname{supp} f \subset \overline{G}\}.$$

Since for $s \geq 0$ there are not elements concentrated at Γ_k ($k = 1, \ldots, \overline{k}$) in $H^{s,p}(\mathbf{R}^n, q)$, the relation

$$H^{s,p}(G,q) \simeq H^{s,p}(G \cup \Gamma_1 \cup \ldots \cup \Gamma_k, q) \quad (s \in \mathbf{R})$$

is true. The norm in $H^{s,p}(G,q)$ ($s \in \mathbf{R}$) is denoted by $||u, G, q||_{s,p}$.

For $s \geq 0$ we denote by $B^{s,p}(\Gamma_k, q)$ ($k = 0, \ldots, \overline{k}$) the Besov space with the norm

$$\langle\langle \varphi, \Gamma_k, q \rangle\rangle_{s,p} = \inf_{u=\varphi|\Gamma_k} ||u, G, q||_{s+i'_k/p,p}.$$

The spaces $B^{s,p}(\Gamma_k, q)$ and $B^{-s,p'}(\Gamma_k, q)$ are dual to each other with respect to the extension $\langle\cdot,\cdot\rangle_{\Gamma_k}$ of the scalar product in $L_2(\Gamma_k)$ (see [R1]).

Let

$$s \in \mathbf{R}, \quad s \neq j + \frac{i'_k}{p} \quad (j = 1, \ldots, \tilde{q}_k,\ k = 0, \ldots, \overline{k},\ \tilde{q}_0 = r). \tag{3.2.20}$$

By $\widetilde{H}^{s,p}(q)$ we denote the completion of $C^\infty(\overline{G})$ in the norm

$$|||u, G, q|||_{s,p} = \Bigg(||u, G, q||^p_{s,p} + \sum_{j=1}^{r} \langle\langle D^{j-1}_\nu u, \Gamma_0, q \rangle\rangle^p_{s-j+1-1/p,p} + \sum_{k=1}^{\overline{k}} \sum_{|\alpha| \leq \tilde{q}_k} \langle\langle D^\alpha_y u, \Gamma_k, q \rangle\rangle^p_{s-|\alpha|-i'_k/p,p} \Bigg)^{1/p}. \tag{3.2.21}$$

The space $\widetilde{H}^{s,p}(q)$ is isometrically equivalent to the closure $\mathcal{H}_0^{s,p}(q)$ of the space of vectors $u \in C^\infty(\overline{G})$ (3.2.10) in the space

$$\mathcal{H}^{s,p}(q) = H^{s,p}(G,q) \times \prod_{j=1}^{r} B^{s-j+1-1/p,p}(\Gamma_0,q)$$
$$\times \prod_{k=1}^{\overline{k}} \prod_{|\alpha|\leq \widetilde{q}_k} B^{s-|\alpha|-i'_k/p,p}(\Gamma_k,q). \quad (3.2.22)$$

Hence, the space $\widetilde{H}^{s,p}(q)$ coincides with the subspace $\mathcal{H}_0^{s,p}(q)$ of the space of vectors

$$u = \Big(u_0, u_1, \dots, u_r, \Big\{u_{\alpha k} : k = 1, \dots, \overline{k};\ |\alpha| \leq \widetilde{q}_k\Big\}\Big) \in \mathcal{H}^{s,p}(q). \quad (3.2.23)$$

In addition, an element $u \in \mathcal{H}^{s,p}(q)$ belongs to the space $\mathcal{H}_0^{s,p}(q) \simeq \widetilde{H}^{s,p}(q)$ if and only if

$$\begin{aligned} u_j &= D_\nu^{j-1} u_0|_{\Gamma_0} \ (\forall j : s - j + 1 - 1/p > 0), \\ u_{\alpha k} &= D_y^{\alpha} u_0|_{\Gamma_k} \ (\forall \alpha, k : s - |\alpha| - i'_k/p > 0) \end{aligned} \quad (3.2.24)$$

(sf. [R1], [R11]). In addition, if $s < 1/p$, then $\mathcal{H}_0^{s,p}(q) = \mathcal{H}^{s,p}(q) = \widetilde{H}^{s,p}(q)$.

By $\widetilde{H}^{s,p,(r)}(q)$ ([R1],[R11]) we denote the completion of $C^\infty(\overline{G})$ in the norm

$$|||u,G,q|||_{s,p,(r)} = \Big(||u,G,q||_{s,p}^p + \sum_{j=1}^{r} \langle\langle D_\nu^{j-1}u, \Gamma_0, q\rangle\rangle_{s-j+1-1/p,p}^p\Big)^{1/p} \quad (3.2.25)$$

The space $\widetilde{H}^{s,p,(r)}(q)$ consists of vectors

$$(u_0, \dots, u_r) \in H^{s,p}(G,q) \times \prod B^{s-j+1-1/p,p}(\Gamma_0,q)$$

such that $u_j = D_\nu^{j-1} u_0|_{\Gamma_0}$ $(\forall j : s-j+1-1/p > 0)$. For exceptional values of s the spaces $\widetilde{H}^{s,p}(q)$ and $\widetilde{H}^{s,p,(r)}(q)$, and norms (3.2.21) and (3.2.25) are defined by complex interpolation.

Theorem 3.2.1. *(sf. [R1], [RSk1], [RSk2], [Gr2], [R11]) For any $s \in \mathbf{R}$ and $p \in (1,\infty)$ the closure $\mathcal{A}(q_1) = A_{s,p}(q_1)$ of the mapping*

$$u \mapsto \Big(Lu|_{\overline{G}}, \Big(D_\nu^{j-1} Lu|_{\Gamma_0}\Big| j : 1 \leq j \leq r - 2m\Big), B_1 u|_{\Gamma_0}, \dots, B_m u|_{\Gamma_0},$$

$$\left(B_{jk}u|_{\Gamma_k} : k = 1, \dots, \overline{k}, j = 1, \dots, m_k\right)\Big) \quad (u \in C^\infty(\overline{G}))$$

acts continuously in the pair of spaces

$$\widetilde{H}^{s,p}(q) \to K^{s,p}(q), \tag{3.2.26}$$

$$K^{s,p}(q) := \widetilde{H}^{s-2m,p,(r-2m)}(q) \times \prod_{j=1}^{m} B^{s-\widetilde{q}_{jo}-1/p,p}(\Gamma_0, q)$$

$$\times \prod_{k=1}^{\overline{k}} \prod_{j=1}^{m_k} B^{s-\widetilde{q}_{jk}-i'_k/p,p}(\Gamma_k, q)$$

If $s_1 \le s$ and $p_1 \le p$, then the operator $\mathcal{A}(q_1) \equiv \mathcal{A}_{s_1,p_1}(q_1)$ is an extension of the operator $\mathcal{A}_{s_2,p_2}(q_1)$ in continuity.

Definition 3.2.1. *An element $u \in \widetilde{H}^{s,p}(q)$ is called a generalized solution of problem (3.2.1)–(3.2.3) if*

$$\mathcal{A}_{s,p}(q_1)u = F = (f, \varphi_{10}, \dots, \varphi_{m0}, \{\varphi_{jk}\}) \in K^{s,p}(q).$$

Theorem 3.2.2. *An element $u \in \widetilde{H}^{s,p}(q)$ is a generalized solution of problem (3.2.1)–(3.2.3) if and only if relations (3.2.12)–(3.2.15) hold.*

3.2.4. In $\overline{G}$ we consider elliptic problem (3.2.1)–(3.2.3) with a parameter. The following statements are true (see [R1]).

I. For any $s \in \mathbf{R}$ and $p \in (1, \infty)$ the closure $A_{s,p}(q_1)$ of the mapping

$$u \mapsto \left(Lu,\, B_1 u|_{\Gamma_0}, \dots, B_m u|_{\Gamma_0}\right) \quad (u \in C^\infty(\overline{G}))$$

acts continuously in the pair of spaces

$$\widetilde{H}^{s,p,(r)}(q) \to K^{s,p,(r)}(q)$$

$$K^{s,p,(r)}(q) := \widetilde{H}^{s-2m,p,(r-2m)}(q) \times \prod_{j=1}^{m} B^{s-\widetilde{q}_{0j}-1/p,p}(\Gamma_0, q).$$

II. The element $\widetilde{u} = (u_0, u_1, \dots, u_r) \in \widetilde{H}^{s,p,(r)}(q)$, is called a generalized solution of problem (3.2.1)–(3.2.2) if $A_{s,p}(q_1)\widetilde{u} = (f, \varphi) \in K^{s,p,(r)}(q)$. Furthermore, $A_{s,p}(q_1)\widetilde{u} = (f, \varphi)$ if and only if equalities (3.2.12)–(3.2.14) hold. Hence, if element $u \in \widetilde{H}^{s,p}(q)$ (3.2.23) is a generalized solution of problem

(3.2.1)–(3.2.3), then the element $\widetilde{u} = (u_0, u_1, \ldots, u_r) \in \widetilde{H}^{s,p,(r)}(q)$ is a generalized solution of problem (3.2.1)–(3.2.2).
III. Let $s \in \mathbf{R}$, and $p \in (1, \infty)$. Then there exists a number $q^0 > 0$ such that the operator $A(q_1)$ esablishes an isomorphism between spaces

$$\widetilde{H}^{s,p,(r)}(q) \to K^{s,p,(r)}(q) \tag{3.2.27}$$

for $|q| \geq q^0$ and $\theta \in [\theta_1, \theta_2]$. There exists a constant $c > 0$ independent of u, q ($|q| \geq q^0$), and $\theta \in [\theta_1, \theta_2]$, such that

$$c^{-1}|||u, G, q|||_{s,p,(r)} \leq |||A(q_1)u|||_{K^{s,p,(r)}(q)} \leq c|||u, G, q|||_{s,p,(r)}$$

$$(u \in \widetilde{H}^{s,p,(r)}(q)).$$

3.2.5. Let us formulate the assumption on the boundary expressions on Γ_k $(k = 1, \ldots, \overline{k})$. Let $u \in \widetilde{H}^{s,p}(q)$ be a generalized solution of problem (3.2.1)–(3.2.3). Then, according to Theorem 3.2.2, the elements $\{u_{\alpha k}\}$ satisfy equalites (3.2.15). For any $k \in \{1, \ldots, \overline{k}\}$ we set

$$\widetilde{T}_{kj\beta}(t, D_t, q_1) = \begin{cases} T_{kj\beta}(t, D_t, q_1) & \text{for } |\beta| \leq \widetilde{q}_{jk} \\ 0 & \text{for } \widetilde{q}_{jk} < |\beta| \leq \widetilde{q}_k \end{cases}$$

(the numbers $\widetilde{q}_k$ are defined by formula (3.2.4)). Then one can rewrite eqalities (3.2.15) in the form

$$B_{jk}u|_{\Gamma_k} \equiv \sum_{|\beta| \leq \widetilde{q}_k} \widetilde{T}_{kj\beta}(t, D_t, q_1)u_{k\beta} = \varphi_{jk} \tag{3.2.28}$$

$$(j = 1, \ldots, m_k;\ k = 1, \ldots, \overline{k}),$$

or, in short,

$$T_k(t, D_t, q_1)U_k = \varphi_k \quad \Big(T_k = (\widetilde{T}_{kj\beta})_{\substack{j=1,\ldots,m_k \\ |\beta| \leq \widetilde{q}_k}};\ \varphi_k = (\varphi_{1k}, \ldots, \varphi_{m_k k})\Big) \tag{3.2.28'}$$

For any k the matrix $(\widetilde{T}_{kj\beta}(t, D_t, q_1))$ has a structure of Douglis–Nirenberg. This means that if we associate the number $s_j = \widetilde{q}_{jk} - \widetilde{q}_k$ with the row with index j, the number $t_\beta = \widetilde{q}_k - |\beta|$ with the column with index β, then we obtain $\operatorname{ord} \widetilde{T}_{kj\beta} \leq s_j + t_\beta$ for $s_j + t_\beta \geq 0$, and $\widetilde{T}_{kj\beta} \equiv 0$ for $s_j + t_\beta < 0$.
We now assume that for any $k \in \{1, \ldots, \overline{k}\}$ system (3.2.28) is a Douglis-Nirenberg elliptic system on Γ_k with a parameter, i.e., the system is square and

$$\det(\widetilde{T}^0_{kj\beta}(t, \sigma, q_1)) \neq 0 \quad \big((\sigma, q) = (\sigma_1, \ldots, \sigma_{i_k}, q) \neq 0,\ t \in \Gamma_k\big),$$

where $\widetilde{T}^0_{kj\beta}$ is the principal symbol of the expression $\widetilde{T}_{kj\beta}(t, D_t, q_1)$.

For each $s \in \mathbf{R}$ and $p \in (1, \infty)$ the mapping

$$T_k(q_1) = T_{k,s,p}(q_1) : U \mapsto T_k(q_1)U : B^{T+s,p}(\Gamma_k, q) \to B^{s-S,p}(\Gamma_k, q), \tag{3.2.29}$$

$$B^{T+s,p}(\Gamma_k, q) := \prod_{j=1}^{m_k} B^{t_j+s,p}(\Gamma_k, q), \quad B^{s-S,p}(\Gamma_k, q) := \prod_{j=1}^{m_k} B^{s-s_j,p}(\Gamma_k, q).$$

is a continuous operator.

Let $s \in \mathbf{R}$ and $p \in (1, \infty)$. Then there exists a number $q^0 > 0$ such that the operator $T_k(q_1)$ $(k = 1, \dots, \overline{k})$ esablishes an isomorphism

$$B^{T+s,p}(\Gamma_k, q) \to B^{s-S,p}(\Gamma_k, q) \tag{3.2.30}$$

for $|q| \geq q^0$ and $\theta \in [\theta_1, \theta_2]$. There exists a constant $c > 0$ independent of u, q $(|q| \geq q^0)$, and $\theta \in [\theta_1, \theta_2]$ such that

$$c^{-1}\langle\langle u, \Gamma_k, q\rangle\rangle_{T+s,p} \leq \langle\langle T_k(q_1)u\rangle\rangle_{B^{s-S,p}(\Gamma_k,q)} \leq c\langle\langle u, \Gamma_k, q\rangle\rangle_{T+s,p}\,.$$

3.2.6.

Theorem 3.2.3. *Let $p \in (1, \infty)$. Then there exists a number $q^0 > 0$ such that, for $|q| \geq q^0$ and $\theta \in [\theta_1, \theta_2]$, the following statements are true:*

(i) for $s < 1/p$ the problem $\mathcal{A}(q_1)u = F \in K^{s,p}(q)$ is unique solvable in $\widetilde{H}^{s,p}(q)$. There exists a constant $c > 0$ independent of u, q $(|q| \geq q^0)$, and $\theta \in [\theta_1, \theta_2]$ such that

$$c^{-1}|||u, G, q|||_{s,p} \leq |||\mathcal{A}(q_1)u|||_{K^{s,p}(q)} \leq c|||u, G, q|||_{s,p} \; (u \in \widetilde{H}^{s,p}(q)). \tag{3.2.31}$$

(ii) for $s > 1/p$ statement (i) holds if and only if the compatibility conditions (see (3.2.34) below) hold.

For the solvability of problem (3.2.1)–(3.2.3) in $\widetilde{H}^{s,p}$ it is nesessary that relations (3.2.25) and (3.2.29) are valid; this condition is also sufficient if, in addition, the compatibility conditions hold for $s > 1/p$.

Proof. If $s < 1/p$, then, by solving problem (3.2.1)–(3.2.2), we find the element

$$\widetilde{u} = A^{-1}(q_1)(f, \varphi) = (u_0, \dots, u_r) \in \widetilde{H}^{s,p,(r)}(q)$$

(see (3.2.27)). The element $\widetilde{u} = (u_0, \dots, u_r)$ gives us the first $r + 1$ components of the solution required. We find the rest of the components by solving the problem

$$T_k(q_1)U_k = \varphi_k \quad (k = 1, \dots, \overline{k}) \tag{3.2.32}$$

(see (3.2.30)). The element $U_k = T_k^{-1}(q_1)\varphi_k$ is the general solution of problem (3.2.32). The vector

$$(\tilde{u}, \{U_k : k = 1, \ldots, \overline{k}\}) \in \widetilde{H}^{s,p}(q) \tag{3.2.33}$$

is the general solution of problem (3.2.1)–(3.2.3). If $s > 1/p$, then $u_0 \in H^{s,p}(G, q)$, and, hence, there exist the traces

$$D_y^\alpha u_0|_{\Gamma_k} \quad (\forall \alpha, k : s - |\alpha| - j_k'/p > 0).$$

In order to the solution obtained above be belonging to $\widetilde{H}^{s,p}(q)$ it is nesessary and sufficient that the compatibility conditions

$$u_{\alpha k} = D_y^\alpha u_0|_{\Gamma_k} \quad (\forall \alpha, k : s - |\alpha| - i_k'/p > 0). \tag{3.2.34}$$

hold. This completes the proof of the theorem. □

3.2.7. Let us consider problem (3.2.1)–(3.2.3) under the conditions $\overline{k} = 1$, $\Gamma_1 = \Gamma$, $i_1 = i$, $i_1' = i'$ and $s < 2m - i'/p'$. In this case it is possible that the element $f_0 \in H^{s-2m,p}(G, q)$ is concentrated at Γ. It turns out that we can add to f_0 the element f_0' concentrated at Γ so that the compatibility conditions hold automatically. The following theorem is true.

Theorem 3.2.4. *Let $1/p < s < 2m - i'/p'$, and $F \in K^{s,p}(q)$. Then one can add to f_0 the element f_0' concentrated at Γ so that for the problem $\mathcal{A}(q_1)u = \widetilde{F}$, $\widetilde{F} = F + (f_0', 0, \ldots, 0)$ the compatibility condition hold automatically, and there exists a number $q^0 > 0$ such that this problem is unique solvable in $\widetilde{H}^{s,p}(q)$ for $|q| \geq q^0$ and $\theta \in [\theta_1, \theta_2]$. There exists a constant $c > 0$ independent of u, q ($|q| \geq q^0$), and $\theta \in [\theta_1, \theta_2]$ such that*

$$c^{-1}|||u, G, q|||_{s,p} \leq |||\mathcal{A}(q_1)u|||_{K^{s,p}(q)} \leq c|||u, G, q|||_{s,p} \; (u \in \widetilde{H}^{s,p}(q)).$$

Note that the element f_0 coincides with the element $f_0 + f_0'$ inside of G.

3.3. Generalizations. Applications

3.3.1.
As it was mentioned in Subsection refsub3.1.1, all the statements with the same proofs remain true for the case where expressions (3.1.5)–(3.1.8) are pseudo differential along ∂G and differential in the directions normal to ∂G.

3.3.2.

The solvability of parabolic Sobolev's problems follows from statement II above. (sf. [AgV], [RR2], [EZh]).

3.3.3.

All the results with the same proofs remain true for the Sobolev problem for Douglis-Nirenberg elliptic systems of order $(T, S) = (t_1, \ldots, t_N, s_1, \ldots, s_N)$ (sf. [R1, Ch.X]).

3.3.4.

Now we show some of possible applications of obtained results. The theorem on complete collection of isomorphisms enables us to construct the Green's function for the Sobolev problem and to study it (sf. [R1, Sec. 7.4]). This theorem gives a possibility to study the local smoothness of the solutions up to Γ_0 and Γ_k, to investigate the strongly degenerated elliptic Sobolev problem, and to study the Sobolev problem in the case where the right-hand sides have arbitrary large power singularities along the manifolds of different dimensionalities (sf. [R1, Ch. 8]).

CHAPTER 4

THE CAUCHY PROBLEM FOR GENERAL HYPERBOLIC SYSTEMS IN THE COMPLETE SCALE OF SOBOLEV TYPE SPACES

4.1. Statement of a Problem. Functional Spaces

Ever since the well-known work of S. L. Sobolev [Sob] generalized functions have frequently been used to study the Cauchy problem for hyperbolic equations (we mention here [Ler], [Går], [Vla], [Hör], the survey [VoG], and the bibliography given there). In this note the Cauchy problem for a system strictly hyperbolic in the Leray–Volevich sense is studied in the complete scale of spaces of Sobolev type depending on real parameters s and τ; s characterizes the order smoothness of a solution in all variables, while τ describes the additional smoothness in the tangential variables. The solutions is the 'more generalized' the smaller s and τ; for sufficiently large s and τ the solution is an ordinary classical solution of the problem under consideration. In [R7] and [R8] such problems were studied for a single equation.

4.1.1. Let $(t,x) = (t,x_1,\dots,t_n) \in \mathbf{R}^{n+1}$, and let $(\sigma,\xi) = (\sigma,\xi_1,\dots,\xi_n)$ be the dual variables. Assume that

$$l = l(D_t,D_x) = (l_{kj}\,(d_t,D_x))_{k,j=1,\dots,N}\,, \tag{4.1.1}$$

where l is a matrix differential expression and the l_{kj} are homogeneous differential expressions of orders $s_k + t_j$ with constant coefficients, with $l_{kj} = 0$ if $s_k + t_j < 0$. Here $D_t = i\partial/\partial t$, $D_x = D_1 \dots D_n$, $D_j = i\partial/\partial x_j$; $s_1,\dots,s_N$ and $t_1,\dots,t_N$ are integers, and

$$t_1 \geq \dots \geq t_N \geq 0 = s_1 \geq \dots \geq s_N.$$

Let $s_1 + \dots + s_N + t_1 + \dots + t_N = r$, and let

$$L(\sigma,\xi) = \det(l(\sigma,\xi)) = \sum_{j+|\alpha|\leq r} a_{j\alpha}\sigma^j\xi^\alpha. \tag{4.1.2}$$

Expression (4.1.1) is said to be *strictly hyperbolic in Leray–Volevich sense* if polynomial (4.1.2) is strictly hyperbolic. This means that the coefficients $a_{r,0,\dots,0}$ of σ^r in (4.1.2) are nonzero, and for each $\xi \in \mathbf{R}^n \setminus \{0\}$ the roots of the equation $L(\sigma, \xi) = 0$ relative to σ are real and distinct.
It is assumed everywhere below that expression (4.1.1) is strictly hyperbolic.

We investigate, first, the solvability of the problem

$$lu = f, \qquad u = (u_1, \dots, u_N), \quad f = (f_1, \dots, f_N) \tag{4.1.3}$$

in $\mathbf{R}^{n+1}$, and, second, the solvability of the Cauchy problem

$$lu = f \quad (\text{ in } \boldsymbol{\Omega}), \qquad D_t^{k-1} u_j \big|_{t=0} = u_{jk} \tag{4.1.4}$$

$$(\forall j : t_j \geq 1,\ k = 1, \dots, t_j)$$

in the half-space $\boldsymbol{\Omega} = \Big\{(t, x) \in \mathbf{R}^{n+1} : t > 0\Big\}$. Moreover, we study the solvability of these problems for the system

$$\Big(l(D_t, D_x) + l'(t, x, D_t, D_x)\Big) u = f \tag{4.1.5}$$

obtained by perturbation of system (4.1.3) by lower-order terms with infinitely smooth coefficients all of whose derivatives are bounded. The solvability of these problems will be established in the complete scale of Sobolev type spaces.

Let us introduce functional spaces for a precise formulation of the problems.

4.1.2. Let s, τ, $\gamma \in \mathbf{R}$. We denote by $H^{s,\tau}(\mathbf{R}^{n-1}, \gamma)$ the space of distributions f with the norm

$$\|f, \mathbf{R}^{n+1}, \gamma\|_{s,\tau} = \left(\int \left(1 + \gamma^2 + \sigma^2 + |\xi|^2\right)^s \right.$$
$$\left. \times \left(1 + \gamma^2 + |\xi|^2\right)^\tau |\widetilde{f}(\sigma, \xi)|^2 \, d\sigma \, d\xi \right)^{1/2}, \tag{4.1.6}$$

where $\widetilde{f}(\sigma, \xi)$ is the Fourier transform of the element f, and the integration goes over the entire space. It is clear that for each fixed $\gamma \in \mathbf{R}$ the norm $\|f, \mathbf{R}^{n+1}, \gamma\|_{s,\tau}$ is equivalent to the norm $\|f, \mathbf{R}^{n+1}, 0\|_{s,\tau} = \|f, \mathbf{R}^{n+1}\|_{s,\tau}$, and the set $H^{s,\tau}(\mathbf{R}^{n+1}, \gamma)$ thus does not depend on γ. However, in this note it is convenient to consider only norms equivalent to (4.1.6) for which the constants in the corresponding two-sided estimates can be chosen not to depend on γ.

For s, τ, $\gamma \in \mathbf{R}$ and $s \geq 0$ we denote by $H^{s,\tau}(\Omega,\gamma)$ the set of restrictions of functions in $H^{s,\tau}(\mathbf{R}^{n+1},\gamma)$ to Ω with the norm of quotient space:

$$\|w,\Omega,\gamma\|_{s,\tau} = \inf \|u,\mathbf{R}^{n+1},\gamma\|_{s,\tau} \quad (s,\ \tau,\ \gamma \in \mathbf{R},\ \ s \geq 0),$$

where the infimum is taken over all functions $u \in H^{s,\tau}(\mathbf{R}^{n+1},\gamma)$ equal to w in Ω. We denote by $H^{-s,-\tau}(\Omega,\gamma)$, s, τ, $\gamma \in \mathbf{R}$, $s \geq 0$, the space dual to $H^{s,\tau}(\Omega,\gamma)$ with respect to the extension $(\cdot,\cdot) = (\cdot,\cdot)_\Omega$ of the scalar product in $L_2(\Omega)$;

$$\|u,\Omega,\gamma\|_{-s,-\tau} = \sup_{v \in H^{s,\tau}(\Omega,\gamma)} |(u,v)|\, \|v,\Omega,\gamma\|_{s,\tau}^{-1}$$

is a norm in $H^{-s,-\tau}(\Omega,\gamma)$, $s \geq 0$.

For $s,\gamma \in \mathbf{R}$, let $H^s(\partial\Omega,\gamma)$ denote the space of distributions g on $\partial\Omega$ with the norm

$$\|g,\partial\Omega,\gamma\|_s = \left(\int_{\partial\Omega} |\widehat{g}(\xi)|^2 \,(1+|\xi|^2+\gamma^2)^s \, d\xi \right)^{1/2} < \infty,$$

where $\widehat{g}(\xi) = (F_{x\to\xi}\, g)(\xi)$ is the Fourier transform of g.

We fix a natural number r and suppose that s, τ, $\gamma \in \mathbf{R}$, $s \neq k+1/2$, $k = 0,\dots,r-1$. We denote by $\widetilde{H}^{s,\tau,(r)}(\Omega,\gamma)$ the completion of $C_0^\infty(\overline{\Omega})$ in the norm

$$|||u,\Omega,\gamma|||_{s,\tau,(r)} = \left(\|u,\Omega,\gamma\|_{s,\tau}^2 + \sum_{j=1}^{r} \|D_t^{j-1}u,\partial\Omega,\gamma\|_{s-j+\tau-1/2}^2 \right)^{1/2}, \quad (4.1.7)$$

where $C_0^\infty(\overline{\Omega})$ denotes the set of restrictions to $\overline{\Omega}$ of functions in $C_0^\infty(\mathbf{R}^{n-1})$. A similar space was introduced by the author in [R10] and studied in [R11] (see also [Ber, Ch. 3, §6.8], and [RS1]).

For $s = k+1/2$ $(k = 0,\dots,r-1)$ the space $\widetilde{H}^{s,\tau,(r)}(\Omega,\gamma)$ and the norm $|||u,\Omega,\gamma|||_{s,\tau,(r)}$ are defined by interpolation.

It follows from (4.1.7) that the closure S of the mapping

$$u \mapsto \left(u\big|_{\overline{\Omega}}, u\big|_{\partial\Omega}, \dots, D_t^{r-1}u\big|_{\partial\Omega} \right) \quad (u \in C_0^\infty(\overline{\Omega}))$$

establishes an isometry between $\widetilde{H}^{s,\tau,(r)}(\Omega,\gamma)$ and a subspace of the direct product

$$H^{s,\tau}(\Omega,\gamma) \times \prod_{j=1}^{r} H^{s+\tau-j+1/2}(\partial\Omega,\gamma).$$

We henceforth agree to identify an element $u \in \widetilde{H}^{s,\tau,(r)}(\Omega,\gamma)$ with the element $Su = (u_0,\dots,u_r)$. We write

$$u = (u_0,\dots,u_r) \in \widetilde{H}^{s,\tau,(r)}(\Omega,\gamma)$$

for each $u \in \widetilde{H}^{s,\tau,(r)}(\Omega,\gamma)$. If $r=0$, then we set

$$\widetilde{H}^{s,\tau,(0)}(\Omega,\gamma) := H^{s,\tau}(\Omega,\gamma), \quad |||u,\Omega,\gamma|||_{s,\tau,(0)} := \|u,\Omega,\gamma\|_{s,\tau}.$$

We further introduce the spaces

$$\mathcal{H}^{s,\tau}(\mathbf{R}^{n+1},\gamma), \qquad \mathcal{H}^{s,\tau}(\Omega,\gamma), \qquad \widetilde{\mathcal{H}}^{s,\tau,(r)}(\Omega,\gamma),$$

and denote the respective norms in these spaces by

$$|u,\mathbf{R}^{n+1},\gamma|_{s,\tau}, \qquad |u,\Omega,\gamma|_{s,\tau}, \qquad |u,\Omega,\gamma|_{s,\tau,(r)}.$$

We set

$$\mathcal{H}^{s,\tau}(\mathbf{R}^{n+1},\gamma) := \left\{u : e^{-\gamma t}u \in H^{s,\tau}(\mathbf{R}^{n+1},\gamma)\right\},$$

$$|u,\mathbf{R}^{n+1},\gamma|_{s,\tau} := \|e^{-\gamma t}u,\mathbf{R}^{n+1},\gamma\|_{s,\tau}\,.$$

If here $\mathbf{R}^{n+1}$ is replaced by Ω, we obtain the definition of $\mathcal{H}^{s,\tau}(\Omega,\gamma)$ and the norm in it. In an entirely similar way,

$$\widetilde{\mathcal{H}}^{s,\tau,(r)}(\Omega,\gamma) := \left\{u : e^{-\gamma t}u \in \widetilde{H}^{s,\tau,(r)}(\Omega,\gamma)\right\},$$

$$|u,\Omega,\gamma|_{s,\tau,(r)} := |||e^{-\gamma t}u,\Omega,\gamma|||_{s,\tau,(r)}\,.$$

4.2. Main Results

In this section we formulate and prove the statements on the Cauchy problem for a system strictly hyperbolic in the Leray–Volevich sense.

4.2.1.

Lemma 4.2.1. *Let*

$$M = M(t,x,D_t,D_x) \quad ((t,x) \in \mathbf{R}^{n-1})$$

be a linear differential expression of order m with infinitely smooth coefficients all of whose derivatives are bounded. Then for any $s,\tau \in \mathbf{R}$ there exists a constant $c>0$ independent of u and γ, such that

$$|Mu,\mathbf{R}^{n+1},\gamma|_{s-m,\tau} \le c|u,\mathbf{R}^{n+1},\gamma|_{s,\tau} \quad (u \in C_0^\infty(\mathbf{R}^{n+1})). \tag{4.2.1}$$

If $m \le r$, then

$$|Mu,\Omega,\gamma|_{s-m,\tau} \le c|u,\Omega,\gamma|_{s,\tau,(r)} \quad (u \in C_0^\infty(\overline{\Omega})), \tag{4.2.2}$$

$$|Mu, \Omega, \gamma|_{s-m,\tau,(r-m)} \leq c|u, \Omega, \gamma|_{s,\tau,(r)} \quad (u \in C_0^\infty(\overline{\Omega})), \tag{4.2.3}$$

and if $m \leq r-1$, *then*

$$|Mu, \Omega, \gamma|_{s-m+\tau-1/2} \leq c|u, \Omega, \gamma|_{s,\tau,(r)} \quad (u \in C_0^\infty(\overline{\Omega})). \tag{4.2.4}$$

Proof. Since

$$M = M(t, x, D_t, D_x) = e^{\gamma t} M(t, x, D_t + i\gamma, D_x)(e^{-\gamma t} u),$$

it follows that

$$\begin{aligned} |Mu, \mathbf{R}^{n-1}, \gamma|_{s-m,\tau} &= \|e^{-\gamma t} M(t, x, D_t, D_x)u, \mathbf{R}^{n+1}, \gamma\|_{s-m,\tau} \\ &= \|M(t, x, D_t + i\gamma, D_x)(e^{-\gamma t} u), \mathbf{R}^{n+1}, \gamma\|_{s-m,\tau} \\ &\leq c\|e^{-\gamma t} u, \mathbf{R}^{n+1}, \gamma\|_{s,\tau} \\ &= c|u, \mathbf{R}^{n+1}, \gamma|_{s,\tau}, \end{aligned}$$

and (4.2.1) is established. Inequalities (4.2.2)–(4.2.4) can be obtained in an entirely similar way (cf. [R11], [RS1], [R7], and [R8]). □

4.2.2. Let us now study hyperbolic systems in $\mathbf{R}^{n+1}$.
It follows from estimate (4.2.1) that for any $s, \tau \in \mathbf{R}$ the closures l and $l+l'$ of the respective mappings $u \mapsto lu$ and $u \mapsto (l+l')u$ $\left(u \in (C_0^\infty(\mathbf{R}^{n+1}, \gamma))^N\right)$ acts continuously in the pair of spaces

$$\mathcal{H}^{T+s,\tau}(\mathbf{R}^{n+1}, \gamma) \quad \to \quad \mathcal{H}^{s-S,\tau}(\mathbf{R}^{n+1}, \gamma), \tag{4.2.5}$$

$$\mathcal{H}^{T+s,\tau}(\mathbf{R}^n, \gamma) \quad := \quad \prod_{j=1}^N \mathcal{H}^{t_j+s,\tau}(\mathbf{R}^{n+1}, \gamma),$$

$$\mathcal{H}^{s-S,\tau}(\mathbf{R}^n, \gamma) \quad := \quad \prod_{k=1}^N \mathcal{H}^{s-s_k,\tau}(\mathbf{R}^{n+1}, \gamma),$$

and

$$\begin{aligned} |lu, \mathbf{R}^{n+1}, \gamma|_{s-S,\tau} &\leq c\,|u, \mathbf{R}^{n+1}, \gamma|_{T+s,\tau}, \\ |(l+l')u, \mathbf{R}^{n+1}, \gamma|_{s-S,\tau} &\leq c\,|u, \mathbf{R}^{n+1}, \gamma|_{T+s,\tau}, \end{aligned} \tag{4.2.6}$$

where $|f, \mathbf{R}^{n+1}, \gamma|_{s-S,\tau}$ and $|u, \mathbf{R}^{n+1}, \gamma|_{T+s,\tau}$ are the respective norms in the spaces of images and pre-images of mapping (4.2.5). In addition, the constant $c > 0$ does not depend on u and γ. The question arises of the invertibility of the operators l and $l + l'$.

Theorem 4.2.1. *For each $f \in \mathcal{H}^{s-S,\tau}(\mathbf{R}^{n+1}, \gamma)$ $(s, \tau, \gamma \in \mathbf{R},\ |\gamma| \geq \gamma_0 > 0)$ there exists one and only one element $u \in \mathcal{H}^{T+s,\tau-1}(\mathbf{R}^{n+1}, \gamma)$ such that $lu = f$. There exists a constant $c > 0$ independent of f, u, and γ $(|\gamma| \geq \gamma_0 > 0)$ such that*

$$|u, \mathbf{R}^{n+1}, \gamma|_{T+s,\tau-1} \leq \frac{c}{|\gamma|} |f, \mathbf{R}^{n+1}, \gamma|_{s-S,\tau}. \tag{4.2.7}$$

If $\operatorname{supp} f \subset \overline{\Omega} = \{(t, x) \in \mathbf{R}^n : t \geq 0\}$ then also $\operatorname{supp} u \subset \overline{\Omega}$.

For the proof, in the equation

$$l(D_t + i\gamma, D_x)(e^{-\gamma t} u) = e^{-\gamma t} f$$

which is equivalent to (4.1.3) we pass to the Fourier transforms and use the following lemma.

Lemma 4.2.2. *There exists a constant $c > 0$ independent of $(\sigma, \gamma, \xi) \in \mathbf{R}^{n+2}$, such that*

$$|L(\sigma + i\gamma, \xi)| \geq c|\gamma| \left(\sigma^2 + \gamma^2 + |\xi|^2\right)^{(r-1)/2} \quad \left((\sigma, \gamma, \xi) \in \mathbf{R}^{n+2}\right).$$

Furthermore, there exists a constant $M > 0$ such that for $(\gamma, \xi) \in \mathbf{R}^{n+1} \backslash \{0\}$ the estimate

$$|L(\sigma + i\gamma, \xi)| \geq c\left(\sigma^2 + \gamma^2 + |\xi|^2\right)^{r/2}$$

holds on the set $|\sigma^2| \geq M(\gamma^2 + |\xi|^2)^{1/2}$.

The last assertion of Theorem (4.2.1) follows from a theorem of Paley–Wiener type.

Comparison of (4.2.6) and (4.2.7) shows that the transition $f \mapsto u$ 'loses one unit of smoothness in the tangential direction'; moreover, the norm of this operator can be estimated in terms of $c\,|\gamma|^{-1}$ and is small for large $|\gamma|$. From Theorem (4.2.1) we therefore obtain the following statement.

Theorem 4.2.2. *There exists a number $\gamma_0 > 0$ such that for $|\gamma| \geq \gamma_0 > 0$ and for any $f \in \mathcal{H}^{s-S,\tau}(\mathbf{R}^{n+1}, \gamma)$ $(s, \tau \in \mathbf{R})$ problem (4.1.5) has one and only one solution $u \in \mathcal{H}^{T+s,\tau-1}(\mathbf{R}^{n+1}, \gamma)$. There exists a constant $c > 0$ independent of f, u, and γ $(|\gamma| \geq \gamma_0)$, such that inequality (4.2.7) holds. If $\operatorname{supp} f \subset \overline{\Omega}$, then also $\operatorname{supp} u \subset \overline{\Omega}$.*

4.2.3. We associate the Cauchy problem (4.1.4) with the operator $A = A_{s,\tau}$ $(s, \tau \in \mathbf{R})$ which is the closure of the mapping

$$u \mapsto \left(lu, \{D^{k-1}u_j \ (j : t_j \geq 1,\ k = 1, \dots, t_j)\}\right) \quad \left(u \in \left(C_0^\infty\left(\overline{\Omega}\right)\right)^N\right)$$

acting continuously from the entire of

$$\widetilde{\mathcal{H}}^{T+s,\tau,(T)}(\Omega, \gamma) := \prod_{1 \leq j \leq N} \widetilde{\mathcal{H}}^{t_j+s,\tau,(t_j)}(\Omega, \gamma)$$

into the space

$$\begin{aligned} \widetilde{K}^{s,\tau} &:= \prod_{1 \leq j \leq N} \widetilde{\mathcal{H}}^{s-s_j,\tau,(-s_j)}(\Omega, \gamma) \times \prod_{j : t_j \geq 1} \prod_{1 \leq k \leq t_j} H^{t_j+s+\tau-k+1/2}(\Omega, \gamma) \\ &= \widetilde{\mathcal{H}}^{s-S,\tau,(-S)}(\Omega, \gamma) \times B^{s,\tau}(\Omega, \gamma) \end{aligned}$$

(see (4.2.3) and (4.2.4)).

We set in (4.1.4) that

$$f = (f_1, \dots, f_N) \in \widetilde{\mathcal{H}}^{s-S,\tau,(-S)}(\Omega, \gamma),$$

$$f_k = (f_{k0}, \dots f_{k,-s_k}) \in \widetilde{\mathcal{H}}^{s-s_k,\tau,(-s_k)}(\Omega, \gamma),$$

$$U = \left\{u_{jk} \ (j : t_j \geq 1,\ k = 1, \dots, t_j)\right\} \in B^{s,\tau}(\Omega, \gamma) \quad (s, \tau \in \mathbf{R}).$$

We call an element

$$u = (u_1, \dots, u_N) \in \widetilde{\mathcal{H}}^{T+s,\tau,(T)}(\Omega, \gamma)$$

satisfying the problem $Au = (f, U)$ a (*generalized*) *solution* of Cauchy problem (4.1.4). A solution of the Cauchy problem for equation (4.1.5) is defined in an entirely similar way.

For solvability of Cauchy problem (4.1.4) it is necessary that f and U be connected by certain compatibility conditions. Indeed, if l_k is the row with index k of the matrix l and $u \in \widetilde{\mathcal{H}}^{T+s,\tau,(T)}(\Omega, \gamma)$ is a solution of problem (4.1.4), then, according to (4.2.5),

$$D_t^{m-1} l_k u\big|_{t=0} = D_t^{m-1} f_k\big|_{t=0} = f_{km} \in H^{s+\tau-s_k-m+1/2}(\partial\Omega, \gamma) \tag{4.2.8}$$

$$\left(k : -s_k \geq 1,\ m = 1, \dots, -s_k\right).$$

The left-hand side of (4.2.8) is completely determined by the vector U of initial data of the Cauchy problem, and the right-hand side is determined by the vector f. Conditions (4.2.8) therefore determine necessary conditions for the solvability of the Cauchy problem.

Theorem 4.2.3. *Let $s, \tau, \gamma \in \mathbf{R}$, $|\gamma| \geq \gamma_0 > 0$, $F = (f, U) \in K^{s,\tau}$, and the certain compatibility conditions are satisfied. Then the Cauchy problem (4.1.4) has one and only one solution $u \in \tilde{H}^{T+s,\tau-1,(T)}(\Omega, \gamma)$. There exists a constant $c > 0$ independent of F, u, and γ ($|\gamma| \geq \gamma_0 > 0$) such that*

$$\|u\|_{\tilde{H}^{T+s,\tau-1,(T)}(\Omega,\gamma)} \leq |\gamma|^{-1}\|F\|_{K^{s,\tau}}. \tag{4.2.9}$$

For the proof, the solution u_0 is extended by zero into the entire space, and Cauchy problem (4.1.4) is reduced to a problem in $\mathbf{R}^{n+1}$ (cf. [R7], [R8], and also [Vla] and [Zhi]). Then we use Theorem 4.2.1.

Let us note that in Theorem 4.2.3 the transition $f \mapsto u$ 'loses one unit of smoothness in the tangential direction'; the norm of this operator does not exceed $c|\gamma|^{-1}$ and is small for large $|\gamma|$. We therefore obtain the following statement.

Theorem 4.2.4. *Let $s, \tau \in \mathbf{R}$. There exists a number $\gamma_0 > 0$ such that for $|\gamma| \geq \gamma_0$ and for each $F = (f, U) \in K^{s,\tau}$ satisfying the compatibility conditions the Cauchy problem for equation (4.1.5) has one and only one solution $u \in \tilde{\mathcal{H}}^{T+s,\tau-1,(T)}(\Omega, \gamma)$. Moreover, estimate (4.2.9) holds.*

4.2.4. It is clear that the compatibility conditions do not arise if $s_1 = \cdots = s_N = 0$. In the general case they cannot be given up. Let us alter the formulation somewhat. In this section we suppose that in (4.1.4)

$$f = f_0 = (f_{10}, \ldots, f_{N0}) \in \prod_{1 \leq k \leq N} \mathcal{H}^{s-s_k,\tau}(\Omega, \gamma) = \mathcal{H}^{s-S,\tau}(\Omega, \gamma),$$

and we seek a solution in $\tilde{H}^{T+s,\tau,(T)}(\Omega, \gamma)$ as before. The possibility of this approach follows easily from inequalities (4.2.2) and (4.2.4). We now determine the elements $f_{km} \in H^{s+\tau-s_k-m+1/2}(\Omega, \gamma)$ from (4.2.8). The conditions

$$(f_{k0}, \ldots, f_{k,-s_k}) \in \tilde{\mathcal{H}}^{s-s_k,\tau,(-s_k)}(\Omega, \gamma) \quad (\forall k: -s_k \geq 1) \tag{4.2.10}$$

are then necessary for the solvability in $\tilde{\mathcal{H}}^{T+s,\tau,(T)}(\Omega, \gamma)$ of Cauchy problem (4.1.4) with $f = f_0 \in \mathcal{H}^{s-S,\tau}(\Omega, \gamma)$. If $|\gamma| \geq \gamma_0 > 0$, then by Theorem 4.2.3 conditions (4.2.10) are also sufficient for the solvability of this problem. Thus, in the formulation considered here conditions (4.2.10) replace compatibility conditions (4.2.8). However, if $s \in \mathbf{R}$ is such that

$$s < s_j + 1/2 \qquad (j = 1, \ldots, N), \tag{4.2.11}$$

then (4.2.10) is satisfied. Therefore, in the class of 'not very smooth functions' the Cauchy problem is always solvable. It is not necessary to require that the compatibility conditions be satisfied. However, the smoothness of such a solution does not always grow with the smoothness of the right hand sides. As soon as (4.2.11) is not satisfied the compatibility conditions are required. An entirely similar assertion holds for the Cauchy problem for equation (4.1.5).

4.2.5. Theorems 4.2.3 and 4.2.4 can be applied to study the Cauchy problem with power singularities in the right hand sides, to construct and to study the Green matrix of the problems in question, and to investigate the Cauchy problem for degenerate hyperbolic systems.

CHAPTER 5

BOUNDARY VALUE AND MIXED PROBLEMS FOR GENERAL HYPERBOLIC SYSTEMS

Boundary value and mixed problems for hyperbolic equations have been studied in classes of sufficiently smooth functions by Sakamoto [Sak], Kreiss [Kre], Agranovich [Agr], Chazarin and Piriou [ChP], Volevich and Gindikin [VoG], and others (see the survey [VIv]). In this chapter we study boundary value and mixed problems for systems strictly hyperbolic in the Leray–Volevich sense in complete scale of spaces of Sobolev type depending on parameters s, $\tau \in \mathbf{R}$; s characterizes the smoothness of the solutions in all the variables, and τ characterizes the additional smoothness with respect to the tangential variables. The smaller s and τ are, the more generalized the solution is; for suffuciently large s and τ the solution is the ordinary classical solution of the problem. The result obtained enable us, in particular, to investigate hyperbolic problems with arbitrary power singularities on the right hand sides, to construct and investigate the Green's matrices of the problems under study, and to investigate the class of degenerate hyperbolic problems for systems of equations.

Earlier Lions, Magenes, Berezanskii, S. Krein, the author, and others studied elliptic problems in scales of spaces of Sobolev type depending on a parameter $s \in \mathbf{R}$ (see [LiM], [Ber], [R6] and the references given there). So called theorems on complete collection of isomorphisms were established in these publications, theorems which have found many applications. For parabolic problems similar theorems were obtained by Zhitarashu [Zh1], and by the author for the Cauchy problem for hyperbolic equations, as well as for boundary value and mixed problems for homogeneous hyperbolic equations ([R7], [R8], [R9]).

5.1. General Strictly Hyperbolic Systems. Statement of Problems

Let $(t,x) = (t,x_1,\ldots,x_n) \in \mathbf{R}^{n+1}$, let $(\sigma,\xi) = (\sigma,\xi_1,\ldots,\xi_n)$ be the dual variables, and let

$$l = l(D_t,D_x) = \big(l_{kj}(D_t,D_x)\big)_{k,j=1,\ldots,N} \tag{5.1.1}$$

be a matrix differential expression, with

$$l_{kj}(D_t, D_x) = \begin{cases} \sum\limits_{p+|\alpha|=s_k+t_j} l^{(k,j)}_{p\alpha} D^p_t D^\alpha_x & \text{for } k,j : s_k + t_j \geq 0, \\ 0 & \text{for } k,j : s_k + t_j < 0. \end{cases} \tag{5.1.2}$$

Here $D_t = i\partial/\partial t$, $D^\alpha_x = D_1^{\alpha_1}, \ldots, D_n^{\alpha_n}$, $D_j = i\partial/\partial x_j$, $|\alpha| = \alpha_1 + \ldots + \alpha_n$, $s_1, \ldots, s_N$, $t_1, \ldots, t_N$ are integers such that

$$t_1 \geq \ldots \geq t_N \geq 0 = s_1 \geq \ldots \geq s_N,$$

and $l^{(k,j)}_{p\alpha}$ are complex numbers. Let

$$L(\sigma, \xi) = \det l(\sigma, \xi) = \sum_{j+|\alpha|=r} a_{j\alpha} \sigma^j \xi^\alpha. \tag{5.1.3}$$

Definition 5.1.1. *The expression (5.1.1) is said to be strictly hyperbolic (in the Leray-Volevich srense) if the polynomial (5.1.3) is strictly hyperbolic: the coefficient $a_{r,0,\ldots,0}$ of σ^r in (1.3) is nonzero, and for each $\xi \in \mathbf{R}\backslash\{0\}$ the roots of the equation $L(\sigma, \xi) = 0$ with respect to σ are real and distinct.*

It will be assumed everywhere below that the expression (5.1.1) is strictly hyperbolic. It will also be assumed that the hyperplane $x_n = 0$ is noncharasteristic with respect to expression (5.1.1), that is, the coefficient $a_{0,\ldots,0,r}$ of ξ_n^r in the polynomial (5.1.3) is nonzero:

$$a_{0,\ldots,0,r} \neq 0. \tag{5.1.4}$$

It follows from the strict hyperbolicity of matrix (5.1.1) that for each $\gamma > 0$ the equation

$$L(\sigma + i\gamma, \xi', \xi_n) = 0 \tag{5.1.5}$$

does not have real roots with respect to ξ_n. Let

$$\zeta_1(\sigma + i\gamma, \xi'), \ldots, \zeta_r(\sigma + i\gamma, \xi') \tag{5.1.6}$$

$$((\sigma + i\gamma, \xi') \neq (0,0),\ \gamma \geq 0)$$

be the roots of (5.1.5). We assume, for definiteness, that, for $\gamma > 0$, the first m roots in (5.1.6) have negative imaginary parts, and the rest of the roots have positive imaginary parts. Of course, m does not depend on $(\sigma + i\gamma, \xi')$. We set

$$\begin{aligned} L(\sigma + i\gamma, \xi', \xi_n) &= L_-(\sigma + i\gamma, \xi', \xi_n) L_+(\sigma + i\gamma, \xi'), \\ L_-(\sigma + i\gamma, \xi', \xi_n) &= \prod_{1 \leq j \leq m} (\xi_n - \zeta_j(\sigma + i\gamma, \xi')). \end{aligned} \tag{5.1.7}$$

In this chapter we investigate the solvability in R^{n+1} of the problem

$$l(D_t, D_x)u = f \qquad (u = (u_1, \ldots, u_N), \quad f = (f_1, \ldots, f_n)), \tag{5.1.8}$$

and also of the problem

$$(l(D_t, D_x) + l'(t, x, D_t, D_x))u = f, \tag{5.1.9}$$

obtained by perturbing the system (5.1.8) by lower terms having having infinitely smooth coefficients with all derivatives bounded. In the half-space

$$G = \{(t, x) = (t, x', x_n) \in \mathbf{R}^{n+1} : x_n > 0\}$$

we study the boundary value problem

$$\begin{gathered} l(D_t, D_x)u = f, \\ (bu)_h = \sum_{j=1}^{N} b_{hj}(D_t, D_x)u\big|_{x_n=0} = \varphi_h \quad (h = 1, \ldots, m), \end{gathered} \tag{5.1.10}$$

where $b = (b_{hj}(D_t, D_x))_{\substack{h=1,\ldots,m \\ j=1,\ldots,N}}$,

$$b_{hj}(D_t, D_x) = \begin{cases} \sum\limits_{k+|\mu|=\sigma_h+t_j} b_{hjk\mu} D_t^k D_x^\mu & \text{for } h, j : \sigma_h + t_j \geq 0 \\ 0 & \text{for } h, j : \sigma_h + t_j < 0, \end{cases} \tag{5.1.11}$$

and $\sigma_1, \ldots, \sigma_m$ are given integers.

Definition 5.1.2. *The problem (5.1.10) is said to be hyperbolic if system (5.1.8) is strictly hyperbolic, condition (5.1.4) holds, the number of the boundary conditions is equal to the number m of roots (5.1.6) with negative imaginary parts, and the Lopatinskii condition L_1 holds:*

L_1: *For each $(\sigma + i\gamma, \xi') \neq (0, 0)$, $\gamma \geq 0$, the rows of the matrix*

$$L(\sigma + i\gamma, \xi', \xi_n)\, b(\sigma + i\gamma, \xi', \xi_n)\, l^{-1}(\sigma + i\gamma, \xi', \xi_n), \tag{5.1.12}$$

whose elements are regarded as polynomials in ξ_n, are linearly independent modulo $L_-(\xi_n) = L_-(\sigma + i\gamma, \xi', \xi_n)$.

In this chapter we study both the hyperbolic problem (5.1.10) and the 'perturbed' problem

$$\begin{gathered} (l(D_t, D_x) + l'(t, x, D_t, D_x))u = f, \\ (b(D_t, D_x) + b'(t, x, D_t, D_x))u\big|_{x_n=0} = \varphi. \end{gathered} \tag{5.1.13}$$

Here

$$b' = (b'_{hj}), \qquad \operatorname{ord} b'_{hj}(t,x,D_t,D_x) \le \sigma_h + t_j - 1 \quad (\sigma_h + t_j \ge 1),$$

and $b'_{hj} = 0$ if $\sigma_h + t_j < 1$;

$$l' = (l'_{jk}(t,x,D_t,D_x)), \qquad \operatorname{ord} l'_{jk} \le s_j + t_k - 1 \quad (s_j + t + k \ge 1),$$

and $l'_{jk} = 0$ if $s_j + t_k < 1$. Problem (5.1.13) is obtained by perturbing problem (5.1.10) by lower terms with infinitely smooth coefficients with all derivatives bounded.

We also mention here that we shall show that if the right hand sides in (5.1.10) and (5.1.13) vanish for $t \le 0$, then the solutions of these problems also vanish for $t \le 0$. Therefore, the solvability theorem for problems (5.1.10) and (5.1.13) yields solvability theorems for the corresponding mixed problems on $G_+ = \{(t,x) \in G : t > 0\}$ with homogeneous (zero) initial data at $t = 0$.

The solvability of all these problems is established in the complete scale of spaces of Sobolev type. For the precise formulation of the problems presented we must introduce some functional spaces.

5.2. Functional Spaces

5.2.1. Let s, τ, $\gamma \in \mathbf{R}$. We denote by $H^{s,\tau}(\mathbf{R}^{n+1},\gamma)$ the space of distributions f with the norm

$$\|f,\mathbf{R}^{n+1},\gamma\|_{s,\tau} = \left(\int (1+\gamma^2+\sigma^2+|\xi|^2)^s (1+\gamma^2+\sigma^2+|\xi'|^2)^\tau |\tilde{f}(\sigma,\xi)|^2 \, d\sigma d\xi\right)^{1/2}. \tag{5.2.1}$$

Here $\tilde{f}$ is the Fourier trensform of the element f,

$$\tilde{f}(\sigma,\xi) = \int f(t,x) \exp i(t\sigma + x\xi)\, dt dx$$

if f is a sufficiently regular function.

It is clear that, for each fixed $\gamma \in \mathbf{R}$, the norm $\|f, R^{n+1},\gamma\|_{s,\tau}$ is equivalent to the norm

$$\|f,\mathbf{R}^{n+1},0\|_{s,\tau} = \|f,\mathbf{R}^{n+1}\|_{s,\tau},$$

and hence the set $H^{s,\tau}(\mathbf{R}^{n+1},\gamma)$ does not depend on γ. However, in this chapter it is convenient to consider only the norms equivalent to (5.2.1)in which the constants in the corresponding two-sided estimates can be chosen to be independent of γ. We note also that

$$\|f, \mathbf{R}^{n+1}, \gamma\|_{s,\tau} \geq |\gamma|^{s_1+\tau_1} \|f, \mathbf{R}^{n+1}, \gamma\|_{s-s_1,\tau-\tau_1}$$
$$(s_1 \geq 0, \tau_1 \geq 0, f \in H^{s,\tau}(\mathbf{R}^{n+1})). \tag{5.2.2}$$

Norms (5.2.1) will often be considered for $\gamma \geq \gamma_0 > 0$. It is clear that then these norms are equivalent to the 'homogeneous' norms obtained by replacing $1+\gamma^2$ by γ^2 in (5.2.1). These homogeneous norms also will be denoted by $\|f, \mathbf{R}^{n+1}, \gamma\|_{s,\tau}$ $(\gamma \geq \gamma_0 > 0)$.

5.2.2. Let $G = \{(t, x', x_n) \in \mathbf{R}^n : x_n > 0\}$, let $\partial G = \{(t,x) : x_n = 0\}$ be the boundary of G, and let s, τ, $\gamma \in \mathbf{R}$, $s \geq 0$. We denote by $H^{s,\tau}(G,\gamma)$ the set of restrictions to G of the functions in $H^{s,\tau}(\mathbf{R}^n, \gamma)$, with the quotient space norm

$$\|w, G, \gamma\|_{s,\tau} = \inf \|u, \mathbf{R}^n, \gamma\|_{s,\tau} \quad (s, \tau, \gamma \in \mathbf{R}, s \geq 0), \tag{5.2.3}$$

where the inf is over all $u \in H^{s,\tau}(\mathbf{R}^n, \gamma)$ equal to w on G. Let $H^{-s,-\tau}(G,\gamma)$ $(s, \tau \in \mathbf{R}, s \geq 0)$ denote the space dual to $H^{s,\tau}(G,\gamma)$ with respect to the extension $(\cdot,\cdot) = (\cdot,\cdot)_G$ of the scalar product in $L_2(G)$, and let

$$\|u, G, \gamma\|_{-s,-\tau} = \sup_{v \in H^{s,\tau}(G,\gamma)} \frac{|(u,v)|}{\|v, G, \gamma\|_{s,\tau}} \tag{5.2.4}$$

be the norm in $H^{-s,-\tau}(G,\gamma)$ $(s \geq 0)$. It is clear that $H^{-s,-\tau}(G,\gamma)$ $(s \geq 0)$ is isometrically equivalent to the subspace $H_{\overline{G}}^{-s,-\tau}(R^{n+1},\gamma)$ of $H^{-s,-\tau}(R^{n+1},\gamma)$ consisting of the elements with support in $\overline{G}$. It follows from (5.2.2) that

$$\|u, G, \gamma\|_{s,\tau} \geq |\gamma|^{s_1+\tau_1} \|u, G, \gamma\|_{s-s_1,\tau-\tau_1}$$
$$(s, \tau, s_1, \tau_1 \in \mathbf{R}, \quad s_1 \geq 0, \tau_1 \geq 0; \quad u \in H^{s,\tau}(G,\gamma)). \tag{5.2.5}$$

5.2.3. An element $w \in H^{s,\tau}(R^{n+1})$ belongs to

$$H_{\overline{G}}^{s,\tau}(R^{n+1}) = \{w \in H^{s,\tau}(R^{n+1}) : \operatorname{supp} w \subset \overline{G}\}$$

if and only if the Fourier transform $\widetilde{w}(\sigma, \xi', \xi_n)$ admits an analytic continuation with respect to ξ_n into the half-plane $\xi_n + i\eta$, $\eta > 0$, with an estimate

$$\int (1+\sigma^2+|\xi'|^2+\xi_n^2+\eta^2)^s (1+\sigma^2+|\xi'|^2)^\tau |\widetilde{w}(\sigma,\xi',\xi_n+i\eta)|^2 \, d\sigma d\xi \leq c,$$

where $c > 0$ does not depend on η (a theorem of Paley–Wiener type [Esk]).

Similarly, $w \in H^{s,\tau}(\mathbf{R}^{n+1})$ belongs to

$$H^{s,\tau}_{\overline{\Omega}}(\mathbf{R}^{n+1}) = \left\{w \in H^{s,\tau}(\mathbf{R}^{n+1}) : \operatorname{supp} w \subset \overline{\Omega} = \{(t,x) \in \mathbf{R}^{n+1} : t \geq 0\}\right\}$$

if and only if the Fourier transform $\widetilde{w}(\sigma, \xi)$ admits an analytic continuation with respect to σ into the half-plane $\sigma + i\eta$, $\eta > 0$ with an estimate

$$\int (1+\sigma^2+\eta^2+|\xi|^2)^s (1+\sigma^2+|\xi'|^2)^\tau |\widetilde{w}(\sigma+i\eta,\xi)|^2 \, d\sigma d\xi \leq c,$$

where $c > 0$ does not depend on η.

5.2.4. Let $\Theta^+(x)$ be the characteristic function of the half-space G, i.e.,

$$\Theta^+(x) = \begin{cases} 1 & \text{if } x_n > 0 \\ 0 & \text{if } x_n < 0. \end{cases}$$

The closure Θ^+ of the mapping $u \mapsto \Theta^+ u$ ($u \in C_0^\infty(\mathbf{R}^n)$) acts continuously in the pair of spaces

$$H^{s,\tau}(\mathbf{R}^{n+1}, \gamma) \to H^{s,\tau}_{\overline{G}}(\mathbf{R}^{n+1}, \gamma) \quad (s, \tau \in \mathbf{R},\ |s| < 1/2)$$

(see [Esk], [RoS1]). Let us extend the action of this operator. If

$$\begin{gathered} w = w_1 + w_2, \\ w_1 \in H^{s,\tau}(R^{n+1}, \gamma)\ (|s| < 1/2), \quad w_2 \in H^{s,\tau}_{CG}(R^{n+1}, \gamma), \end{gathered} \tag{5.2.6}$$

$$H^{s,\tau}_{CG}(R^{n+1}, \gamma) = \{w \in H^{s_1,\tau}(R^{n+1}, \gamma) : \operatorname{supp} w \subset \mathbf{R}^{n+1} \backslash G\}, \quad (s_1 \leq s),$$

then we set

$$\Theta^+ w = \Theta^+ w_1 \in H^{s,\tau}_{\overline{G}}(\mathbf{R}^{n+1}, \gamma). \tag{5.2.7}$$

5.2.5. For any s, $\gamma \in \mathbf{R}$ let

$$\Lambda^s_\pm = \Lambda^s_\pm(\gamma) = F^{-1}\langle\xi\rangle^s_\pm F \quad \langle\xi\rangle_\pm = \xi_n \pm i\sqrt{1+\gamma^2+\sigma^2+|\xi'|^2}, \tag{5.2.8}$$

where F and F^{-1} are the direct and inverse Fourier transformations, and $\langle\xi\rangle^s_\pm = \exp s \ln\langle\xi\rangle_\pm$, the branch of the logarithm being chosen so that $\operatorname{Im}\ln\langle\xi\rangle_\pm \to 0$ as $\xi_n \to +\infty$. Since $|\langle\xi\rangle_\pm|^2 = 1+\gamma^2+\sigma^2+|\xi|^2$, it follows that $\Lambda^s_\pm$ realizes an isometry between $H^{s_1,\tau}(R^{n+1},\gamma)$ and $H^{s_1-s,\tau}(\mathbf{R}^n,\gamma)$. But since the function $\langle\xi\rangle^s_+$ admits an analytic continuation with respect to ξ_n into the upper half-plane $\xi_n + i\eta$, $\eta > 0$, it follows from 5.2.3 that Λ^s_+ maps $H^{s_1,\tau}_{\overline{G}}(\mathbf{R}^{n+1},\gamma)$ isometrically onto $H^{s_1-s,\tau}_{\overline{G}}(R^{n+1},\gamma)$; similarly, Λ^s_- maps $H^{s_1,\tau}_{CG}(\mathbf{R}^{n+1},\gamma)$ isometrically onto

$$H^{s_1-s,\tau}_{CG}(\mathbf{R}^{n+1},\gamma) \qquad (s, s_1, \tau \in \mathbf{R},\ CG = \mathbf{R}^{n+1}\backslash G).$$

If now $w_1 \in H^{s_1,\tau}(\mathbf{R}^{n+1},\gamma)$, $w_2 \in H^{s_2,\tau}(\mathbf{R}^{n+1},\gamma)$ $(s_1 < s_2)$, and $w_2 - w_1 \in H^{s_2,\tau}_{CG}(\mathbf{R}^{n+1},\gamma)$, then

$$\Lambda_-^{s_1}(w_1 - w_2) \in H^{s_2-s_1,\tau}_{CG}(\mathbf{R}^{n+1},\gamma),$$

and hence

$$\Theta^+\Lambda_-^{s_1}w_1 = \Theta^+\Lambda_-^{s_1}w_2$$

by (5.2.7). From this it follows that if w belongs to $H^{s,\tau}(G,\gamma)$ $(s \geq 0)$, and $w_1 \in H^{s_1,\tau}(\mathbf{R}^{n+1},\gamma)$ $(s_1 \leq s)$ is an arbitrary extension of w, then $\Theta^+\Lambda_-^s w_1$ does not depend on the way of the extension.

Consequently, the operator $w \mapsto \Theta^+\Lambda_-^s w_1$ realizes a one-to-one cotinuous (hence also inversely continuous) mapping acting in the pair of spaces

$$H^{s,\tau}(\mathbf{R}^{n+1},\gamma) \to H^{0,\tau}_{\overline{G}}(\mathbf{R}^{n+1},\gamma).$$

Therefore, the norm $\|w, G, \gamma\|_{s,\tau}$ is equivalent to the norm

$$\|\Theta^+\Lambda_-^s w_+, \mathbf{R}^{n+1}, \gamma\|_{0,\tau}.$$

The following norms are also equivalent to each other:

$$\|w, G, \gamma\|_{s,\tau} \simeq \|\Theta^+\Lambda_-^{s-\eta} w_+, \mathbf{R}^{n+1}, \gamma\|_{\eta,\tau}$$
$$(\gamma, s, \tau \in \mathbf{R}, s \geq 0, |\eta| < 1/2) \qquad (5.2.9)$$

(here w_+ is the extension of w by zero to $\mathbf{R}^{n+1}\backslash G$).

We remark that, if $|\gamma| \geq \gamma_0 > 0$, then it is convenient to replace the operator $\Lambda_\pm^s(\gamma)$ by the homogeneous operator $F^{-1}(\xi_n \pm i\sqrt{\gamma^2 + \sigma^2 + |\xi'|^2}F)$, which will also be denoted by $\Lambda_\pm^s(\gamma)$. Here the corresponding norms are replaced by equivalent norms, and the constants in the two-sided estimates will not depend on γ $(|\gamma| \geq \gamma_0 > 0)$.

5.2.6.

The space $H^s(\partial G,\gamma)$ $(s,\gamma \in \mathbf{R})$ is the space of distributions g such that

$$\|g, \partial G, \gamma\|_s = \left(\int_{\partial G} |\widehat{g}(\sigma,\xi')|^2(1+\gamma^2+\sigma^2+|\xi'|^2)^s\, d\sigma d\xi'\right)^{1/2} < \infty. \quad (5.2.10)$$

Here $\widehat{g}(\sigma,\xi') = F'_{(t,x')\mapsto(\sigma,\xi')}g$ is the Fourier transform of the element g. If $s, \tau \in \mathbf{R}$ and $s > 1/2$, then the closure of the mapping

$$u(t,x',x_n) \mapsto u(t,x',0) \quad (u \in C_0^\infty(\overline{G}))$$

acts continuously in the pair of spaces

$$H^{s,\tau}(G,\gamma) \to H^{s+t-1/2}(\partial G,\gamma)$$

(here and below, $C_0^\infty(\overline{G})$ is the set of restrictions to $\overline{G}$ of functions in $C_0^\infty(\mathbf{R}^{n+1})$).

5.2.7.
We fix a natural number r and assume that $s, \tau, \gamma \in \mathbf{R}$, $s \neq k + 1/2$ $(k = 0, \ldots, r-1)$. Let $\widetilde{H}^{s,\tau,(r)}(G,\gamma)$ benote the completion of $C_0^\infty(\overline{G})$ in the norm

$$|||u, G, \gamma|||_{s,\tau,(r)} = \Big(\|u, G, \gamma\|_{s,\tau}^2 + \sum_{j=1}^{r} \|D_n^{j-1} u, \partial G, \gamma\|_{s-j+\tau-1/2}^2 \Big)^{1/2}. \quad (5.2.11)$$

A similar space was introduced by the author in [R10] and studied in [R11] (see also [R1], [Ber, Ch. 3, §6.8], and [RS1]). For $s = k+1/2$ $(k = 0, \ldots, r-1)$ the space $\widetilde{H}^{s,\tau,(r)}(G,\gamma)$ and norm $|||u, G, \gamma|||_{s,\tau,(r)}$ are defined by interpolation. It follows from (5.2.11) that the closure S of the mapping

$$u \mapsto (u|_{\overline{G}}, u|_{\partial G}, \ldots, D_n^{r-1} u|_{\partial G}) \quad (u \in C_0^\infty(\overline{G}))$$

establishes an isometry between $\widetilde{H}^{s,\tau,(r)}(G,\gamma)$ and a subspace of the direct product

$$H^{s,\tau}(G,\gamma) \times \prod_{j=1}^{r} H^{s+\tau-j+1/2}(\partial G, \gamma).$$

In addition, if $s < 1/2$ then $S\widetilde{H}^{s,\tau,(r)}(G,\gamma)$ coincides with the whole of this product space, and if $s > 1/2$, $s \neq k + 1/2$ $(k = 0, \ldots, r-1)$, then

$$\begin{aligned} &S\widetilde{H}^{s,\tau,(r)}(G,\gamma) \\ &= \Big\{ (u_0, \ldots, u_r) : u_0 \in H^{s,\tau,(r)}(G,\gamma); \\ &\qquad u_j = D_n^{j-1} u_0|_{\partial G} \in H^{s+\tau-j+1/2}(\partial G, \gamma) \text{ if } s - j + 1/2 > 0; \\ &\qquad u_j \text{ is an arbitrary element in } H^{s+\tau-j+1/2}(\partial G, \gamma) \text{ if } s - j + 1/2 < 0 \Big\} \end{aligned}$$

(sf. [R11]). Below we shall identify an element $u \in \widetilde{H}^{s,\tau,(r)}(G,\gamma)$ with the element $Su = (u_0, \ldots, u_r)$. We write $u = (u_0, \ldots, u_r) \in \widetilde{H}^{s,\tau,(r)}(G,\gamma)$ for each $u \in \widetilde{H}^{s,\tau,(r)}(G,\gamma)$. Also, let $u|_{\overline{G}} := u_0$, $D_n^{j-1} u|_{\partial G} := u_j$ $(j = 1, \ldots, r)$. Finally, for $r = 0$ we set

$$\widetilde{H}^{s,\tau,(r)}(G,\gamma) := H^{s,\tau}(G,\gamma), \quad |||u, G, \gamma|||_{s,\tau,(r)} := \|u, G, \gamma\|_{s,\tau}.$$

Let

$$M = M(t, x, \gamma, D_t, D_x) = \sum_{j=0}^{m} M_j(t, x, \gamma, D_t, D_{x'}) D_n^j$$

$$= \sum_{j+|\alpha|+k\leq m} \gamma^k a_{j\alpha k}(t,x) D_x^\alpha D_t^j. \tag{5.2.12}$$

be the differential expression of order m that is polynomially dependent on the real parameter $\gamma \in \mathbf{R}$, $a_{j\alpha k} \in C^\infty\,(\mathbf{R}^{n+1})$, with all the derivatives of the coefficients bounded. Then, for any $s, \tau \in \mathbf{R}$, the estimate

$$\|Mu, G, \gamma\|_{s-m,\tau} \leq c_1 |||u, G, \gamma|||_{s,\tau,(r)} \quad (u \in C_0^\infty(\overline{G})) \tag{5.2.13}$$

holds for $m \leq r$, and the estimate

$$\|Mu, \partial G, \gamma\|_{s-m+\tau-1/2} \leq c_2 |||u, G, \gamma|||_{s,\tau,(r)} \quad (u \in C_0^\infty(\overline{G})) \tag{5.2.14}$$

holds for $m \leq r-1$. Here the positive constants c_1 and c_2 do not depend on u and γ ([R1, Ch.9], [RS1], [R10], [R11], [Ber]). It follows directly from (5.2.13), (5.2.14), and the definition of the norm in (5.2.11) that for $m \leq r$

$$\|Mu, G, \gamma\|_{s-m,\tau,(r-m)} \leq c |||u, G, \gamma|||_{s,\tau,(r)} \quad (u \in C_0^\infty(\overline{G})),$$

where $c > 0$ is independent of u and γ.

From (5.2.13)–(5.2.7) we obtain:
a) the closure of the mapping $u \mapsto Mu|_{\overline{G}}$ $(u \in C_0^\infty(\overline{G}))$ for $m \leq r$ acts continuously from the whole of $\widetilde{H}^{s,\tau,(r)}(G,\gamma)$ into $H^{s-m,\tau}(G,\gamma)$;
b) the closure of the mapping $u \mapsto Mu|_{\partial G}$ $(u \in C_0^\infty(\overline{G}))$ for $m \leq r-1$ acts continuously from the whole of $\widetilde{H}^{s,\tau,(r)}(G,\gamma)$ into $H^{s+\tau-m-1/2}(\partial G,\gamma)$;
c) the closure of the mapping $u \mapsto Mu$ $(u \in C_0^\infty(\overline{G}))$ for $m \leq r$ acts continuously from the whole of $\widetilde{H}^{s,\tau,(r)}(G,\gamma)$ into $\widetilde{H}^{s-m,\tau,(r-m)}(G,\gamma)$.

Action of differential expressions to elements in $\widetilde{H}^{s,\tau,(r)}(G,\gamma)$ can be understood also in another (weak) sense ([R1], [R11], [RS1]). Integrating by parts, we obtain

$$(Mw, v) = (w, M^+v) - i\sum_{j=1}^{m}\sum_{k=1}^{j}\langle D_n^{k-1}w, D_n^{j-k}M_j^+v\rangle \quad (w, v \in C_0^\infty(\overline{G})). \tag{5.2.15}$$

Here and below, $(\cdot,\cdot)$ and $\langle\cdot,\cdot\rangle$ stand for the scalar products (or their extensions) in $L_2(G)$ and $L_2(\partial G)$, respectively; and M^+ and M_j^+ are the expressions formally adjoint to M and M_j. Formula (5.2.15) can be written in the form

$$(Mw)_+ = Mw_+ - i\sum_{j=1}^{m}\sum_{k=1}^{j} M_j D_n^{j-k}(D_n^{k-1}w|_{x_n=0} \times \delta(x_n)), \tag{5.2.16}$$

where f_+ will stand for the extension of f by zero to $\mathbf{R}^{n+1}\backslash\overline{G}$.

In the set of expressions $M = \sum_{0\le j\le m} M_j D_n^j$ of form (5.2.12) we introduce the operator J:

$$JM = \begin{cases} 0, & \text{if } m = 0 \\ \sum\limits_{1\le j\le m} M_j D_n^{j-1}, & \text{if } m \ge 1. \end{cases}$$

Then (5.2.16) can be written in the form

$$(Mw)_+ = Mw_+ - i\sum_{k=1}^{m}(J^k M)(D_n^{k-1}w\big|_{t=0} \times \delta(x_n)). \qquad (5.2.17)$$

If $u = (u_0, \dots, u_r) \in \widetilde{H}^{s,\tau,(r)}(G,\gamma)$ then, by passing to the limit in (5.2.15), we can show that $Mu|_{\overline{G}} = f \in H^{s-m,\tau}(G,\gamma)$ if and only if

$$(f, v) = (u_0, M^+v) - i\sum_{j=1}^{m}\sum_{k=1}^{j}\langle u_k, D_n^{j-k}M_j^+ v\rangle \quad (v \in C_0^\infty(\overline{G})). \qquad (5.2.18)$$

From this relation it follows

$$(Mu)_+ = Mu_{0+} - i\sum_{k=1}^{m}(J^k M)(u_k \times \delta(x_n)). \qquad (5.2.19)$$

Formula (5.2.19) gives a rule for computing $Mu|_{\overline{G}}$ for each $u = (u_0, \dots, u_r) \in \widetilde{H}^{s,\tau,(r)}(G,\gamma)$. We recall (see 5.2.2) that if $s - m < 1/2$ then the space $H^{s-m,\tau}(G,\gamma)$ is isometrically equivalent to $H_{\overline{G}}^{s-m,\tau}(\mathbf{R}^n,\gamma)$, and, thus, in this case we can assume that $Mu|_{\overline{G}} = (Mu)_+$. Similarly, if $s < 1/2$ then we can assume that $u_0 = u_{0+}$.

By passing to the limit we also obtain that if $u = (u_0, \dots, u_r) \in \widetilde{H}^{s,\tau,(r)}(G,\gamma)$ then

$$Mu|_{\partial G} = \sum_{j=0}^{m} M_j(t, x', \gamma, D_t, D_{x'})u_{j+1} \in H^{s+\tau-m-1/2}(\partial G,\gamma). \qquad (5.2.20)$$

5.2.8. We next introduce the spaces

$$\mathcal{H}^{s,\tau}(\mathbf{R}^{n+1},\gamma), \quad \mathcal{H}^{s,\tau}(G,\gamma), \quad \mathcal{H}^{s}(\partial G,\gamma), \quad \widetilde{\mathcal{H}}^{s,\tau,(r)}(G,\gamma);$$

we denote the norms in these spaces by

$$|u, \mathbf{R}^{n+1}, \gamma|_{s,\tau}, \quad |u, G, \gamma|_{s,\tau}, \quad |u, G, \gamma|_{s,\tau}, \quad |u, \partial G, \gamma|_{s}, \quad |u, G, \gamma|_{s,\tau,(r)},$$

respectively.
Let

$$\begin{aligned} \mathcal{H}^{s,\tau}(\mathbf{R}^{n+1},\gamma) &:= \{u : e^{-\gamma t}u \in H^{s,\tau}(\mathbf{R}^{n+1},\gamma)\}, \\ |u,\mathbf{R}^{n+1}|_{s,\tau} &:= \|e^{-\gamma t}u,\mathbf{R}^{n+1},\gamma\|_{s,\tau}. \end{aligned} \tag{5.2.21}$$

Replacing $\mathbf{R}^{n+1}$ by G in (5.2.21), we obtain the definition of the space $\mathcal{H}^{s,\tau}(G,\gamma)$ and the norm in it.

Similarly, let

$$\begin{aligned} \mathcal{H}^{s}(\partial G,\gamma) &:= \{u : e^{-\gamma t}u \in H^{s}(\partial G,\gamma)\}, \\ |u,\partial G,\gamma|_{s} &:= \|e^{-\gamma t}u,\partial G,\gamma\|_{s}; \end{aligned} \tag{5.2.22}$$

$$\begin{aligned} \widetilde{\mathcal{H}}^{s,\tau,(r)}(\partial G,\gamma) &:= \{u : e^{-\gamma t}u \in \widetilde{H}^{s,\tau,(r)}(G,\gamma)\}, \\ |u,G,\gamma|_{s} &:= |||e^{-\gamma t}u,G,\gamma|||_{s,\tau,(r)}. \end{aligned} \tag{5.2.23}$$

We have the following result.

Lemma 5.2.1. *Let $M = M(t,x,D_t,D_x)$ be a linear differential expression of order m with infinitely smooth coefficients with all derivatives bounded. Then for any $s,\tau \in \mathbf{R}$ there exists a constant $c > 0$ independent of u and γ such that*

$$|Mu,\mathbf{R}^{n+1},\gamma|_{s-m,\tau} \le c|u,\mathbf{R}^{n+1},\gamma|_{s,\tau} \quad (u \in C_0^\infty(\mathbf{R}^{n+1})); \tag{5.2.24}$$

if $m \le r$ then

$$|Mu,G,\gamma|_{s-m,\tau} \le c|u,G,\gamma|_{s,\tau,(r)} \quad (u \in C_0^\infty(\overline{G})); \tag{5.2.25}$$

if $m \le r-1$ then

$$|Mu,\partial G,\gamma|_{s+\tau-m-1/2} \le c|u,G,\gamma|_{s,\tau,(r)} \quad (u \in C_0^\infty(\overline{G})); \tag{5.2.26}$$

if $m \le r$ then

$$|Mu,G,\gamma|_{s-m,\tau,(r-m)} \le c|u,G,\gamma|_{s,\tau,(r)} \quad (u \in C_0^\infty(\overline{G})). \tag{5.2.27}$$

Proof. Since

$$M(t,x,D_t,D_x)u = e^{\gamma t}M(t,x,D_t+i\gamma,D_x)(e^{-\gamma t}u), \tag{5.2.28}$$

we get from (5.2.22), (5.2.23) and (5.2.13) that

$$\begin{aligned} |Mu,G,\gamma|_{s-m,\tau} &= \|e^{-\gamma t}Mu,G,\gamma\|_{s-m,\tau} \\ &= \|M(t,x,D_t+i\gamma,D_x)(e^{-\gamma t}uG,\gamma\|_{s-m,\tau} \\ &\le c|||e^{-\gamma t}u,G,\gamma|||_{s,\tau,(r)} \\ &= c|u,G,\gamma|_{s,\tau,(r)}, \end{aligned}$$

and inequality (5.2.25) is established. The remaining estimates are proved similarly. □

5.2.9. The case of k-dimentional parameter γ
In 5.2.1–5.2.8 we studied spaces and norms depending on the real parameter γ. In all these considerations one can assume that

$$\gamma = (\gamma_1,\dots,\gamma_k) \in \mathbf{R}^k, \quad |\gamma|^2 = \gamma_1^2+\dots+\gamma_k^2,$$

where k is any natural number. The formulation of the assertions and their proofs are unchanged for this case. In the arguments in 5.2.8 it is then necessary to replace $e^{-\gamma t}$, for example, by $e^{-\gamma_k t}$.

5.3. Hyperbolic Systems in $\mathbf{R}^{n+1}$

5.3.1. It follows from Lemma 5.2.1 that for any $s,\tau,\in\mathbf{R}$ the closure l of the mapping $u\mapsto lu$ $(u\in(C_0^\infty(\overline{G}))^N)$ (see (5.1.1)) acts continuously in the pair of spaces

$$\mathcal{H}^{T+s,\tau}(\mathbf{R}^{n+1},\gamma)\to\mathcal{H}^{s-S,\tau}(\mathbf{R}^{n+1},\gamma), \tag{5.3.1}$$

$$\mathcal{H}^{T+s,\tau}(\mathbf{R}^{n+1},\gamma) := \prod_{j=1}^{N}\mathcal{H}^{t_j+s,\tau}(\mathbf{R}^{n+1},\gamma),$$

$$\mathcal{H}^{s-S,\tau}(\mathbf{R}^{n+1},\gamma) := \prod_{k=1}^{N}\mathcal{H}^{s-s_k,\tau}(\mathbf{R}^{n+1},\gamma),$$

and

$$|lu,\mathbf{R}^{n+1},\gamma|_{s-S,\tau}\le c|u,\mathbf{R}^{n+1},\gamma|_{T+s,\tau}, \tag{5.3.2}$$

where $|f,\mathbf{R}^{n+1},\gamma|_{s-S,\tau}$ and $|u,\mathbf{R}^{n+1},\gamma|_{T+s,\tau}$ are the respective norms in the spaces of images and pre-images of mapping (5.3.1);the constant $c>0$

does not depend on u and γ. The question arises of the invertibility of the operator l.

Theorem 5.3.1. *Suppose that expression (5.1.1) is strictly hyperbolic, $s, \tau, \gamma \in \mathbf{R}$, $\gamma \geq \gamma_0 > 0$, and (5.1.4) holds. Then for each $f \in \mathcal{H}^{s-S,\tau+1}(\mathbf{R}^{n+1}, \gamma)$ there exists one and only one element $u \in \mathcal{H}^{T+s,\tau}(\mathbf{R}^{n+1}, \gamma)$ such that $lu = f$. There exists a constant $c > 0$ independent of f, u, and γ ($\gamma \geq \gamma_0 > 0$) such that*

$$|u, \mathbf{R}^{n+1}, \gamma|_{T+s,\tau} \leq \frac{c}{\gamma} |f, \mathbf{R}^{n+1}, \gamma|_{s-S,\tau+1}. \tag{5.3.3}$$

If $\operatorname{supp} f \subset \overline{\Omega} = \{(t, x) \in \mathbf{R}^{n+1} : t \geq 0\}$ *then also* $\operatorname{supp} u \subset \overline{\Omega}$.

The proof is based on the following lemma.

Lemma 5.3.1. *Let expression (5.1.1) be strictly hyperbolic, and $L(\sigma, \xi) = \det l(\sigma, \xi)$. Then there exists a constant $c > 0$ independent of $(\sigma, \gamma, \xi) \in \mathbf{R}^{n+2}$ such that*

$$|L(\sigma + i\gamma, \xi)| \geq c|\gamma|(\sigma^2 + \gamma^2 + |\xi|^2)^{(r-1)/2} \quad ((\sigma, \gamma, \xi) \in \mathbf{R}^{n+2}). \tag{5.3.4}$$

There exists a constant $M > 0$ such that for $(\gamma, \xi) \in \mathbf{R}^{n+1} \backslash \{0\}$ the estimate

$$|L(\sigma + i\gamma, \xi)| \geq c(\sigma^2 + \gamma^2 + |\xi|^2)^{r/2}. \tag{5.3.5}$$

holds on the set $|\sigma| > M(\gamma^2 + |\xi|^2)^{1/2}$.

If (5.1.4) holds then the constant $M > 0$ can be chosen so that inequality (5.3.5) holds also on the set $|\xi_n| > M(\sigma^2 + \gamma^2 + |\xi'|^2)^{1/2}$.

Proof. Let $\sigma_1(\xi), \ldots, \sigma_r(\xi)$ be the roots σ of the equation $L(\sigma, \xi) = 0$. From the strict hyperbolicity of l it follows that for each $\xi \neq 0$ these roots are real and distinct. By the Theorem in implicit function, $\sigma_j(\xi)$ ($j = 1, \ldots, r$) is a continuous function of $\xi \in \mathbf{R}^n \backslash \{0\}$. It is also clear that the mapping $\xi \mapsto \sigma_j(\xi)$ is a homogeneous function of the first degree and that (see (5.1.3))

$$L(\sigma, \xi) = a_{r,0,\ldots,0}(\sigma - \sigma_1(\xi)) \cdots (\sigma - \sigma_r(\xi)). \tag{5.3.6}$$

Denote by T_j ($j = 1, \ldots, r$) the manifold along which the cone $\sigma = \sigma_j(\xi)$ intersects the sphere $\sigma^2 + |\xi|^2 = 1$ in $\mathbf{R}^{n+1}$. The manifolds $T_1, \ldots, T_r$ are at posistive distances from each other; since $a_{r,0,\ldots,0} \neq 0$, no manifold T_j contains the point $(1, 0, \ldots, 0)$. If (5.1.4) holds then no T_j contains the point $(0, \ldots, 0, 1)$ either.

It follows from (5.3.6) that

$$|L(\sigma+i\gamma,\xi| = |a_{r,0,\dots,0}| \prod_{1\le j\le r} |\sigma+i\gamma-\sigma_j(\xi)|.$$

We consider the function $\psi = |L(\sigma+i\gamma,\xi)|$ on the unit sphere S^{n+1}: $\sigma^2+\gamma^2+|\xi|^2=1$ in $\mathbf{R}^{n+2}$. The manifolds $T_1,\dots,T_r$ lie on the equator $\gamma=0$ of this sphere. Let Γ_j be a sufficiently small neighborhood in S^{n+1} of T_j ($j=1,\dots,r$) such that $\overline{\Gamma}_j\cap\overline{\Gamma}_k=\emptyset$ ($j\neq k$) and $\overline{\Gamma}_1\cup\dots\cup\overline{\Gamma}_r$ does not contain the point $(1,0,\dots,0)$ (and also does not contain the point $(0,\dots,0,1)$ if (5.1.4) holds). The domain $\Gamma_j\subset S^{n+1}$ determines a cone K_j in $\mathbf{R}^{n+2}$ with vertex at the origin of coordinates; let $\overline{K}_j$ be the closure of K_j ($j=1,\dots,r$).

On the compact set $S^{n+1}\backslash(\Gamma_1\cup\dots\cup\Gamma_r)$ the function $\psi=|L(\sigma+i\gamma,\xi)|$ is nonzero and continuous. Therefore, $\psi\ge c_1>0$ on this set. And since ψ is a homogeneous function, it follows that

$$|L(\sigma+i\gamma,\xi)|\ge c_1(\sigma^2+\gamma^2+|\xi|^2)^{r/2} \tag{5.3.7}$$

$$(\forall(\sigma,\gamma,\xi)\in\mathbf{R}^{n+2}\backslash(K_1\cup\dots\cup K_r)).$$

Similarly, on the set $\overline{\Gamma}_j$, we have

$$|a_{r,0,\dots,0}|\prod_{k\neq j}|\sigma+i\gamma-\sigma_k(\xi)|\ge c_2>0,\quad |\sigma+i\gamma-\sigma_j(\xi)|\ge|\gamma|.$$

From this, $|L(\sigma+i\gamma,\xi)|\ge c_2|\gamma|(\sigma^2+\gamma^2+|\xi|^2)^{(r-1)/2}$ in $\overline{K}_j$ ($j=1,\dots,r$), and estimate (5.3.4) is established. The last conclusions of the lemma follow from (5.3.7) and the fact that the point $(1,0,\dots,0)$, as well as the point $(0,0,\dots,1)$ if (5.1.4) holds, does not belong to the set $\bigcup_{1\le j\le r}\overline{\Gamma}_j$. This copletes the proof of the lemma. □

Now let us prove Theorem 5.3.1. It follows from (5.2.28) that the equation (5.1.8) is equivalent to the equation

$$l(D_t+i\gamma,D_x)(e^{-\gamma t}u)=e^{-\gamma t}f. \tag{5.3.8}$$

In (5.3.8) we pass to Fourier transforms and use the fact that for $\gamma\ge\gamma_0>0$ the matrix $l(\sigma+i\gamma,\xi)$ is non-singular. Then

$$\widetilde{e^{-\gamma t}u}=l^{-1}(\sigma+i\gamma,\xi)\widetilde{e^{-\gamma t}f}. \tag{5.3.9}$$

Denoting by $L_{kj}(\sigma+i\gamma,\xi)$ the cofactor of the element $l_{kj}(\sigma+i\gamma,\xi)$ of the matrix $l(\sigma+i\gamma,\xi)$, we now rewrite (5.3.9) in the form

$$\widetilde{e^{-\gamma t}u_j}=L^{-1}(\sigma+i\gamma,\xi)\sum_{k=1}^{N}L_{kj}(\sigma+i\gamma,\xi)\widetilde{e^{-\gamma t}f_k}\quad (j=1,\dots,N). \tag{5.3.10}$$

Since

$$|L_{kj}(\sigma + i\gamma, \xi)| \le c_1(\sigma^2 + \gamma^2 + |\xi|^2)^{r-s_k-t_j},$$

we get from (5.3.10) that

$$\left|\widetilde{e^{-\gamma t}u_j}\right|^2(\sigma^2 + |\xi|^2 + \gamma^2)^{t_j+s}$$

$$\le c|L(\sigma + i\gamma, \xi)|^{-2}\sum_{k=1}^{N}(\sigma^2 + \gamma^2 + |\xi|^2)^{r+s-s_k}\left|\widetilde{e^{-\gamma t}f_k}\right|^2. \qquad (5.3.11)$$

Integrating (5.3.11) with respect to ξ_n, we use estimate (5.3.5) for $|\xi_n| > M(\sigma^2 + \gamma^2 + |\xi'|^2)^{1/2}$ and estimate (5.3.4) for $|\xi_n| \le M(\sigma^2 + \gamma^2 + |\xi'|^2)^{1/2}$. We obtain the estimate

$$\int_{-\infty}^{+\infty}\left|\widetilde{e^{-\gamma t}u_j}\right|^2(\sigma^2 + |\xi|^2 + \gamma^2)^{t_j+s}\, d\xi_n$$

$$\le \frac{c_1}{\gamma^2}\int_{-\infty}^{+\infty}\sum_{k=1}^{N}(\sigma^2 + \gamma^2 + |\xi|^2)^{s-s_k}(\gamma^2 + \sigma^2 + |\xi'|^2)\left|\widetilde{e^{-\gamma t}f_k}\right|^2 d\xi_n \qquad (5.3.12)$$

$$(j = 1, \ldots, N).$$

Multiplying the left and right hand sides of (5.3.12) by $(\sigma^2 + \gamma^2 + |\xi'|^2)^{\tau}$ and integrating with respect to (σ, ξ'), we obtain the required estimate (5.3.3). the last assertion of the theorem now follows from a theorem of Paley–Wiener type, since

$$\widetilde{e^{-\gamma t}g}(\sigma, \xi) = \tilde{g}(\sigma + i\gamma, \xi).$$

We remark that a comparison of formulas (5.3.2) and (5.3.3) shows that the transition $f \longmapsto u$ involves the 'loss a unit of smoothness in the tangential direction'. Such a circumstance is typical for hyperbolic problems (see [Sak], [Kre], [Agr], [ChP], [VoG]). It will be encountered also in the investigation of boundary value and mixed problems.

5.3.2. We now consider the problem (5.1.9) in $\mathbf{R}^{n+1}$.

Theorem 5.3.2. *Under the conditions of Theorem 5.3.1, there exists a number $\gamma_0 > 0$ such that for each $\gamma \ge \gamma_0$ system (5.1.9) with any $f \in \mathcal{H}^{s-S,\tau+1}(\mathbf{R}^{n+1}, \gamma)$ has one and only one solution $u \in \mathcal{H}^{T+s,\tau}(\mathbf{R}^{n+1}, \gamma)$. There exists a constant $c > 0$ independent of f, u, and γ $(\gamma \ge \gamma_0)$ such that estimate (5.3.3) holds. If $\operatorname{supp} f \subset \overline{\Omega}$ then also $\operatorname{supp} u \subset \overline{\Omega}$.*

Proof. According to Theorem 5.3.1, the system (5.1.9) is equivalent to the equation

$$u + l^{-1}l'u = l^{-1}f. \tag{5.3.13}$$

Lemma 5.2.1 yield the estimate

$$|l'u, \mathbf{R}^{n+1}, \gamma|_{s-S,\tau+1} \le |l'u, \mathbf{R}^{n+1}, \gamma|_{s-S+1,\tau} \le c_1|u, \mathbf{R}^{n+1}, \gamma|_{T+s,\tau},$$

where $c_1 > 0$ is independent of u and γ. But then by Theorem 5.3.1, the norm of the operator

$$l^{-1}l' : \mathcal{H}^{T+s,\tau}(\mathbf{R}^{n+1}, \gamma) \mapsto \mathcal{H}^{T+s,\tau}(\mathbf{R}^{n+1}, \gamma)$$

does not exceed $c_1 c\gamma^{-1}$. Choosing $\gamma_0 > 0$ so that $c_1 c\gamma_0^{-1} < 1/2$, we obtain that for $\gamma \ge \gamma_0$ equation (5.3.13) is uniquely solvable in $\mathcal{H}^{T+s,\tau}(\mathbf{R}^{n+1}, \gamma)$. Further,

$$\begin{aligned} |u, \mathbf{R}^{n+1}, \gamma|_{T+s,\tau} &\le 2|l^{-1}f, \mathbf{R}^{n+1}, \gamma|_{T+s,\tau} \\ &\le 2c\gamma^{-1}|f, \mathbf{R}^{n+1}, \gamma|_{s-S,\tau+1}. \end{aligned}$$

To obtain the last assertion of the theorem we must replace the spaces in the arguments by the corresponding subspaces with supports in $\overline{\Omega}$ and take into account that $\operatorname{supp} f \subset \overline{\Omega}$ implies that $\operatorname{supp} l^{-1}f \subset \overline{\Omega}$. □

5.4. Solvability of Boundary Value and Mixed Hyperbolic Problems in the Complete scale Sobolev Type Spaces

5.4.1. We cosider hyperbolic problem (5.1.10) (see Definition 5.1.2). Let $s, \tau, \gamma \in \mathbf{R}$, $\gamma \ge \gamma_0 > 0$, and let

$$æ = \max\{0, \sigma_1 + 1, \dots, \sigma_m + 1\}. \tag{5.4.1}$$

It follows immediately from Lemma 5.2.1 that the closure $A = (l, b)$ of the mapping $u \mapsto (lu, bu)$ $\left(u \in (C_0^\infty(\overline{G}))^N\right)$ acts continuously from the whole space

$$\widetilde{\mathcal{H}}^{T+s,\tau,(T+æ)}(G, \gamma) := \prod_{j=1}^{N} \widetilde{\mathcal{H}}^{t_j+s,\tau,(t_j+æ)}(G, \gamma) \tag{5.4.2}$$

into the space

$$\begin{aligned} K_{s,\tau} &:= \widetilde{\mathcal{H}}^{s-S,\tau,(æ-S)}(G, \gamma) \times \mathcal{H}^{s-\sigma-1/2+\tau}(\partial G, \gamma) \\ &:= \prod_{j=1}^{N} \widetilde{\mathcal{H}}^{s-s_j,\tau,(æ-s_j)}(G, \gamma) \times \prod_{h=1}^{N} \mathcal{H}^{s-\sigma_h-1/2+\tau}(\partial G, \gamma). \end{aligned} \tag{5.4.3}$$

An element $u \in \widetilde{\mathcal{H}}^{T+s,\tau,(T+æ)}(G,\gamma)$ such that $Au = F = (f,\varphi) \in K_{s\tau}$ is called a (generalized) solution of problem (5.1.10). If $s \geq æ$, then a generalized solution is an ordinary solution of this problem.
In the absolutely similar way we define a (generalized) solution of problem (5.1.13).

The question on the solvability of problem (5.1.10) in complete scale of spaces (5.4.2) is characterized by the next theorem.

Theorem 5.4.1. *Assume that problem (5.1.10) is hyperbolic,*

$$F = (f,\varphi) \in K_{s,\tau},$$

$$f = (f_1,\dots,f_N), \quad f_j = (f_{j0}, f_{j1},\dots,f_{j,æ-s_j}) \in \widetilde{\mathcal{H}}^{s-s_j,\tau,(æ-s_j)}(G,\gamma)$$
$$(s,\tau,\gamma \in \mathbf{R}, \quad \gamma \geq \gamma_0 > 0),$$

and

$$\begin{aligned}
&f_{j0} \in \mathcal{H}^{s-s_j,\tau+1}(G,\gamma) && (j = 1,\dots,N),\\
&f_{jk} \in \mathcal{H}^{s-s_j-k+1+\tau}(\partial G,\gamma) && (\forall j : æ - s_j \geq 1; k = 1,\dots,æ - s_j), \qquad (5.4.4)\\
&\varphi_h \in \mathcal{H}^{s-\sigma_h+\tau}(\partial G,\gamma) && (h = 1,\dots,m).
\end{aligned}$$

Then problem (5.1.10) has one and only one solution

$$u = (u_1,\dots,u_N) \in \widetilde{\mathcal{H}}^{T+s,\tau,(æ+T)}(G,\gamma),$$

$$u_j = (u_{j0},\dots,u_{j,æ+t_j}) \quad (j = 1,\dots,N),$$
$$u_{j0} \in \mathcal{H}^{t_j+s,\tau}(G,\gamma), \quad u_{jk} \in \mathcal{H}^{t_j+s+\tau-k+1}(\partial G,\gamma)$$
$$(j = 1,\dots,N; k = 1,\dots,t_j + æ),$$

and the estimate

$$\sum_{j=1}^{N}\Big(\gamma|u_{j0},G,\gamma|^2_{t_j+s,\tau} + \sum_{k=1}^{t_j+æ} |u_{jk},\partial G,\gamma|^2_{t_j+s+\tau-k+1}\Big)$$
$$\leq c\Big(\gamma^{-1}\sum_{j=1}^{N}|f_{j0},G,\gamma|^2_{s-s_j,\tau+1}$$
$$+ \sum_{j:æ-s_j\geq 1}\sum_{k=1}^{æ-s_j} |f_{jk},\partial G,\gamma|^2_{s-s_j+\tau-k+1} + \sum_{h=1}^{m}|\varphi_h,\partial G,\gamma|^2_{s-\sigma_h+\tau}\Big)$$
(5.4.5)

holds with a constant $c > 0$ independent of u, f, φ, and γ ($\gamma \geq \gamma_0 > 0$). If $\operatorname{supp} F \subset \overline{G}_+$ then also $\operatorname{supp} u \subset \overline{G}_+$.

This theorem will be proved in Section 5.6; in Section 5.5 we shall give various equivalent forms of the Lopatinskii condition to be used in the proof. In this section we present some corollaries to Theorem 5.4.1.

Corollary 5.4.1. *Suppose that problem (5.1.10) is hyperbolic, $s, \tau, \gamma \in \mathbf{R}$, $\gamma \geq \gamma_0 > 0$, and $F \in K_{s,\tau+1}$. Then problem (5.1.10) has one and only one solution $u \in \widetilde{\mathcal{H}}^{T+s,\tau,(æ+T)}(G,\gamma)$, and it satisfies the estimate*

$$\|u\|_{\widetilde{\mathcal{H}}^{T+s,\tau,(æ+T)}(G,\gamma)} \leq \frac{c}{\gamma}\|F\|_{K_{s,\tau+1}} \tag{5.4.6}$$

with a constant $c > 0$ independent of u, F, and γ ($\gamma \geq \gamma_0$). If $\operatorname{supp} F \subset \overline{G}_+$ then also $\operatorname{supp} u \subset \overline{G}_+$.

Indeed, the inclusion $F \in K_{s,\tau+1}$ implies the validity of inclusions (5.4.4), and, hence, the validity of (5.4.5). Estimate (5.4.6) follows from (5.4.5) and the inequalities of type (5.2.2), (5.2.5) for ∂G.

5.4.2. Let us now investigate the solvability of problem (5.1.13).

Theorem 5.4.2. *Suppose that problem (5.1.10) is hyperbolic, and $s, \tau \in \mathbf{R}$. Then there exists a number $\gamma_0 > 0$ such that for each $\gamma \geq \gamma_0$ problem (5.1.13) has one and only one solution*

$$u \in \widetilde{\mathcal{H}}^{T+s,\tau,(T+æ)}(G,\gamma)$$

for any

$$F = (f, \varphi) \in K_{s,\tau+1}.$$

There exists a constant $c > 0$ independent of u, F, and γ ($\gamma \geq \gamma_0$) such that estimate (5.4.6) holds. If $\operatorname{supp} F \subset \overline{G}_+$ then also $\operatorname{supp} u \subset \overline{G}_+$.

Proof. The assertion of Theorem 5.4.2 is derived from Corollary 5.4.1 in exactly the same way as Theorem 5.3.2 is derived from Theorem 5.3.1. Indeed, let $A' = (l', b')$ be the operator determined by the lower terms of problem (5.1.13). Then the problem can be written in the form

$$(A + A')u = F \in K_{s,\tau+1}. \tag{5.4.7}$$

According to Corollary 5.4.1, this problem is equivalent to the problem

$$u + A^{-1}A'u = A^{-1}F \in \widetilde{\mathcal{H}}^{T+s,\tau,(T+æ)}(G,\gamma). \tag{5.4.8}$$

It follows from (5.4.6) and Lemma 5.2.1 that

$$\|A^{-1}A'u\|_{\widetilde{\mathcal{H}}^{T+s,\tau,(T+\text{æ})}(G,\gamma)} \le \frac{c_1}{\gamma}\|A'u\|_{K_{s,\tau+1}} \le \frac{c_1}{\gamma}\|u\|_{\widetilde{\mathcal{H}}^{T+s,\tau,(T+\text{æ})}(G,\gamma)}.$$

Therefore, the norm of the operator $A^{-1}A'$ on $\widetilde{\mathcal{H}}^{T+s,\tau,(T+\text{æ})}(G,\gamma)$ does not exceed $c_1\gamma^{-1}$. If we choose $\gamma_0 > 0$ so that $c_1\gamma_0^{-1} \le 1/2$ then for $\gamma \ge \gamma_0$ we obtain the solvability of equation (5.4.8) and the validity of estimate (5.4.6).

To obtain the last assertion of the theorem we must consider equation (5.4.8) in the subspace of the corresponding space that consists of the elements with supports in $\overline{G}_+$. □

By similar arguments it can be shown that if the right hand side $F = (f, \varphi)$ of problem (5.1.13) satisfies relations (5.4.4) then the solution u satisfies the inequality (5.4.5) with a constant $c > 0$ independent of u, F, and γ $(\gamma \ge \gamma_0)$.

5.5. Lopatinskii Condition

5.5.1. We consider hyperbolic problem (5.1.10). By using the identity (5.2.28) it is possible to write the problem in the form

$$l(D_t + i\gamma, D_x)(e^{-\gamma t}u) = e^{-\gamma t}f \quad (\text{in } G),$$

$$b_h(D_t + i\gamma, D_x)(e^{-\gamma t}u) := \sum_{j=1}^{N} b_{hj}(D_t + i\gamma, D_x)(e^{-\gamma t}u_j)\big|_{x_n=0} = e^{-\gamma t}\varphi_h \tag{5.5.1}$$

$$(h = 1, \dots, m).$$

Let us give different forms of the Lopatinskii condition. Using the partial Fourier transform $(t, x') \to (\sigma, \xi')$ in (5.5.1) with $f = 0$, on the semi-axis $x_n > 0$ we obtain the following problem for a system of ordinary differential equations

$$l(\sigma + i\gamma, \xi', D_n)\widehat{e^{-\gamma t}u}(\sigma, \xi', x_n) = 0 \quad (x_n > 0), \tag{5.5.2}$$

$$b_h(\sigma + i\gamma, \xi', D_n)(\widehat{e^{-\gamma t}u})(\sigma, \xi', x_n)\big|_{x_n=0} = (\widehat{e^{-\gamma t}u})(\sigma, \xi') \tag{5.5.3}$$

$$(h = 1, \dots, m).$$

Since the expression l is strictly hyperbolic, for any $\gamma > 0$ equation (5.1.5) does not have real ξ_n–roots. Therefore, for $\gamma > 0$ all the ξ_n–roots (5.1.6) of equation (5.1.5) are not real. We recall that the first m roots in (5.1.6) have negative imaginary parts, and the rest have positive imaginary parts.

Let

$$\mathfrak{M}^+ = \mathfrak{M}^+(\sigma + i\gamma, \xi') \qquad ((\sigma + i\gamma, \xi') \neq (0,0), \gamma \geq 0)$$

denote the m-dimensional space of solutions of system (5.5.2) determined by the first m roots in (5.1.6); for $\gamma > 0$ the space $\mathfrak{M}^+(\sigma + i\gamma, \xi')$ consists of all the stable (that is, decreasing as $x_n \to +\infty$) solutions of system (5.5.2). The Lopatinskii condition can now be formulated as follows.

L_2. For each $(\sigma + i\gamma, \xi') \neq (0,0)$ with $\gamma \geq 0$ the problem

$$\begin{gathered} l(\sigma + i\gamma, \xi', D_n)V(x_n) = 0, \\ b_h(\sigma + i\gamma, \xi', D_n)V(x_n)\big|_{x_n=0} = a_h (h = 1, \dots, m) \end{gathered} \tag{5.5.4}$$

is uniquely solvable in $\mathfrak{M}^+ = \mathfrak{M}^+(\sigma + i\gamma, \xi')$; in other words, problem (5.5.4) with $a_1 = \dots = a_m = 0$ has only the trivial solution in $\mathfrak{M}^+$.

We establish the equivalence of L_1 and L_2 in the next subsection; here we give other equivalent forms of the Lopatinskii condition (sf. [Vol], where similar assertions are proved for elliptic systems).

Let

$$w_1(\sigma + i\gamma, \xi', x_n), \dots, w_m(\sigma + i\gamma, \xi', x_n)$$

be a basis in $\mathfrak{M}^+$, and let $w(\sigma + i\gamma, \xi', x_n)$ be the matrix whose columns are the vectors $w_1, \dots, w_m$. Each element in $\mathfrak{M}^+$ has the form

$$\alpha_1 w_1(\sigma + i\gamma, \xi', x_n) + \dots + \alpha_m w_m(\sigma + i\gamma, \xi', x_n);$$

therefore, the unique solvability in $\mathfrak{M}^+$ of problem (5.5.4) is equivalent to the unique solvability of the system of linear equations

$$\sum_{k=1}^{m} \alpha_k b_n(\sigma + i\gamma, \xi', D_n) w_k(\sigma + i\gamma, \xi', x_n)\big|_{x_n=0} = a_n \quad (h = 1, \dots, m). \tag{5.5.5}$$

Therefore, the condition L_2 is equivalent to the following condition.

L_3. For each $(\sigma + i\gamma, \xi') \neq (0,0)$ with $\gamma \geq 0$,

$$\det(b((\sigma + i\gamma, \xi', D_n) w(\sigma + i\gamma, \xi', x_n)\big|_{x_n=0}) \neq 0. \tag{5.5.6}$$

This immediately implies that the condition L_2 is equivalent to the following condition

L_4. Problem (5.5.4) is solvable in $\mathfrak{M}^+$ for any complex numbers $a_1, \dots, a_m$.

5.5.2. We now show that L_2 is equivalent to L_1 (see Definition 5.1.2). Assume first that the matrix $l(D_t, D_x)$ is diagonal. Then the problem (5.5.4) has the form

$$l_{jj}(\sigma+i\gamma,\xi',D_n)V_j(x_n)=0 \quad (j=1,\ldots,N,\ x_n>0), \tag{5.5.7}$$

$$b_h(\sigma+i\gamma,\xi',D_n)V(x_n)\big|_{x_n=0}=a_h \quad (h=1,\ldots,m), \tag{5.5.8}$$

with

$$l_{jj}(\sigma+i\gamma,\xi',D_n)=l_{jj}^+(\sigma+i\gamma,\xi',D_n)l_{jj}^-(\sigma+i\gamma,\xi',D_n),$$

where the roots ξ_n of the equation $l_{jj}^{\pm}(\sigma+i\gamma,\xi',D_h)$ have positive (negative) imaginary parts for $\gamma>0$. Now

$$\mathfrak{M}^+(\sigma+i\gamma,\xi')=\{V=(V_1,\ldots,V_N):$$

$$l_{jj}^-(\sigma+i\gamma,\xi',D_n)V_j=0 \quad (j=1,\ldots,N)\}, \tag{5.5.9}$$

$$\mathfrak{M}^+=\mathfrak{M}_1^+\oplus\ldots\oplus\mathfrak{M}_N^+,\quad \mathfrak{M}_j^+=\{V_j: l_{jj}^-(\sigma+i\gamma,\xi',D_n)V_j=0\}.$$

The condition L_1 means that for each $(\sigma+i\gamma,\xi')\neq(0,0)$ with $\gamma\geq 0$ the equalities

$$\sum_{j=1}^m c_j b_{jk}(\sigma+i\gamma,\xi',\xi_n)L_{kk}(\sigma+i\gamma,\xi',\xi_n)=L_-(\sigma+i\gamma,\xi',\xi_n)P_k(\xi_n), \tag{5.5.10}$$

$$(k=1,\ldots,N),$$

are possible only for $c_1=\ldots=c_m=0$, where L_{kk} are the cofactors of the elements l_{kk} in the determinant $L(\sigma+i\gamma,\xi',\xi_n)$, the expressions $P_k(\xi_n)$ are the polynomials, and $\{c_j\}$ are complex numbers. From the other hand,

$$L_{kk}(\sigma+i\gamma,\xi',\xi_n)=\frac{L(\sigma+i\gamma,\xi',\xi_n)}{l_{kk}(\sigma+i\gamma,\xi',\xi_n)};$$

therefore (5.5.10) can be written in the form

$$\sum_{j=1}^m c_j b_{jk}(\sigma+i\gamma,\xi',\xi_n)=\frac{l_{kk}(\sigma+i\gamma,\xi',\xi_n)P_k(\xi_n)}{L_+(\sigma+i\gamma,\xi',\xi_n)},$$

and since $l_{kk}=l_{kk}^+l_{kk}^-$, it follows that (5.5.10) is equivalent to the relation

$$\sum_{j=1}^m c_j b_{jk}(\sigma+i\gamma,\xi',\xi_n)=\frac{l_{kk}^-(\sigma+i\gamma,\xi',\xi_n)P_k(\xi_n)}{\prod_{i\neq k} l_{ii}^+(\sigma+i\gamma,\xi',\xi_n)} \quad (k=1,\ldots,N). \tag{5.5.11}$$

The left hand side of (5.5.11) is a polynomial in ξ_n, and, hence, the right hand side is also a polynomial. For $\gamma > 0$ the polynomials l^-_{kk} and l^+_{ii} are mutually prime. Therefore, $P_k(\xi_n)$ is divisible by $\prod_{i\neq k} l^+_{ii}(\sigma + i\gamma, \xi', \xi_n)$, and the coefficients of quotient are continuous functions of $(\sigma + i\gamma, \xi')$. This implies that relation (5.5.11) can be written in the form

$$\sum_{j=1}^{m} c_j b_{jk}(\sigma+i\gamma,\xi',\xi_n) = p_k(\xi_n) l^-_{kk}(\sigma+i\gamma,\xi',\xi_n) \quad (k=1,\dots,N). \quad (5.5.12)$$

Thus, in the case under consideration, the condition L_1 can be formulated as follows: for each $(\sigma + i\gamma, \xi') \neq (0,0)$ with $\gamma \geq 0$ equalities (5.5.12) are possoble only for $c_1 = \dots = c_m = 0$, where $p_k(\xi_n)$ are polynomials, and $\{c_j\}$ are complex comstants.

Let us prove that L_2 implies L_1. Assume that relation (5.12) holds and, say, $c_1 \neq 0$. Since the space $\mathfrak{M}^+(\sigma + i\gamma, \xi')$ is m-dimentional, there exists a nonzero vector $V = (V_1, \dots, V_N) \in \mathfrak{M}^+(\sigma + i\gamma, \xi')$ such that

$$\sum_{k=1}^{N} b_{jk}(\sigma + i\gamma, \xi', D_n) V_k(x_n)\big|_{x_n=0} = 0 \quad (j = 2, \dots, m). \qquad (5.5.13)$$

Then it follows from (5.5.12) that (5.5.13) holds also for $j = 1$. Then $V \in \mathfrak{M}^+$ is a nonzero solution of problem (5.5.7) – (5.5.8) with $a_1 = \dots = a_m = 0$, and this contradicts to L_2. The assertion $L_2 \Rightarrow L_1$ is proved.

Let us prove the converse. Suppose that L_1 holds, that is (see (5.5.12)), the rows

$$(b_{j1}(\sigma + i\gamma, \xi', \xi_n), \dots, b_{jN}(\sigma + i\gamma, \xi', \xi_n)) \quad (j = 1, \dots, m) \qquad (5.5.14)$$

are linearly independent modulo

$$(l^-_{11}(\sigma + i\gamma, \xi', \xi_n), \dots, l^-_{NN}(\sigma + i\gamma, \xi', \xi_n)). \qquad (5.5.15)$$

Let $b'_{jk}(\sigma + i\gamma, \xi', \xi_n)$ be the remainder after division of the polynomial $b_{jk}(\xi_n) = b_{jk}(\sigma + i\gamma, \xi', \xi_n)$ by $l^-_{kk}(\xi_n) = l^-_{kk}(\sigma + i\gamma, \xi', \xi_n)$. Then the rows

$$(b'_{j1}(\xi_n), \dots, b'_{jN}(\xi_n)) \quad (j = 1, \dots, m) \qquad (5.5.16)$$

are linearly independent.

Let us show that the problem

$$l^-_{jj}(\sigma + i\gamma, \xi', D_n) V_j(x_n) = 0 \quad (j = 1, \dots, N), \qquad (5.5.17)$$

$$\sum_{k=1}^{N} b'_{hk}(D_n) V_k(x_n)\big|_{x_n=0} = 0 \quad (h = 1, \dots, m), \qquad (5.5.18)$$

has only the trivial solution. Indeed, let $m_j^- = \operatorname{ord} l_j^-(\xi_n)$. Then $\sum m_j^- = m$, and

$$\operatorname{ord} b'_{hk}(D_n) \le m_k^- - 1 \quad (k = 1, \ldots, N;\ h = 1, \ldots, m);$$

$$b'_{hk}(\sigma + i\gamma, \xi', D_n) = \sum_{1 \le p \le m_k^-} r_{hkp}(\sigma + i\gamma, \xi') D_n^{p-1}$$

$$(k = 1, \ldots, N;\ h = 1, \ldots, m).$$

It follows from the linear independence of rows (5.5.16) that the rank of the square matrix

$$(r_{h;(k,p)})(\sigma + i\gamma, \xi') \quad (h = 1, \ldots, m; k = 1, \ldots, N; p = 1, \ldots, m_k^-) \tag{5.5.19}$$

is equal to m. Here h is the row index, and

$$(k, p) \quad (k = 1, \ldots, N;\ p = 1, \ldots, m_k^-)$$

is the column index. Since $m_1^- + \ldots + m_N^- = m$, there are also m columns. Then, from equalities (5.5.18), we can deduce that $(D_n^{p-1} V_k)(0) = 0$, and we can see that problem (5.5.17)–(5.5.18) is equivalent to the problem

$$\begin{gathered} l_{jj}^-(\sigma + i\gamma, \xi', D_n) V_j = 0 \quad (j = 1, \ldots, N), \\ D_n^{t-1} V_j \big|_{x_n = 0} = 0 \quad (t = 1, \ldots, m_j^-). \end{gathered} \tag{5.5.20}$$

Since for each j Cauchy problem (5.5.20) has only the trivial solution, problem (5.5.17)–(5.5.18) has only the trivial solution. Then problem (5.5.4) with $a_1 = \ldots = a_m = 0$ has in $\mathfrak{M}^+$ only the trivial solution. Thus, the assertion $L_1 \Rightarrow L_2$ is established in the case under consideration when the matrix $l(\sigma + i\gamma, \xi', D_n)$ is diagonal.

Remark 5.5.1. If l is diagonal matrix then problem (5.5.4) in $\mathfrak{M}^+$ is equivalent to equations (5.5.17) and the boundary conditions

$$\sum_{k=1}^{N} b'_{hk}(\sigma + i\gamma, \xi', D_n) V(x_n) \big|_{x_n = 0} = a_h \quad (h = 1, \ldots, m). \tag{5.5.21}$$

Since matrix (5.5.19) is non-singular and the elements of it depend continuously on $(\sigma + i\gamma, \xi') \ne (0, 0)$ with $\gamma \ge 0$, equalities (5.5.21) implies that

$$(D_n^{p-1} V_j)(0) \quad (j = 1, \ldots, N; p = 1, \ldots, m_j^-) \tag{5.5.22}$$

can be expressed linearly in terms of the constants $\{a_h\}$. Furthermore, the coefficients of a_h are continuous and homogeneous functions of $(\sigma + i\gamma, \xi') \ne$

$(0,0)$, $\gamma \geq 0$. If we then use equalities (5.5.17), we can successively express the higher derivatives $(D_n^p V_j)(0), (D_n^{p+1} V_j)(0), \ldots$ (linearly) in terms of $\{a_h\}$. The coefficients of a_h will as before be continuous homogeneous functions of $(\sigma + i\gamma, \xi') \neq (0,0)$, $\gamma \geq 0$. Therefore, the solvability in $\mathfrak{M}^+(\sigma + i\gamma, \xi')$ of problem (5.5.7)–(5.5.8) is equivalent to the solvability of the corresponding Cauchy problem for system (5.5.7).

Suppose now that the matrix $l(D_n) = l(\sigma + i\gamma, \xi', D_n)$ is not diagonal. Then the matrix $l(D_n)$ can be reduced to diagonal form by means of elementary transformations (see [Vol]): there exist polynomial matrices $P(D_n)$ and $Q(D_n)$ such that

$$\det P(\xi_n) = c \neq 0, \quad \det Q(\xi_n) = \text{const} \neq 0, \tag{5.5.23}$$

$$P(D_n) l(D_n) Q(D_n) = \Lambda(D_n), \tag{5.5.24}$$

where $\Lambda(D_n)$ is a diagonal matrix. In addition, the inverse matrices $P^{-1}(D_n)$ and $Q^{-1}(D_n)$ exist, and their elements are polynomials in D_n. It is clear that

$$l(D_n) = P^{-1}(D_n) \Lambda(D_n) Q^{-1}(D_n). \tag{5.5.25}$$

Then problem (5.5.4) is equivalent to the problem

$$\begin{aligned} &\Lambda(D_n) Q^{-1}(D_n) V(x_n) = 0, \\ &b_h(D_n) V(x_n)\big|_{x_n=0} = a_h \quad (h = 1, \ldots, m). \end{aligned} \tag{5.5.26}$$

Let $Q^{-1}(D_n)V = U$, $V = QU$, and

$$\Lambda(D_n) = \begin{pmatrix} \Lambda_{11}(D_n) & \ldots & 0 \\ \vdots & \ddots & \vdots \\ 0 & \ldots & \Lambda_{NN}(D_n) \end{pmatrix}, \quad \Lambda_{jj}(D_n) = \Lambda_{jj}^{+}(D_n) \Lambda_{jj}^{-}(D_n),$$

where for $\gamma > 0$ the roots of $\Lambda_{jj}^{\pm}(\xi_n)$ lie above (below) the real axis. Then problem (5.5.4) in $\mathfrak{M}^+$ is equivalent to the problem

$$\Lambda_{jj}(D_n) U_j = 0 \quad (j = 1, \ldots, N), \tag{5.5.27}$$

$$b_h(QU)(x_n)\big|_{x_n=0} = a_h \quad (h = 1, \ldots, m). \tag{5.5.28}$$

Equalities (5.5.12) take the form

$$\sum_{j=1}^{m} c_j \sum_{h=1}^{N} b_{jh}(\xi_n) Q_{hk}(\xi_n) = p_k(\xi_n) \Lambda_{kk}^{-}(\xi_n) \qquad (k = 1, \ldots, N). \tag{5.5.29}$$

Since the matrices $P(\xi_n)$ and $Q(\xi_n)$ are non-singular, the condition L_1 can be formulated as follows: for each $(\sigma + i\gamma, \xi') \neq (0,0)$ with $\gamma \geq 0$ equalities (5.5.29) are possible only for $c_1 = \ldots = c_m = 0$, where the expressions $p_k(\xi_n)$ are polynomials and $\{c_j\}$ are complex constants.

We prove that L_2 implies L_1. Assume that (5.5.29) holds, and, say, $c_1 \neq 0$. Since the space $\mathfrak{M}^+$ is m-dimentional, there exists a nonzero vector $U = (U_1, \ldots, U_N) \in \mathfrak{M}^+$ such that

$$\sum_{k=1}^{N}\Big(\sum_{h=1}^{N} b_{jh}(D_n)Q_{hk}(D_n)\Big)U_k(x_n)\Big|_{x_n=0} = 0 \quad (j = 2, \ldots, m). \tag{5.5.30}$$

It follows from (5.5.29) that equality (5.5.30) is valid also for $j = 1$, and problem (5.5.27)–(5.5.28) has a non-trivial solution, that is impossible. Thus, it is proved the assertion $L_2 \Rightarrow L_1$.

Let us prove the converse. Suppose that L_1 holds, that is, the rows

$$\Big(\sum_{h=1}^{N} b_{jh}(\xi_n)Q_{hk}(\xi_n) : k = 1, \ldots, N\Big) \quad (j = 1, \ldots, m) \tag{5.5.31}$$

are linearly independent modulo

$$(\Lambda_{11}^-(\xi_n), \ldots, \Lambda_{NN}^-(\xi_n)). \tag{5.5.32}$$

Again replacing the elements of rows (5.5.31) by their remainders after division by $\Lambda_{kk}^-(\xi_n)$, we obtain a problem of type (5.5.17)–(5.5.18) which again has only the trivial solution. Then problem (5.5.27)–(5.5.28) has only the trivial solution. The implication $L_1 \Rightarrow L_2$ is thus established.

The assertion expressed in Remark 3.5.1 remains valid: If the Lopatinskii condition holds then problem (5.5.4) is solvable (uniquely) for any $a_1, \ldots, a_m$. Furthermore,

$$(D_n^{p-1}V(0) : p = 1, \ldots, t_j + æ,\ æ \geq 0 \text{ an integer};\ j = 1, \ldots, N)$$

can be expressed linearly in terms of $a_1, \ldots, a_m$. The coefficients of the a_h will be continuous homogeneous functions of $(\sigma + i\gamma, \xi') \neq (0,0)$, $\gamma \geq 0$. Problem (5.5.4) in $\mathfrak{M}^+(\sigma + i\gamma, \xi')$ is thus equivalent to the corresponding Cauchy problem for the equation $l(\sigma + i\gamma, \xi', D_n)V = 0$.

5.5.3. In this subsection we give another form of the Lopatinskii condition.

We represent the expressions $l_{kj}(D_t, D_x)$ in (5.1.2) in the form

$$l_{kj}(D_t, D_x) = \sum_{\nu=1}^{s_k+t_j+1} l^{kj\nu}(D_t, D_{x'})D_n^{\nu-1} \quad (\forall k, j : s_k + t_j \geq 0), \tag{5.5.33}$$

and the boundary expressions $b_{hj}(D_t, D_x)$ in (5.1.11) in the form

$$b_{hj}(D_t, D_x) = \sum_{\nu=1}^{\sigma_h+t_j+1} b^{hj\nu}(D_t, D_{x'})D_n^{\nu-1} \quad (\forall h, j : \sigma_h + t_j \geq 0). \quad (5.5.34)$$

Denote by V_+ the extension of $V \in \mathfrak{M}^+ = \mathfrak{M}^+(\sigma + i\gamma, \xi')$ by zero to the semi-axis $x_n < 0$ and set:

$$D_n^{k-1}V_j\big|_{x_n=0} = V_{jk} \quad (\forall j : t_j \geq 1,\ k = 1, \ldots, t_j). \quad (5.5.35)$$

Then (5.2.16) and (5.2.17) yield that one can rewrite the system

$$l(\sigma + i\gamma, \xi', D_n)V(x_n) = 0 \quad (x_n > 0)$$

in the form

$$\sum_{k=1}^{n} l_{jk}(\sigma + i\gamma, \xi', D_n)V_{k+}$$

$$= i \sum_{k:s_j+t_k\geq 1} \sum_{1\leq p\leq s_j+t_k} J^p(l_{jk}(\sigma + i\gamma, \xi', D_n))(V_{kp} \times \delta(x_n)) \quad (5.5.36)$$

$$(j = 1, \ldots, N),$$

where

$$J^p(l_{kj}(\sigma + i\gamma, \xi', D_n)) = \sum_{\nu=p+1}^{s_k+t_j+1} l^{kj\nu}(D_t, D_{x'})D_n^{\nu-p-1} \quad (1 \leq p \leq s_k + t_j). \quad (5.5.37)$$

Boundary conditions (5.5.4) can be written in the form

$$\sum_{j:\sigma_h+t_j\geq 0} \sum_{1\leq\nu\leq\sigma_h+t_j+1} b^{hj\nu}(\sigma + i\gamma, \xi')V_{j\nu} = a_h \quad (h = 1, \ldots, m). \quad (5.5.38)$$

It is clear that the condition L_2 is equivalent to the unique solvability in $\mathfrak{M}^+(\sigma + i\gamma, \xi')$ of problem (5.5.36), (5.5.38), (5.5.35) for each $(\sigma + i\gamma, \xi') \neq (0, 0)$, $\gamma \geq 0$. Furtermore, conditions (5.5.38) can be replaced here by conditions of form (5.5.21) equivalent to them.

Passing to Fourier transforms $x_n \to \xi_n$ in (5.5.36), we obtain

$$\sum_{k=1}^{N} l_{jk}(\sigma + i\gamma, \xi', \xi_n)\widetilde{V}_{k+}(\xi_n)$$

$$= i \sum_{k:s_j+t_k\geq 1} \sum_{1\leq p\leq s_j+t_k} J^p(l_{jk}(\sigma + i\gamma, \xi', \xi_n))V_{kp} \quad (5.5.39)$$

$$(j = 1, \ldots, N),$$

Since the support of V_+ lies on the half-line $x_n \geq 0$, the element $\widetilde{V}_+(\xi_n)$ admits an analytic continuation $\widetilde{V}_+(\zeta)$ into the half-plane $\zeta = \xi_n + i\eta$, $\eta > 0$. Let $L_{jk} = (\sigma + i\gamma, \xi', \zeta)$ be the cofactor of the element $l_{jk}(\sigma + i\gamma, \xi', \zeta)$ of the matrix $l(\sigma + i\gamma, \xi', \zeta)$. Then

$$l^{-1} = L^{-1}(\sigma + i\gamma, \xi', \zeta)(L_{kj}(\sigma + i\gamma, \xi', \zeta))$$

is the matrix inverse to the matrix $l = (l_{jk}(\sigma + i\gamma, \xi', \zeta))$. Taking into account (5.1.7), we obtain from (5.5.39) that

$$L_-(\sigma+i\gamma, \xi', \zeta)\widetilde{V}_+(\sigma+i\gamma, \xi', \zeta)$$

$$= i\sum_{r=1}^{N} L_+^{-1}(\sigma+i\gamma, \xi', \zeta)$$

$$\times \left(L_{rj}(\sigma + i\gamma, \xi', \zeta) \sum_{k:s_r+t_k\geq 1} \sum_{p=1}^{s_r+t_k} (J^p l_{rk})(\sigma + i\gamma, \xi', \zeta) V_{kp}(\sigma + i\gamma, \xi', \zeta) \right) \tag{5.5.40}$$

$$(j = 1, \ldots, N).$$

Here the left hand side is an analytic function in the upper ζ-half-plane. The right hand side (for $\gamma > 0$) has $r - m$ poles there. This imposes certain conditions: the singular points of the right hand side of (5.5.40) are removable.

We confine ourselves first to the case of a diagonal matrix $l(\sigma + i\gamma, \xi', D_n)$. Then (5.5.39) takes the form

$$l_{jj}(\sigma+i\gamma, \xi', \xi_n)\widetilde{V}_{j+}(\sigma+i\gamma, \xi', \xi_n)$$

$$= i \sum_{1\leq p\leq s_j+t_j} J^p(l_{jj}(\sigma + i\gamma, \xi', \xi_n)) V_{jp}. \tag{5.5.41}$$

If $s_j + t_j = 0$, then the right hand side of (5.5.41) is equal to 0, and $l_{jj} = \text{const} \neq 0$. Then $V_{j+} = 0$. We pass in (5.5.41) to the analytic continuation into the half-plane $\zeta = \xi_n + i\eta$, $\eta > 0$ and take into account that $l_{jj}(\zeta) = l_{jj}^+(\zeta) l_{jj}^-(\zeta)$, where the roots ζ of the polynomial $l_{jj}^+(\zeta)(l_{jj}^-(\zeta))$ lie above (below) the real axis for $\gamma > 0$. We obtain that

$$l_{jj}^-(\sigma + i\gamma, \xi', \zeta)\widetilde{V}_+(\sigma + i\gamma, \xi', \zeta)$$

$$= i(l_{jj}^+(\sigma + i\gamma, \xi', \zeta))^{-1} \sum_{p=1}^{s_j+t_j} J^p(l_{jj}^-(\sigma + i\gamma, \xi', \zeta))V_{jp}. \tag{5.5.42}$$

Here for $\gamma > 0$ the left hand side is an analytic function in the upper ζ-half-plane. Therefore, ζ-zeros of the polynomial $l^+_{jj}(\sigma + i\gamma, \xi', \zeta)$ are removable singular points of the right hand side, and, hence, the polynomial

$$\sum_{1 \le p \le s_j + t_j} J^p(l_{jj}(\zeta))V_{jp}$$

is divisible by the polynomial $l^+_{jj}(\zeta)$. After division we obtain from (5.5.42) that

$$l^-_{jj}(\sigma + i\gamma, \xi', \zeta)\widetilde{V}_{j+}(\sigma + i\gamma, \xi', \zeta)$$
$$= i \sum_{1 \le p \le n^-_j} J^p(l^-_{jj}(\sigma + i\gamma, \xi'\zeta))V_{jp} \quad (j : s_j + t_j \ge 1). \quad (5.5.43)$$

Here n^-_j is the degree of the polynomial $l^-_{jj}(\zeta)$. It is clear that equality (5.5.43) can be obtained directly in a way analogous to the way formula (5.5.41) was obtained. Suppose that the polynomial of degree at most $n^+_j - 1$

$$r_{jp}(\zeta) = r_{jp}(\sigma + i\gamma, \xi', \zeta) \quad (j : s_j + t_j \ge 1; \quad 1 \le p \le s_j + t_j) \quad (5.5.44)$$

is the remainder after division of $J^p(l_{jj}(\sigma + i\gamma, \xi', \zeta))$ by $l^+_{jj}(\sigma + i\gamma, \xi', \zeta)$. Then for $V \in \mathfrak{M}^+$, $V_{jp} = D^{p-1}_n V_j\big|_{x_n=0}$, we have the equalities

$$\sum_{1 \le p \le s_j + t_j} r_{jp}(\sigma + i\gamma, \xi', \zeta)V_{jp} = 0 \quad (j : s_j + t_j \ge 1), \quad (5.5.45)$$

where

$$r_{jp}(\sigma + i\gamma, \xi', \zeta) = \sum_{1 \le k \le n^+_j} r_{jpk}(\sigma + i\gamma, \xi')\zeta^{k-1}. \quad (5.5.46)$$

From (5.5.45) and (5.5.46),

$$\sum_{1 \le p \le s_j + t_j} r_{jkp}(\sigma + i\gamma, \xi')V_{jp} = 0 \quad (1 \le k \le n^+_j, \ j : n^+_j \ge 1). \quad (5.5.47)$$

Relations (5.5.47) form a linear system of $\sum n^+_j = r - m$ equations in the unknowns $\{V_{jp}\}$. We adjoin to it the system (5.5.21). This gives us a linear system of r equations in r unknowns:

$$\{V_{jp} : \quad j : s_j + t_j \ge 1; \quad 1 \le p \le s_j + t_j\}. \quad (5.5.48)$$

The Lopatinskii condition can be written in the following form.

L$_5$. For any $(\sigma + i\gamma, \xi') \ne 0$ $(\gamma \ge 0)$ the determinant $\Delta(\sigma + i\gamma, \xi')$ of system (5.5.21), (5.5.47) iz nonzero:

$$\Delta(\sigma + i\gamma, \xi') \ne 0 \quad ((\sigma + i\gamma, \xi') \ne (0, 0), \quad \gamma \ge 0). \quad (5.5.49)$$

Indeed, suppose that (5.5.2) and (5.5.3) hold and the condition L_2 fails for $(\sigma_0 + i\gamma_0, \xi'_0) \neq 0$, $\gamma_0 \geq 0$. Then problem (5.5.4) has a nonzero solution $V^0 \in \mathfrak{M}^+(\sigma_0 + i\gamma_0, \xi'_0)$. Then $\{V^0_{jp}\}$ is a nonzero solution of homogeneous system (5.5.38) (with $a_h = 0$), (5.5.47), and this contradicts (5.5.49). The implication $L_5 \Rightarrow L_2$ is established.

Conversely, let

$$\Delta(\sigma_0 + i\gamma_0, \xi'_0) = 0 \quad ((\sigma_0 + i\gamma_0, \xi'_0) \neq (0,0),\ \gamma_0 \geq 0).$$

Then homogeneous system (5.5.21) (with $a_h = 0$), (5.5.47) has a nonzero solution $\{V_{jp} : j : s_j + t_j \geq 1; p = 1, \ldots, s_j + t_j\}$. Let W_j $(j : n_j^- \geq 1)$ be a solution of the Cauchy problem

$$l_{jj}^-(\sigma_0 + i\gamma_0, \xi'_0, D_n)W_j = 0, \quad D_n^{h-1}W_j\big|_{x_n=0} = V_{jh} \quad (h = 1, \ldots, n_j^-) \tag{5.5.50}$$

on the semi-axis. It is clear that $W = (W_1, \ldots, W_N) \in \mathfrak{M}^+(\sigma_0 + i\gamma_0, \xi'_0)$. Let W_+ denote the extension of W by zero to the semi-axis $x_n < 0$. Then for $\widetilde{W}_{j+}$ we have relations (5.5.43) with $\widetilde{V}_{j+}$ replaced by $\widetilde{W}_{j+}$, and, hence, also relations (5.5.42) and (5.5.41). From this,

$$D_n^{h-1}W_j\big|_{x_n=0} = V_{jh} \quad (j : s_j + t_j \geq 1,\ h = 1, \ldots, s_j + t_j),$$

and $W(\sigma_0 + i\gamma_0, \xi'_0, x_n) \in \mathfrak{M}^+$ is a nonzero solution of problem (5.5.7), (5.5.8) (with $a_1 = \ldots = a_m = 0$), which contradicts L_2.

Thus, if l is a diagonal matrix, then the conditions L_1 and L_5 are equivalent to each other. If the system l is not diagonal then we can modify the above arguments correspondingly. We can argue just as in the proof that L_1 and L_2 are equivalent to each other. In Subsection 5.5.7 we obtain an analogie of the condition L_5 for problem (5.5.4) without passing to a diagonal system.

5.5.4. In $\mathbf{R}^n$ we consider the hyperbolic system

$$l(D_t, D_x)u = \Phi \in \mathcal{H}^{s-S,\tau+1}(\mathbf{R}^{n+1}, \gamma), \tag{5.5.51}$$

and let $\gamma \geq \gamma_0 > 0$. By Theorem 5.3.1, this problem has a unique solution $u \in \mathcal{H}^{T+s,\tau}(\mathbf{R}^{n+1}, \gamma)$, which satisfies the estimate

$$|u, \mathbf{R}^{n+1}|_{T+s,\tau} \leq \frac{c}{\gamma}|\Phi, R^{n+1}, \gamma|_{s-S,\tau+1}, \tag{5.5.52}$$

where $c > 0$ is independent of Φ, u, and γ $(\gamma \geq \gamma_0 > 0)$. The question arises of conditions under which the inclusion $\operatorname{supp}\Phi \subset \overline{G}$ implies that $\operatorname{supp} u \subset \overline{G}$.

Lemma 5.5.1. *Under the conditions of Theorem 5.3.1, let $u \in \mathcal{H}^{T+s,\tau}(\mathbf{R}^{n+1},\gamma)$ be a solution of problem (5.5.41), $s,\tau \in \mathbf{R}$. Then the inclusion* $\operatorname{supp}\Phi \subset \overline{G}$ *implies the inclusion* $\operatorname{supp} u \subset \overline{G}$ *if and only if the equalities*

$$\sum_{1\le k\le N}\int_{-\infty}^{\infty} L^{-1}(\sigma+i\gamma,\xi',\xi_n)(\xi_n+i\sqrt{\sigma^2+\gamma^2+|\xi'|^2})^{t_j+s-(r-m)}$$

$$\times L_{kj}(\sigma+i\gamma,\xi',\xi_n)\widetilde{\Phi}_k(\sigma+i\gamma,\xi',\xi_n)\xi_n^\nu\psi(\xi_n)\,d\xi_n=0 \qquad (5.5.53)$$

$$(\nu=0,\ldots,r-m-1;\quad j=1,\ldots,N).$$

hold for almost all $(\sigma+i\gamma,\xi')$. Here $L_{kj}(\sigma+i\gamma,\xi',\xi_n)$ are the cofactors of the elements $l_{kj}(\sigma+i\gamma,\xi',\xi_n)$ of the matrix $l(\sigma+i\gamma,\xi',\xi_n)$, and $\psi(\xi_n)=\psi(\sigma+i\gamma,\xi',\xi_n)$ is any continuous function admitting an analytic continuation with respect to ζ_n into the half-plane $\zeta=\xi_n+i\eta$, $\eta>0$ and such that $\psi(\zeta)$ is bounded and nonvanishing for $\eta>0$ (in particular, $\psi\equiv 1$), and

$$\widetilde{\Phi}_r(\sigma+i\gamma,\xi',\xi_n)=F(e^{-\gamma t}\Phi_r)(\sigma,\xi).$$

Proof. (cf. [RS1], Lemma 3). It follows from (5.2.28) that system (5.5.51) is equivalent to system (5.3.8) with Φ instead of f. Then we obtain from (5.3.9) and (5.3.10) that

$$\widetilde{e^{-\gamma t}u_j}\langle\xi\rangle_+^{t_j+s-(r-m)}\psi(\xi_n)\xi_n^\nu$$

$$=\sum_{r=1}^{N}(\sigma+i\gamma,\xi')\langle\xi\rangle_+^{t_j+s-(r-m)}$$

$$\times L_{rj}(\sigma+i\gamma,\xi',\xi_n)\,\widetilde{\Phi}_r(\sigma+i\gamma,\xi',\xi_n)\,\xi_n^\nu\,\psi(\xi_n) \qquad (5.5.54)$$

$$\left(j=1,\ldots,N;\quad \nu=0,\ldots,r-m-1;\quad \langle\xi\rangle_+=\xi_n+i\sqrt{\sigma^2+\gamma^2+|\xi'|^2}\right).$$

Suppose that $\operatorname{supp}\Phi\subset\overline{G}$ and $\operatorname{supp} u\subset\overline{G}$. Then the functions $\widetilde{\Phi}_r(\sigma+i\gamma,\xi',\xi_n)$ and $\widetilde{e^{-\gamma t}u_j}=\tilde{u}_j(\sigma+i\gamma,\xi',\xi_n)$ admit analytic continuations with respect to ξ_n into the half-plane $\zeta=\xi_n+i\eta$, $\eta>0$. In this case the function

$$\tilde{u}_j(\sigma+i\gamma,\xi',\zeta)(\zeta+i\sqrt{\sigma^2+\gamma^2+|\xi'|^2})^{t_j+s-(r-m)}\psi(\zeta)\zeta^\nu \qquad (5.5.55)$$

is analytic in the half-plane $\eta>0$. Due to the Cauchy theorem, (5.5.54) gives us after analytic continuation that

$$\sum_{k=1}^{N}\int_{\Gamma} L^{-1}(\sigma+i\gamma,\xi',\zeta)(\zeta+i\sqrt{\sigma^2+\gamma^2+|\xi'|^2})^{t_j+s-(r-m)}$$

$$\times L_{kj}(\sigma + i\gamma, \xi', \zeta)\zeta^{\nu}\psi(\zeta)\,d\zeta = 0 \qquad (5.5.56)$$

$$(j = 1, \ldots, N; \quad \nu = 0, \ldots, r - m - 1),$$

where $\Gamma = \Gamma(\sigma + i\gamma, \xi')$ is a contour in the upper ζ-half-plane that encompasses all the zeros of the polynomial $L_+(\zeta) = L_+(\sigma + i\gamma, \xi', \zeta)$. It can be assumed that Γ consists of an interval $[-\rho, \rho]$ of the real axis and the semicircle $|\zeta| = \rho$, $\Im\zeta > 0$, where $\rho > 0$ is a sufficiently large number. Passing to the limit as $\rho \to +\infty$ in (5.5.56), we obtain required equality (5.5.53), and the necessity is proved.

To prove sufficiency these arguments should be reserved. It follows from (5.5.56) and (5.5.54) that the left hand side of (5.5.54) admits an analytic continuation (5.5.55) into the half-space $\eta > 0$. By a theorem of Paley-Wiener type (see 5.2.3), we then have that $\operatorname{supp} u \subset \overline{G}$ (cf. [RS1], Lemma 3).

In particular, if we set

$$\psi(\zeta) = L_-(\sigma + i\gamma, \xi', \zeta)(\zeta + i\sqrt{\sigma^2 + \gamma^2 + |\xi'|^2})^{-m}, \qquad (5.5.57)$$

then for the solvability of problem (5.5.51) with $\operatorname{supp} \Phi \subset \overline{G}$ in the space

$$\mathcal{H}_+^{T+s,\tau}(R^{n+1}, \gamma) = \{u \in \mathcal{H}^{T+s,\tau}(\mathbf{R}^n, \gamma) : \operatorname{supp} u \subset \overline{G}\} \qquad (5.5.58)$$

we can write a necessary and sufficient condition in the form

$$\sum_{k=1}^{N} \int_{-\infty}^{\infty} \langle\xi\rangle_+^{t_j+s-r} L_{kj}(\sigma + i\gamma, \xi', \xi_n)$$

$$\times\, \widetilde{\Phi}_k(\sigma + i\gamma, \xi', \xi_n)\xi_n^{\nu} L_+^{-1}(\sigma + i\gamma, \xi', \xi_n)\,d\xi_n = 0 \qquad (5.5.59)$$

$$(j = 1, \ldots, N; \quad \nu = 0, \ldots, r - m - 1; \quad \langle\xi\rangle_+ = \xi_n + i\sqrt{\sigma^2 + \gamma^2 + |\xi'|^2}).$$

□

Remark 5.5.2. Applying the tangential Fourier transformation $F'_{t,x' \to \sigma,\xi'}$ to system (5.3.8) with Φ instead of f, we obtain the following system on the axis $x_n \in \mathbf{R}$:

$$l(\sigma + i\gamma, \xi', D_n)\widehat{u}(\sigma + i\gamma, \xi', x_n) = \widehat{\Phi}(\sigma + i\gamma, \xi', x_n).$$

After multiplication by $(\sigma^2 + \gamma^2 + |\xi'|^2)^{\tau}$ and replacement f by Φ, inequalities (5.3.12) can be interpreted as solvability of this system with $\widehat{\Phi} \in \mathcal{H}^{s-S,\tau+1}(\mathbf{R}, (\sigma, \xi', \gamma))$ in $\mathcal{H}^{T+s,\tau}(R, (\sigma, \xi', \gamma))$ on the axis $x_n \in R$ (here (σ, ξ', γ) is a parameter, and $\gamma \geq \gamma_0 > 0$ (see Subsection 5.2.9)). It follows from the proof of Lemma 5.5.1 that condition (5.5.53) is necessary and sufficient for supp $\widehat{\Phi} \subset \{x_n \geq 0\}$ to imply that supp $\widehat{u} \subset \{x_n \geq 0\}$.

5.5.5. We now consider in G hyperbolic boundary value problem (5.1.10). Using identity (5.2.28), we write this problem in the form (5.5.1). Then with the help of (5.2.16) and (5.2.17) (see also (5.5.36) and (5.5.38)) we write it in the form

$$\sum_{k=1}^{N} l_{jk}(D_t+i\gamma, D_{x'}, D_n)(e^{-\gamma t}u_{k0})_+$$

$$= i \sum_{k:s_j+t_k\geq 1} \sum_{1\leq p\leq s_j+t_k} J^p(l_{jk}(D_t+i\gamma, D_{x'}, D_n)(e^{-\gamma t}u_{k,p}\times\delta(x_n)) + (e^{-\gamma t}f_{j0})_+ \tag{5.5.60}$$

$$(j = 1, \ldots, N),$$

$$\sum_{j:\sigma_h+t_j\geq 0} \sum_{\nu=1}^{\sigma_h+t_j+1} b^{hj\nu}(D_t+i\gamma, D_{x'})(e^{-\gamma t}u_{j\nu}) = e^{-\gamma t}\varphi_h \tag{5.5.61}$$

$$(h = 1, \ldots, m).$$

Here

$$f_{j0} = f_j; \quad u_{kp} = D_n^{p-1}u_k\big|_{x_n=0} \quad (k = 1, \ldots, N; \quad 1 \leq p \leq t_k); \tag{5.5.62}$$

g_+ is the continuation of the function g by zero outside G. As in Subsection 5.4.1, let

$$æ = \max\{0, \sigma_1+1, \ldots, \sigma_m+1\}. \tag{5.5.63}$$

It follows directly from Lemma 5.2.1 that for any $s, \tau \in \mathbf{R}$ the closure l of the mapping $u \mapsto lu \quad (u \in (C_0^\infty(\overline{G}))^N)$ acts continuously in the pair of spaces

$$\widetilde{\mathcal{H}}^{T+s,\tau,(T+æ)}(G,\gamma) \longrightarrow \widetilde{\mathcal{H}}^{s-S,\tau,(æ-S)}(G,\gamma), \tag{5.5.64}$$

$$\widetilde{\mathcal{H}}^{T+s,\tau,(T+æ)}(G,\gamma) := \prod_{1\leq j\leq N} \widetilde{\mathcal{H}}^{t_j+s,\tau,(t_j+æ)}(G,\gamma),$$

$$\widetilde{\mathcal{H}}^{s-S,\tau,(æ-S)}(G,\gamma) := \prod_{1\leq j\leq N} \widetilde{\mathcal{H}}^{s-S,\tau,(æ-s)}(G,\gamma),$$

and the closure b of the mapping

$$u \mapsto bu|_{x_n=0} \qquad (u \in (C_0^\infty(\overline{G}))^N)$$

acts continuously from the whole of $\widetilde{\mathcal{H}}^{T+s,\tau,(T+æ)}(G,\gamma)$ into the space

$$\mathcal{H}^{s-\sigma+\tau-1/2}(\partial G,\gamma) := \prod_{1\le h\le m} \mathcal{H}^{s-\sigma_h+\tau-1/2}(\partial G,\gamma). \tag{5.5.65}$$

We recall (see Subsections 5.2.7, 5.2.8)that if

$$\begin{gathered} u = (u_1,\dots,u_N) \in \widetilde{\mathcal{H}}^{T+s,\tau,(T+æ)}(G,\gamma), \\ u_j = (u_{j0},\dots,u_{j,æ+t_j}) \in \widetilde{\mathcal{H}}^{t_j+s,\tau,(t_j+æ)}, \end{gathered} \tag{5.5.66}$$

$$f = (f_1,\dots,f_N) \in \widetilde{\mathcal{H}}^{s-S,\tau+1,(æ-S)}(G,\gamma), \quad f_j = (f_{0j},\dots,f_{j,æ-s_j}), \tag{5.5.67}$$

then the equality $lu = f$ means that equalities (5.5.60) hold, and, moreover,

$$D_n^{p-1}(l_k u)\big|_{x_n=0} = f_{kp} \quad (k : æ - s_k \ge 1;\ 1 \le p \le æ - s_k),$$

or, in more detail (see (5.5.33)),

$$\sum_{j=1}^{N} \sum_{\nu=1}^{s_k+t_j+1} l^{kj\nu}(D_t + i\gamma, D_{x'})(e^{-\gamma t} u_{j,\nu+p-1}) = e^{-\gamma t} f_{kp} \tag{5.5.68}$$

$$(k : æ - s_k \ge 1; \quad 1 \le p \le æ - s_k),$$

and the equality $bu = \varphi$ means that equalities (5.5.61) hold.
Thus, an element $u \in \widetilde{\mathcal{H}}^{T+s,\tau,(T+æ)}(G,\gamma)$ is a solution of problem (5.1.10) with $f \in \widetilde{\mathcal{H}}^{s-S,\tau+1,(æ-S)}(G,\gamma)$ and $\varphi \in \widetilde{\mathcal{H}}^{s-\sigma+\tau}(\partial G,\gamma)$ if equalities (5.5.60), (5.5.68), and (5.5.61) hold.

5.5.6. System (5.5.60) is considered in the whole space $\mathbf{R}^{n+1}$, and both the support of the right hand side and the support of the solution belong to $\overline{G}$. This enables us to use Lemma 5.5.1. We obtain from (5.5.59) the following condition for solvability of system (5.5.60) in $\mathcal{H}_+^{T+s,\tau}(R^{n+1},\gamma)$:

$$\begin{aligned} &\sum_{k=1}^{N} i \int_{-\infty}^{\infty} \langle\xi\rangle_+^{t_j+s-r} L_{kj}(\sigma+i\gamma,\xi',\xi_n) L_+^{-1}(\sigma+i\gamma,\xi',\xi_n) \\ &\qquad \times \sum_{\nu: s_k+t_\nu \ge 1} \sum_{1\le p\le s_k+t_\nu} J^p(l_{k\nu}(\sigma+i\gamma,\xi',\xi_n)\widehat{u}_{\nu p}(\sigma+i\gamma,\xi'))\xi_n^\mu \, d\xi_n \\ &= -\sum_{k=1}^{N} \int_{-\infty}^{\infty} \langle\xi\rangle_+^{t_j+s-r} L_{kj}(\sigma+i\gamma,\xi',\xi_n) L_+^{-1}(\sigma+i\gamma,\xi',\xi_n) \\ &\qquad \times \xi_n^\mu \widetilde{f}_{k0+}(\sigma+i\gamma,\xi',\xi_n)\, d\xi_n \end{aligned} \tag{5.5.69}$$

$$(j = 1, \ldots, N, \quad \mu = 0, \ldots, r - m - 1).$$

Let $\widehat{f}^{j\mu}(\sigma + i\gamma, \xi')$ denote the right hand side of (5.5.69), and let $c_{j\mu\nu p}(\sigma + i\gamma, \xi')$ denote the coefficient of $\widehat{u}_{\nu p}$ on the left hand side. Then (5.5.69) can be written in the form

$$\sum_{\nu : t_\nu \geq 1} \sum_{p=1}^{t_\nu} c_{j\mu\nu p}(\sigma + i\gamma, \xi') \widehat{u}_{\nu p}(\sigma + i\gamma, \xi') = \widehat{f}^{j\mu}(\sigma + i\gamma, \xi') \tag{5.5.70}$$

$$(j = 1, \ldots, N; \quad \mu = 0, \ldots, r - m - 1).$$

Lemma 5.5.2. *For each $(\sigma + i\gamma, \xi')$, $\gamma > 0$, the $N(r-m)$ equations (5.5.70) include $r - m$ equations that are linearly independent, and the rest can be expressed linearly in terms of them.*

Proof. Suppose that $\Phi \in (C_0^\infty(R^{n+1}))^N$, and $\operatorname{supp} \Phi \subset \overline{G}$ in Lemma 5.5.1. Then for solvability of system (5.5.51) in $\mathcal{H}_+^{T+s,\tau}(R^{n+1}, \gamma)$ it is necessary and sufficient that equalities (5.5.53) (and, hence, also (5.5.59)) hold for all $(\sigma + i\gamma, \xi')$, $\gamma > 0$. We first replace the integration in (5.5.59) by integration along a contour $\Gamma = \Gamma(\sigma + i\gamma)$ located in the upper ζ-half-plane and encompassing all the ζ-roots of the polynomial

$$L_+(\zeta) = L_+(\sigma + i\gamma, \xi', \zeta).$$

We obtain equalities (5.5.56), in which ψ has the form (5.5.57). We next take into account that

$$\widetilde{\Phi}_k(\sigma + i\gamma, \xi', \zeta) = \frac{1}{\sqrt{2\pi}} \int_0^\infty \widehat{\Phi}_k(\sigma + i\gamma, \xi', x_n) e^{i\zeta x_n} \, dx_n \quad (k = 1, \ldots, N);$$

here $\widetilde{\Phi}_k(\sigma + i\gamma, \xi', \zeta) = \widetilde{\Phi}_k(\zeta)$ is an analytic continuation into the upper half-plane of the function $\widetilde{\Phi}_k(\sigma + i\gamma, \xi', \xi_n) = \widetilde{\Phi}_k(\xi_n)$, and $\widehat{\Phi}_k(\sigma + i\gamma, \xi', x_n)$ is the partial Fourier transform of the function $e^{-\gamma t}\Phi_k(t, x', x_n)$. Then

$$\int_0^\infty \Big(\sum_{k=1}^N \int_\Gamma \frac{(\zeta + i\sqrt{\sigma^2 + \gamma^2 + |\xi'|^2})^{t_j + s - r}}{L_+(\sigma + i\gamma, \xi', \zeta)} \zeta^\nu L_{kj}(\sigma + i\gamma, \xi', \zeta) e^{i\zeta x_n} \Big) \, d\zeta$$

$$\times \widehat{\Phi}_k(\sigma + i\gamma, \xi', x_n) \, dx_n = 0 \tag{5.5.71}$$

$$(j = 1, \ldots, N, \quad \nu = 0, \ldots, r - m - 1)$$

(see (5.5.59)).

Equalities (5.5.71) mean othogonality in $(L_2(R_+))^N$ of the vector

$$\widehat{\Phi}_1(\sigma+i\gamma,\xi',x_n) = (\widehat{\Phi}(\sigma+i\gamma,\xi',x_n),\dots,\widehat{\Phi}_N(\sigma+i\gamma,\xi',x_n))$$

to the columns of the matrix

$$\int_{\Gamma_-} e^{-i\zeta x_n}(l^+(\zeta))^{-1}\overline{L_-(\zeta)}$$

$$\times\begin{pmatrix} (\zeta - i\sqrt{\sigma^2+\gamma^2+|\xi'|^2})^{t_1+s-r} & & \\ & \ddots & \\ & & (\zeta - i\sqrt{\sigma^2+\gamma^2+|\xi'|^2})^{t_N+s-r} \end{pmatrix}$$

$$\times (E,\zeta E,\dots,\zeta^{r-m-1}E)\,d\zeta. \tag{5.5.72}$$

Here $l^+(\zeta)$ is the matrix formally adjoint to the matrix

$$l(\zeta) = l(\sigma+i\gamma,\xi',\zeta) = (l_{jk}(\zeta)) : l^+(\zeta) = (\overline{l_{kj}(\zeta)}),$$

Γ_- is the contour complex conjugate to the contour Γ, and Γ_- encompasses all the ζ-roots of the equation $\det l^+(\zeta) = 0$ lying in the lower half-plane. It can be verified directly that the columns of matrix (5.5.72) are stable (that is, decreasing as $x_n \to +\infty$) solutions of the problem

$$L^+(D_n)V(x_n) = 0 \tag{5.5.73}$$

on the half-line $\mathbf{R}_+ : 0 \le x_n < \infty$. From the other hand,

$$(l(\sigma+i\gamma,\xi',D_n)\widehat{u}(\sigma+i\gamma,\xi',x_n),V(x_n))_{L_2(R_+))^N}$$

$$= (\widehat{u}(x_n),l^+(D_n)V)_{(L_2(R_+))^N} = 0$$

if V is a stable solution of equation (5.5.73),

$$u_j \in \mathcal{H}^{t_j}(\mathbf{R}^n,\gamma) \quad (j=1,\dots,N),$$

and $\operatorname{supp} u \in \overline{G}$. Therefore, for the solvability of the problem $lu = \Phi$ in $\mathcal{H}^T_+(R^{n+1},\gamma)$ it is necessary that the vector $\widehat{\Phi}(\sigma+i\gamma,\xi',x_n)$ be ortogonal in $(L_2(R_+))^N$ to all the stable solutions of equation (5.5.73). Therefore, each stable solution of problem (5.5.73) is a linear combination of the columns of matrix (5.5.72). However, it follows from the hyperbolicity of problem (5.1.10) that for each $(\sigma+i\gamma,\xi')$ the space of stable solutions of system (5.5.73) is $(r-m)$-dimentional. Thus, among $N(r-m)$ conditions (5.5.70) there are $(r-m)$ conditions that are linearly independent, and the rest can be expressed linearly in terms of them.

In what follows, when we refer to conditions (5.5.70) we shall always assume that at the point $(\sigma+i\gamma,\xi')$ under consideration (and, hence, also in some neighborhood of this point) only $r-m$ linearly independent relations remain of relations (5.5.70), and the others have been thrown out. We also remark that if $(\sigma+i\gamma,\xi')\neq(0,0)$, $\gamma=0$, then the rank of the system (5.5.70) cannot exceed $r-m$ (otherwise it would exeed $r-m$ also in a neighborhood of this point).

We explain the meaning of equalities (5.5.70) with $\widehat{f}^{j\mu}=0$, hat is, (5.5.69) with $\widetilde{f}_{k0+}=0$. Let V_+ be the continuation of

$$V\in\mathfrak{M}^+=\mathfrak{M}^+(\sigma+i\gamma,\xi')$$

by zero to the semi-axis $x_n<0$. We shall use formulas (5.5.35) and (5.5.36). Let

$$L_{jq}(D_n)=L_{jq}(\sigma+i\gamma,\xi',D_n)$$

be the cofactor of the element $l_{jq}(\sigma+i\gamma,\xi',D_n)$ of the matrix $l(\sigma+i\gamma,\xi',D_n)$. Applying to each equality (5.5.36) the operator $L_{jq}(D_n)$ and summing over j, we obtain by (5.1.7) that

$$L_+(\sigma+i\gamma,\xi',D_n)L_-(\sigma+i\gamma,\xi',D_n)V_{q+}(x_n)$$
$$=i\sum_{j=1}^N l_{jq}(\sigma+i\gamma,\xi',D_n)\sum_{k:s_j+t_k\geq1}\ \sum_{1\leq p\leq s_j+t_k}J^p(l_{jk}(\sigma+i\gamma,\xi',D_n))V_{kp}\times\delta(x_n)$$
$$(q=1,\dots,N).$$

Let us pass to Fourier transforms here. After analytic continuation into the half-plane $\zeta=\xi_n+i\eta$, $\eta>0$, we obtain

$$L_-(\sigma+i\gamma,\xi',\zeta)\widetilde{V}_q(\zeta)$$
$$=\Big(i\sum_{j=1}^N\ \sum_{k:s_j+t_k\geq1}\ \sum_{1\leq p\leq s_j+t_k}L_{jq}(\sigma+i\gamma,\xi',\zeta)$$
$$\times J^p(l_{jk}(\sigma+i\gamma,\xi',\zeta))V_{kp}\Big)L_+^{-1}(\sigma+i\gamma,\xi',\zeta)\qquad(q=1,\dots,N).$$

Here the left hand side is an analytic function in the upper half-plane, and for $\gamma>0$ the right hand side has $r-m$ singular points in the upper half-plane. These singular points must be removable, of course. With the help of the division algorithm we find that

$$L_{jq}(\sigma+i\gamma,\xi',\zeta)J^pl_{jk}(\sigma+i\gamma,\xi',\zeta)$$
$$=\ P_{jqpk}(\sigma+i\gamma,\xi',\zeta)L_+(\sigma+i\gamma,\xi',\zeta)+r_{jqpk}(\sigma+i\gamma,\xi',\zeta),$$

where the degree of the polynomial

$$r_{jqpk}(\zeta) = r_{qkpk}(\sigma + i\gamma, \xi', \zeta)$$

is less than $r - m$.

The conditions

$$\sum_{j=1}^{N} \sum_{k:s_j+t_k\geq 1} \sum_{1\leq p\leq s_j+t_k} r_{jqpk}(\sigma + i\gamma, \xi', \zeta)V_{kp} = 0 \qquad (q = 1, \ldots, N)$$

are necessary and sufficient for the corresponding singular points to be removable. If here we equate the coefficients of like powers of ζ to zero, we obtain equalities (5.5.70) with $\widetilde{f}^{j\mu} = 0$ and V_{kp} instead of $\widehat{u}_{kp}$. From this, it follows also that the coefficients

$$c_{j\mu\nu p}(\sigma + i\gamma, \xi')$$

are continuous homogeneous functions of $(\sigma + i\gamma, \xi') \neq (0, 0)$, $\gamma \geq 0$, and that equalities (5.5.70) make sense also for $\gamma = 0$. □

5.5.7. Let us pass to Fourier transforms in (5.5.61) and (5.5.68). We obtain

$$\sum_{j:\sigma_h+t_j\geq 0} \sum_{1\leq\nu\leq\sigma_h+t_j+1} b^{hj\nu}(\sigma + i\gamma, \xi')\widehat{u}_{j\nu}(\sigma + i\gamma, \xi') = \widehat{\varphi}_h(\sigma + i\gamma, \xi') \quad (5.5.74)$$

$$(h = 1, \ldots, m),$$

$$\sum_{j=1}^{N} \sum_{\beta=p}^{s_k+t_j+p} l^{k,j,\beta-p+1}(\sigma + i\gamma, \xi')\widehat{u}_{j\beta}(\sigma + i\gamma, \xi') = \widehat{f}_{kp}(\sigma + i\gamma, \xi') \quad (5.5.75)$$

$$(k : æ - s_k \geq 1, \quad 1 \leq p \leq æ - s_k).$$

Equations (5.5.74), (5.5.75), and (5.5.70) form a linear system of

$$m + Næ - (s_1 + \ldots + s_N) + r - m = t_1 + \ldots + t_N + Næ$$

equations in the same number of unknowns

$$\{\widehat{u}_{j\nu} | j : æ + t_j \geq 1, 1 \leq \nu \leq t_j + æ\}.$$

Denote by $\Delta(\sigma + i\gamma, \xi')$ the determinant of this system.

Lemma 5.5.3. *Suppose that problem (5.1.10) is hyperbolic. Then*

$$\Delta(\sigma + i\gamma, \xi') \neq 0 \qquad (\forall(\sigma + i\gamma, \xi') \neq (0, 0) \quad \gamma \geq 0). \qquad (5.5.76)$$

Proof. Assume the contrary. Suppose that $\Delta(\sigma+i\gamma,\xi')=0$ for some $(\sigma+i\gamma,\xi')\neq(0,0)$, $\gamma\geq 0$ (if $\gamma=0$ then it must to be assumed that $\Delta(\sigma+i\gamma,\xi')=0$ for any $r-m$ equalities in (5.5.70) and all the equalities in (5.5.74) and (5.5.75)). Then homogeneous system (5.5.74), (5.5.75), (5.5.70) with

$$\widehat{\varphi}_h=0 \quad (h=1,\dots,m), \qquad \widehat{f}_{jk}=0 \quad (j:æ-s_j\geq 1,\ 1\leq k\leq æ-s_j)$$

has a nonzero solution

$$\{\widehat{u}_{j\nu}(\sigma+i\gamma,\xi')\,|\,j:t_j+æ\geq 1,\ 1\leq k\leq æ-s_j\}.$$

Let us note that not all the elements

$$\widehat{u}_{jk} \quad (j:t_j\geq 1;\ k=1,\dots,t_j)$$

are equal to zero. Indeed, otherwise we would find successively from system (5.5.75) with $\widehat{f}_{kp}=0$ that all the elements

$$\widehat{u}_{j\nu} \quad (j:t_j+æ\geq 1,\ 1\leq \nu\leq t_j+æ)$$

are equal to zero, which contradicts the assumption. Therefore, the solution

$$V(x_n)=V(\sigma+i\gamma,\xi',x_n)\in\mathfrak{M}^+(\sigma+i\gamma,\xi')$$

of the Cauchy problem

$$\begin{aligned} &l(\sigma+i\gamma,\xi',D_n)V=0 \quad (x_n>0),\\ &D_n^{k-1}V_j\big|_{x_n=0}=\widehat{u}_{jk} \quad (j:t_j\geq 1;\quad k=1,\dots,t_j) \end{aligned} \tag{5.5.77}$$

is also nonzero. Then the element

$$V\in\mathfrak{M}^+(\sigma+i\gamma,\xi')$$

is a nonzero solution of problem (5.5.4) with $a_1=\dots=a_m=0$, which contradicts the Lopatinskii condition. Consequently, to conclude the proof it remains to see that Cauchy problem (5.5.77) is solvable in $\mathfrak{M}^+(\sigma+i\gamma,\xi')$.

Let V_+ be the extension of V by zero to the semi-axis $x_n<0$. Then problem (5.5.77) reduces to system (5.5.36) with $V_{kp}=u_{kp}$. Let

$$L_{jq}(\sigma+i\gamma,\xi',D_n)=L_{jq}(D_n)$$

be the cofactor of the element $l_{jq}(\sigma+i\gamma,\xi',D_n)$ of the matrix $l(\sigma+i\gamma,\xi',D_n)$. Applying to each equality in (5.5.36) the operator $L_{jq}(D_n)$ and summing over j, we obtain as at the end of Subsection 5.5.6 that

$L(\sigma+i\gamma,\xi',D_n)V_{q+}$

$$= i\sum_{j=1}^{N} L_{jq}(D_n) \sum_{k:s_j+t_k\geq 1} \sum_{1\leq p\leq s_j+t_k} J^p(l_{jk}(\sigma+i\gamma,\xi',D_n)) \times \widehat{u}_{kp} \times \delta(x_n)$$

$$(q=1,\dots,N).$$

Let us pass to Fourier transforms. After analytic continuation to the half-plane $\zeta = \xi_n + i\eta$, $\eta > 0$, we obtain

$$L(\sigma+i\gamma,\xi',\zeta)\widetilde{V}_{q+}(\sigma+i\gamma,\xi',\zeta)$$

$$= i\sum_{j=1}^{N} L_{jq}(\sigma+i\gamma,\xi',\zeta) \sum_{k:s_j+t_k\geq 1} \sum_{1\leq p\leq s_j+t_k} J^p(l_{jk}(\sigma+i\gamma,\xi',\zeta))\widehat{u}_{kp} \quad (5.5.78)$$

$$(q=1,\dots,N).$$

But, by (5.1.7),

$$L(\sigma+i\gamma,\xi',\zeta) = L_+(\sigma+i\gamma,\xi',\zeta)L_-(\sigma+i\gamma,\xi',\zeta), \quad (5.5.79)$$

and it follows from equalities (5.5.70) with $\widehat{f}^{j\mu} = 0$, that is, from (5.5.69) with $\widetilde{f}_{k0+} = 0$, that the right hand side of (5.5.79), as a polynomial in ζ, is divisible by the polynomial

$$L_+(\zeta) = L_+(\sigma+i\gamma,\xi',\zeta).$$

Thus, by (5.5.78),

$$L_-(\sigma+i\gamma,\xi',\zeta)\widetilde{V}_q(\sigma+i\gamma,\xi',\zeta) = i\sum_{k:s_j+t_k\geq 1} \sum_{1\leq p\leq s_j+t_k} P_{jqkp}(\sigma+i\gamma/\xi',\zeta)\widehat{u}_{kp} \quad (5.5.80)$$

$$(q=1,\dots,N);$$

here the polynomial

$$P_{jqkp}(\zeta) = P_{jqkp}(\sigma+i\gamma,\xi',\zeta)$$

is a homogeneous function of degree $m - t_q + t_k - p$; this polynomial is the quotient upon division of the polynomial

$$L_{jq}(\sigma+i\gamma,\xi',\zeta)J^p(l_{jk}(\sigma+i\gamma,\xi',\zeta))$$

by $L_+(\zeta) = L_+(\sigma+i\gamma,\xi',\zeta)$. With the help of the division algorothm it is easy to see that the coefficients of the powers of ζ in this polynomial are homogeneous and continuous functions of $(\sigma+i\gamma,\xi')$. From (5.5.80) we obtain

$$L_-(\sigma+i\gamma,\xi',D_n)V_{q+}(x_n)$$

$$= i \sum_{j=1}^{N} \sum_{k:s_j+t_k \geq 1} \sum_{p=1}^{s_j+t_k} P_{jqkp}(\sigma + i\gamma, \xi', D_n)\hat{u}_{kp} \times \delta(x_n) \quad (5.5.81)$$

$$(q = 1, \ldots, N).$$

Thus, the problem of solvability in $\mathfrak{M}^+(\sigma + i\gamma, \xi')$ of Cauchy problem (5.5.77) has been reduced to problem (5.5.81) on the whole axis. The solvability of the latter follows from the solvability of the Cauchy problems

$$L_-(\sigma + i\gamma, \xi', D_n)V_q(x_n) = 0 \quad (x_n > 0),$$

$$D_n^{k-1}V_q(x_n)\big|_{x_n=0} = \hat{u}_{qk} \quad (k = 1, \ldots, m) \qquad (5.5.82)$$

$$(q = 1, \ldots, N).$$

Moreover, if $t_q + æ < m$ then it is necessary to find the missing values of $\hat{u}_{qk}$ from equalities (5.5.68) with $f_{kp} = 0$ and $p = æ - s_k + 1, æ - s_k + 2, \ldots$, and so on. The solvability in $\mathfrak{M}^+(\sigma + i\gamma, \xi')$ of Cauchy problem (5.5.77) has thus been estblished, and Lemma 5.5.3 is proved. □

It is easy to see that the condition L_5:

$$\Delta(\sigma + i\gamma, \xi') \neq 0 \quad (\forall(\sigma + i\gamma, \xi') \neq (0, 0), \gamma \geq 0)$$

is equivalent to the Lopatinskii condition. For the case of a diagonal matrix this was proved in Subsection 5.5.3. In the general case the implication $\mathrm{L}_2 \Rightarrow \mathrm{L}_5$ was established in Lemma 5.5.3. The converse implication can be proved just as for a diagonal matrix.

5.6. Proof of Theorem 5.4.1

5.6.1.

Let us consider the hyperbolic problem (5.1.10), rewritten in the form (5.5.60), (5.5.61), (5.5.68). Then we obtain linear system (5.5.74), (5.5.75), (5.5.70) for determining of

$$\{\hat{u}_{j\nu} | j : t_j + æ \geq 1, \nu = 1, \ldots, t_j + æ\};$$

in addition, the determinant $\Delta(\sigma + i\gamma, \xi')$ of this system is nonzero because problem (5.1.10) is hyperbolic. This enables us to express and estimate the elements $\{u_{j\nu}\}$ in terms of $\{\varphi_h\}$, $\{f_{jk}\}$, and $\{f^{j\mu}\}$. We first estimate $f^{j\mu}$ in terms of $f_0 = (f_{10}, \ldots, f_{N0})$. To do this we first present the following estimate of Cheztain and Piriou (see [ChP], Proposition 10.1), and then we estimate $f^{j\mu}$ in terms of f_0.

Lemma 5.6.1. *[ChP] There exists a constant $c > 0$ independent of $(\sigma + i\gamma, \xi')$, $\gamma > 0$, such that*

$$\int_{-\infty}^{+\infty} \left| \Theta^+ F^{-1}_{\xi_n \to x_n} \frac{\xi^{h-1}}{L_+(\sigma + i\gamma, \xi', \xi_n)} \right|^2 dx_n \leq \frac{c}{\gamma}(\sigma^2 + \gamma^2 + |\xi'|^2)^{-(r-m)+h} \quad (5.6.1)$$

$$(h = 1, \ldots, r - m).$$

Here $r - m$ is the degree of $L_+(\xi_n)$, and Θ^+ is the operator introduced in 5.2.4

Corollary 5.6.1. *Estimate (5.6.1) is true also for each natural number $h > r - m$.*

Indeed, separating out the integral part with the help of the division algorithm, we obtain that

$$\xi_n^{h-1}(L_+(\sigma + i\gamma, \xi', \xi_n))^{-1} = P(\xi_n) + r(\xi_n)(L_+(\sigma + i\gamma, \xi', \xi_n))^{-1}.$$

Here $P(\xi_n)$ is a polynomial of degree $h - 1 - (r - m)$, and

$$r(\xi_n) = \sum_{1 \leq j \leq r-m} r_j \xi_n^{j-1}, \quad r_j = r_j(\sigma + i\gamma, \xi')$$

are homogeneous functions of orders $h - (r - m) - j$. But $F^{-1}_{\xi_n \to x_n} P(\xi_n)$ is a linear combination of derivatives of the Dirac measure concentrated at the point $x_n = 0$, and hence

$$\Theta^+ F^{-1}_{\xi_n \to x_n} P(\xi_n) = 0, \quad \Theta^+ F^{-1}(\xi_n^{h-1}/L_+(\xi_n) = \Theta^+ F^{-1}(r(\xi_n)/L_+(\xi_n)),$$

and the case $h > (r - m)$ reduces to the case already treated in Lemma 5.6.1.

Lemma 5.6.2. *Let*

$$f_0 = (f_{10}, \ldots, f_{N0}) \in \mathcal{H}^{s-S,\tau+1}(G, \gamma) = \prod_{1 \leq k \leq N} \mathcal{H}^{s-s_k,\tau+1}(G, \gamma),$$

and

$$\widehat{e^{-\gamma t} f^{j\mu}}(\sigma, \xi') = \widehat{f}^{j\mu}(\sigma + i\gamma, \xi')$$

$$= -\sum_{k=1}^{N} \int_{-\infty}^{+\infty} \frac{(\xi_n + i\sqrt{\sigma^2 + \gamma^2 + |\xi'|^2})^{t_j+s-r}}{L_+(\sigma + i\gamma, \xi', \xi_n)}$$

$$\times L_{kj}(\sigma + i\gamma, \xi', \xi_n) \xi_n^{\mu} \widetilde{f}_{k0}(\sigma + i\gamma, \xi', \xi_n)\, d\xi_n \quad (5.6.2)$$

$$(j = 1,\dots,N;\quad \mu = 0,\dots,r-m-1).$$

Then

$$f^{j\mu} \in \mathcal{H}^{r-m-\mu+\tau}(\partial G,\gamma),$$

and

$$|f^{j\mu},\partial G,\gamma|^{\alpha}_{r-m-\mu+\tau} \le \frac{c}{\gamma}\|f_0\|^2_{\mathcal{H}^{s-S,\tau+1}(G,\gamma)} \tag{5.6.3}$$

$$(j = 1,\dots,N;\quad \mu = 0,\dots,r-m-1).$$

Proof. If $s - s_k < 0$ then $\mathcal{H}^{s-s_k,\tau+1}(G,\gamma)$ is isometrically equivalent to the subspace

$$\mathcal{H}^{s-s_k,\tau+1}_{\overline{G}}(R^{n+1},\gamma) = \{g \in \mathcal{H}^{s-s_k,\tau}(R^{n+1},\gamma):\quad \operatorname{supp} g \subset \overline{G}\}$$

of the space $\mathcal{H}^{s-s_k,\tau+1}(\mathbf{R}^n,\gamma)$. Therefore, in this case we can assume that

$$f_{k0} = f_{k0+},$$

and

$$|F_{k0},G,\gamma|^2_{s-s_k,\tau+1}$$

$$= \int |\widetilde{f}_{k0+}(\sigma+i\gamma,\xi',\xi_n)|^2(\sigma^2+\gamma^2+|\xi|^2)^{s-s_k}(\sigma^2+\gamma^2+|\xi'|^2)^{\tau+1}\,d\sigma d\xi. \tag{5.6.4}$$

By a theorem of Paley–Wiener type (see 5.2.3), the element

$$\widetilde{f}_{k0+}(\sigma+i\gamma,\xi',\xi_n)$$

can be continuated analytically with respect to ξ_n into the upper half-plane $\zeta = \xi_n + i\eta$, $\eta > 0$. Then the function

$$\widetilde{f}_{k0+}(\sigma+i\gamma,\xi',\xi_n)\langle\xi\rangle_+^{s-s_k}L_{kj}(\sigma+i\gamma,\xi',\xi_n)\langle\xi\rangle_+^{t_j+s_k-r}$$

$$(\langle\xi\rangle_+ = \xi_n + i\sqrt{\sigma^2+\gamma^2+|\xi'|^2})$$

(see (5.2.8)) also can be continuated analytically into the upper half-plane, and, by a theorem of Paley–Wiener type,

$$\operatorname{supp} F^{-1}(\widetilde{f}_{k0+}(\sigma+i\gamma,\xi',\xi_n)\langle\xi\rangle_+^{s-s_k}L_{kj}(\sigma+i\gamma,\xi',\xi_n)\langle\xi\rangle_+^{-(r-t_j-s_k)}) \subset \overline{G}.$$

Then by Parseval's equality and the Cauchy–Bunyakovskii–Schwarz inequality we obtain

$$\Big|\int_{-\infty}^{\infty}(\widetilde{f}_{k0+}(\sigma+i\gamma,\xi',\xi_n)\langle\xi\rangle_+^{s-s_k}L_{kj}(\sigma+i\gamma,\xi',\xi_n)\langle\xi\rangle_+^{-(r-t_j-s_k)})$$

$$\times (\xi_n^\mu L_+^{-1}(\sigma + i\gamma, \xi', \xi_n))\, d\xi_n \Big|^2$$

$$= \Big| \sqrt{2\pi} \int_0^\infty F^{-1}_{\xi_n \to x_n}(\ldots) F^{-1}_{\xi_n \to x_n} \left(\frac{\xi_n^\mu}{L_+(\xi_n)} \right) dx_n \Big|^2$$

$$\leq \int_{-\infty}^{\infty} |\tilde{f}_{k0+}(\sigma + i\gamma, \xi', \xi_n) \langle \xi \rangle_+^{s-s_k} L_{kj}(\sigma + i\gamma, \xi', \xi_n) \langle \xi \rangle_+^{-(r-t_j-s_k)}|^2\, d\xi_n$$

$$\times \int_{-\infty}^{\infty} \left| \Theta^+ F^{-1}_{\xi_n \to x_n} \frac{\xi_n^\mu}{L_+(\sigma + i\gamma, \xi', \xi_n)} \right|^2 dx_n$$

$$\leq \frac{c}{\gamma} \int_{-\infty}^{\infty} |\tilde{f}_{k0+}(\sigma + i\gamma, \xi', \xi_n)(\sigma^2 + \gamma^2 + |\xi|^2)^{s-s_k} (\sigma^2 + \gamma^2 + |\xi'|^2)^{\mu-(r-m)+1}\, d\xi_n. \tag{5.6.5}$$

To obtain the last estimate we employ Lemma 5.6.1, Corollary 5.6.1, and the obvious estimate

$$|L_{kj}(\sigma + i\gamma, \xi', \xi_n) \langle \xi \rangle_+^{t_j+s_k-r}| \leq c_1.$$

If $s - s_k > 0$ and

$$f_{k0} \in \mathcal{H}^{s-s_k, \tau+1}(G, \gamma)$$

then, because of the continuity of the imbedding

$$\mathcal{H}^{s-s_k, \tau+1}(G, \gamma) \subset \mathcal{H}^{0, s-s_k+\tau+1}(G, \gamma) \tag{5.6.6}$$

we can repeat calculations (5.6.5) with $s - s_k$ replaced by 0, and τ by $s - s_k + \tau$. As a result we obtain from (5.6.5) and (5.6.2) required estimate (5.6.3). Indeed, if s is an integer such that $s - s_k > 0$, then

$$\Big| \int_{-\infty}^{\infty} \left(\tilde{f}_{k0+}(\sigma + i\gamma, \xi', \xi_n) \frac{L_{kj}(\sigma + i\gamma, \xi', \xi_n)}{\langle \xi \rangle^{r-t_j-s_k}} \right) \left(\frac{\xi_n^\mu \langle \xi \rangle_+^{s-s_k}}{L_+(\sigma + i\gamma, \xi', \xi_n)} \right) d\xi_n \Big|^2$$

$$= \Big| \sqrt{2\pi} \int_0^\infty F^{-1}_{\xi_n \to x_n}(\ldots) F^{-1}_{\xi_n \to x_n}(\ldots)\, dx_n \Big|^2$$

$$\leq c_1 \int_{-\infty}^{\infty} |\tilde{f}_{k0+}(\sigma + i\gamma, \xi', \xi_n)|^2\, d\xi_n \int_{-\infty}^{\infty} \left| \Theta^+ F^{-1}_{\xi_n \to x_n} \frac{\xi_n^\mu \langle \xi \rangle_+^{s-s_k}}{L_+(\sigma + i\gamma, \xi', \xi_n)} \right|^2 dx_n$$

$$\leq \frac{c_2}{\gamma} (\sigma^2 + \gamma^2 + |\xi'|^2)^{\mu-(r-m)+s-s_j+1} \int_{-\infty}^{\infty} |\tilde{f}_{k0+}(\sigma + i\gamma, \xi', \xi_n)|^2\, d\xi_n.$$

As a result we obtain the estimate

$$|F^{j\mu}, \partial G, \gamma|_{r-m-\mu+\tau}$$
$$\leq \frac{c}{\gamma}\Big(\sum_{k:s-s_k\leq 0} |f_{k0}, G, \gamma|^2_{s-s_k,\tau+1} + + \sum_{k:s-s_k>0} |f_{k0}, G, \gamma|^2_{0,s-s_k+\tau+1}\Big)$$
$$\leq \frac{c}{\gamma}\sum_{k=1}^{N} |f_{k0}, G, \gamma|_{s-s_k,\tau+1},$$

and required estimate (5.6.3) is established for integral s. For nonintegral s it follows from this by the interpolation theorem. □

5.6.2. In the system (5.5.74), (5.5.75), (5.5.70) we now express $\{\widehat{u}_{j\nu}\}$ in terms of the right hand sides. We remark that the coefficient $b^{hj\nu}(\sigma+i\gamma,\xi')$ of $\widehat{u}_{j\nu}$ in system (5.5.74) is a continuous homogeneous function of degree $\sigma_h + t_j + 1 - \nu$. Therefore,

$$|b^{hj\nu}(\sigma+i\gamma,\xi')| \leq c(|\sigma|+|\gamma|+|\xi'|)^{\sigma_h+t_j+1-\nu}; \tag{5.6.7}$$

similarly, the coefficient $l^{k,j,\beta-p+1}(\sigma+i\gamma,\xi')$ of $\widehat{u}_{j\beta}$ in system (5.5.75) is a continuous homogeneous function of degree $s_k + t_j - \beta + p$. Therefore,

$$|l^{k,j,\beta-p+1}(\sigma+i\gamma,\xi')| \leq c(|\sigma|+|\gamma|+|\xi'|)^{s_k+t_j-\beta+p}$$
$$(j : æ - s_j \geq 1, \quad 1 \leq p \leq æ - s_j). \tag{5.6.8}$$

The coefficient $c_{j\mu\nu p}(\sigma + i\gamma,\xi')$ of $\widehat{u}_{\nu p}$ in system (5.5.70) is a continuous homogeneous function of degree $s - (r-m) + t_\nu - p + 1 + \mu$. Therefore,

$$|c_{j\mu\nu p}(\sigma+i\gamma,\xi')| \leq c(|\sigma|+|\gamma|+|\xi'|)^{s-(r-m)+t_\nu-p+1+\mu}$$
$$(j = 1,\dots,N,\ \mu = 0,\dots,r-m-1,\ \nu : t_\nu \geq 1,\ p = 1,\dots,t_\nu). \tag{5.6.9}$$

According to Lemma 5.5.2, for each $(\sigma + i\gamma,\xi')$, $\gamma > 0$, the rank of the matrix A_1 of coefficients of the function $\widehat{u}_{\nu p}$ in system (5.5.70) is equal to $r-m$. Since the rank cannot exceed $r-m$ even for $\gamma \geq 0$, this and Lemma 5.5.3 give us that for each $(\sigma + i\gamma,\xi') \neq (0,0)$, $\gamma \geq 0$, this rank is equal to $r-m$.

Let

$$S_+^n = \{(\sigma+i\gamma,\xi') : \sigma^2+\gamma^2+|\xi'|^2 = 1,\ \gamma \geq 0\}.$$

For each point $(\sigma_0,\gamma_0,\xi_0')$ there exists a nonzero minor of A_1 of order $r-m$. By continuity, this minor is nonzero in some neighborhood $U(\sigma_0,\gamma_0,\xi_0')$ of this point in S_+^n. It can be assumed without loss of generality that the minor iz nonzero also in the closure $\overline{U}$ of U in S_+^n. Let $U_1, \dots, U_k$ be a finite covering of S_+^n by the indicated neighborhoods. For each neighborhood U_j we single out $r - m$ equations in (5.5.70) that contain the minor under consideration. This system of $r - m$ equations will be considered in the cone

$$V_j = \{(\sigma + i\gamma, \xi') \in \mathbf{R}^{n+1} : (\sigma + i\gamma, \xi')/|(\sigma + i\gamma, \xi')| \in U_j.$$

In referring below in this subsection to (5.5.70) for $(\sigma + i\gamma, \xi') \in V_j$ we shall always assume that we are working only with the indicated $r - m$ equations.

System (5.5.74), (5.5.75), (5.5.70) has the structure of a Leray–Volevich system: if we associate the number $t_j - \nu + 1$ with the column of coefficients of the elements $\widehat{u}_{j\nu}$, the number σ_h with the row with index h in system (5.5.74), the number $s_k + p - 1$ with the row with index (k, p) in system (5.5.75), and the number $s - (r - m) + \mu$ with the row with index (j, μ) in the system (5.5.70), then, in view of (5.6.7)–(5.6.9) the order of homogeneity of the coefficient of $\widehat{u}_{j\nu}$, which lies at the intersection of the indicated rows and columns, is equal to the sum of the corresponding numbers. Therefore, the determinant

$$\Delta(\sigma + i\gamma, \xi') \quad ((\sigma + i\gamma, \xi') \in V_j)$$

is a homogeneous function of order M_j, and

$$c^{-1}(\sigma^2 + \gamma^2 + |\xi'|^2)^{M_j/2} \le |\Delta(\sigma + i\gamma, \xi')| \le c(\sigma^2 + \gamma^2 + |\xi'|^2)^{M_j/2}.$$

Therefore, if t_{jk} is the order of the element $a_{jk}(\sigma + i\gamma, \xi')$, and $A_{jk}(\sigma + i\gamma, \xi')$ is the cofactor of this element, then

$$\left| \frac{A_{jk}(\sigma + i\gamma, \xi')}{\Delta(\sigma + i\gamma, \xi')} \right| \le c\, (\sigma^2 + \gamma^2 + |\xi'|^2)^{-t_{jk}/2},$$

where $c > 0$ does not depend on $(\sigma + i\gamma, \xi') \in V_j$.

Then the solutions $\{\widehat{u}_{j\nu}(\sigma + i\gamma, \xi')\}$ of system (5.5.74), (5.5.75), (5.5.70) satisfy the estimate

$$|\widehat{u}_{j\nu}(\sigma + i\gamma, \xi')|^2$$

$$\le c_1 \Big(\sum_{h=1}^{m} |\widehat{\varphi}_h(\sigma + i\gamma, \xi')|^2 (\sigma^2 + \gamma^2 + |\xi'|^2)^{-(t_j - \nu + 1 + \sigma_h)}$$

$$+ \sum_{k : æ - s_k \ge 1} \sum_{p=1}^{æ - s_k} |\widehat{f}_{kp}(\sigma + i\gamma, \xi')|^2 (\sigma^2 + \gamma^2 + |\xi'|^2)^{-(s_k + p + t_j - \nu)} +$$

$$+\sum_{j=1}^{N}\sum_{\mu=0}^{r-m-1}|\hat{f}^{j\mu}(\sigma+i\gamma,\xi')|^2(\sigma^2+\gamma^2+|\xi'|^2)^{-(t_j-\nu+1+s-(r-m)+\mu)}\Big) \tag{5.6.10}$$

$$(j=1,\ldots,N;\quad \nu=1,\ldots,t_j+æ).$$

Multiplying the left and right hand sides of (5.6.10) by the term $(\sigma^2+\gamma^2+|\xi'|^2)^{t_j-\nu+1+\tau+s}$, integrating with respect to $(\sigma,\xi')\in\mathbf{R}^{n+1}$, and using the estimate (5.6.3), we obtain the following estimate:

$$|u_{j\nu},\partial G,\gamma|^2_{t_j+s-\nu+1+\tau}$$

$$\leq c\Big(\sum_{h=1}^{m}|\varphi_h,\partial G,\gamma|^2_{s+\tau-\sigma_h}+\sum_{j:æ-s_j\geq 1}\sum_{p=1}^{æ-s_j}|f_{jp},\partial G,\gamma)|^2_{s-s_j-p+1+\tau}$$

$$+\frac{1}{\gamma}\sum_{k=1}^{N}|f_{k0},G,\gamma|^2_{s-s_k,\tau+1}\Big) \tag{5.6.11}$$

$$(j=1,\ldots,N,\quad \nu=1,\ldots,t_j+æ,\quad \gamma\geq\gamma_0>0).$$

Thus, we have proved

Lemma 5.6.3. *Suppose that*

$$f\in\widetilde{\mathcal{H}}^{s-S,\tau+1,(-S+æ)}(G,\gamma),$$

$$\varphi_h\in\mathcal{H}^{s-\sigma_h+\tau}(\partial G,\gamma)\quad(h=1,\ldots,m),$$

$$\gamma\geq\gamma_0>0,$$

and problem (5.1.10) is hyperbolic. Then there exist elements u_0 and

$$u_{j\nu}\in\mathcal{H}^{t_j+s+\tau-\nu+1}(\partial G,\gamma)\quad(j:t_j+æ\geq 1,\quad \nu=1,\ldots,t_j+æ) \tag{5.6.12}$$

such that equalities (5.5.60), (5.5.61), (5.5.68) (or equalities (5.5.74), (5.5.75), (5.5.70)) are valid. Elements (5.6.12) are uniquely determined by these equalities, and estimates (5.6.11) hold with a constant $c>0$ independent of f, φ, and γ ($\gamma\geq\gamma_0>0$). If

$$\operatorname{supp} f\subset\overline{G}_+=\{(t,x)\in\overline{G}:t\geq 0\},$$

and

$$\operatorname{supp}\varphi\subset(\partial G)_+=\{(t,x)\in\partial G:t\geq 0\},$$

then also $\operatorname{supp}U_0\subset\overline{G}_+$ and $\operatorname{supp}u_{j\nu}\subset(\partial G)_+$.

It is clear that elements (5.6.12) satisfy equalities (5.5.61) and (5.5.68); and if we substitute these elements into (5.5.60) then we obtain a system in $\mathbf{R}^{n+1}$ for determining

$$\{(e^{-\gamma t}u_{k0})_+ : k = 1, \dots, N\}.$$

By Theorem 5.3.1, this system has a solution, and by Lemma 5.5.1, the support of the solution belongs to $\overline{G}$.

5.6.3. Let us investigate in greater detail system (5.5.60). As already mentioned, its solvability follows from Theorem 5.3.1. However, the use of Theorem 5.3.1 gives somewhat underestimated results on the regularity of u_0. To see this we first consider the case of $t_1 + s < 1/2$. Then it follows from (5.6.12) that

$$u_{j\nu} \times \delta(x_n) \in \mathcal{H}^{t_j+s-\nu, \tau+1/2}(G, \gamma) \quad (t_1 + s < 1/2);$$

$$|u_{j\nu} \times \delta(x_n), G, \gamma|_{t_j+s-\nu, \tau+1/2} \leq |u_{j\nu}, \partial G, \gamma|_{t_j+s-\nu+1+\tau} \tag{5.6.13}$$

$$(j : t_j + æ \geq 1; \quad \nu = 1, \dots, t_j + æ).$$

Indeed, suppose that

$$v(t, x', x_n) \in C_0^\infty(\mathbf{R}^n).$$

Then

$$\begin{aligned} &|(u_{j\nu} \times \delta(x_n), v(t, x', x_n)| \\ &\quad = |\langle u_{j\nu}, v(t, x', 0)\rangle| \\ &\quad \leq |u_{j\nu}, \partial G, \gamma|_{t_j+s-\nu+\tau+1} |v(t, x', 0), \partial G, \gamma|_{-(t_j+s-\nu)-(\tau+1)} \\ &\quad \leq |u_{j\nu}, \partial G, \gamma|_{t_j+s-\nu+\tau+1} |v(t, x', x_n), G, \gamma|_{-t_j-s+\nu, -\tau-1/2}, \end{aligned} \tag{5.6.14}$$

and estimate (5.6.13) follows. This implies that

$$g_j = i \sum_{k: s_j+t_k \geq 1} \sum_{p=1}^{s_j+t_k} J^p(l_{jk}(D_t + i\gamma, D_x))(e^{-\gamma t}u_{kp}) \times \delta(x_n)$$

$$\in H^{s-s_j, \tau+1/2}(R^{n+1}, \gamma) \qquad (j = 1, \dots, N),$$

while we have that $(e^{-\gamma t} f_{j0})_+ \in H^{s-s_j, \tau+1}(\mathbf{R}^{n+1}, \gamma)$. Therefore,

$$g_j + (e^{-\gamma t} f_{j0})_+ \in H^{s-s_j, \tau+1/2}(\mathbf{R}^{n+1}, \gamma),$$

and thus

$$(e^{-\gamma t}u_{k0})_+ \in H_{\overline{G}}^{t_k+s, \tau-1/2}(\mathbf{R}^{n+1}, \gamma) \quad (k = 1, \dots, N)$$

in this case by Theorem 5.3.1, while (as will be shown below) it is actually true that

$$(e^{-\gamma t}u_{k0})_+ \in H_{\overline{G}}^{t_k+s,\tau}(\mathbf{R}^{n+1},\gamma) \quad (k=1,\ldots,N).$$

In view of the similar arguments we can deduce from Lemma 5.6.3 and Theorem 5.3.1 that if

$$f \in \widetilde{\mathcal{H}}^{s-S,\tau+1,(T+æ)}(G,\gamma), \quad \varphi \in \prod_{1\le h\le m} \mathcal{H}^{s-\sigma_h+\tau}(\partial G,\gamma),$$

$$s,\tau\in\mathbf{R}, \quad \gamma\ge\gamma_0>0,$$

and problem (5.1.10) is hyperbolic, then there exists one and only one solution

$$u=(u_1,\ldots,u_N)\in\widetilde{\mathcal{H}}^{T+s,\tau-1/2,(T+æ)}(G,\gamma)$$

of this problem, and this solution depends continuously on (f,φ). We shall not develop these arguments in detail because, as will be shown below,

$$u\in\widetilde{\mathcal{H}}^{T+s,\tau,(T+æ)}(G,\gamma).$$

Furthermore, in the next subsection we go through such arguments in detail for the case $f=0$. In the last case it is possible to avoid an additional loss of smoothness.

5.6.4. Let us consider problem (5.5.60), (5.5.61), (5.5.68) with $f=0$ abd $\gamma\ge\gamma_0>0$. Using Lemma 5.6.3, we find the elements $\{u_{j\nu}\}$, and then we pass in (5.5.60) (with $f=0$) to Fourier transforms and obverse equalities (5.5.78), (5.5.80). We obtain, that

$$\widetilde{u}_{q0+}(\sigma+i\gamma,\xi',\xi_n)$$

$$=\Big(i\sum_{j=1}^{N}\sum_{k:s_j+t_k\ge1}\sum_{1\le p\le s_j+t_k}P_{jqkp}(\sigma+i\gamma,\xi',\xi_n)\widehat{u}_{kp}\times L_-^{-1}(\sigma+i\gamma,\xi',\xi_n) \tag{5.6.15}$$

$$(q=1,\ldots,N).$$

Suppose first that $t_q+s>0$, s is an integer, and $\tau\in\mathbf{R}$. Then by (5.2.9) we have that

$$|u_{q0},G,\gamma|^2_{t_q+s,\tau}$$

$$\le c_1\|\Theta^+F^{-1}_{\xi_n\to x_n}\langle\xi\rangle_-^{t_q+s}\widetilde{u}_{q0+}(\sigma+i\gamma,\xi',\xi_n),\mathbf{R}^{n+1},\gamma\ |^2_{0,\tau}$$

$$\leq c_2 \Bigg\| \Theta^+ F^{-1}_{\xi_n \to x_n} \langle\xi\rangle_-^{t_q+s}$$

$$\times \sum_{j=1}^{N} \sum_{k:s_j+t_k\geq 1} \sum_{1\leq p\leq s_j+t_k} P_{jqkp}(\sigma+i\gamma,\xi',\xi_n)\, L_-^{-1}(\sigma+i\gamma,\xi',\xi_n)$$

$$\times\, \widehat{u}_{kp}(\sigma+i\gamma,\xi')(\sigma^2+\gamma^2+|\xi'|^2)^{\tau/2} \Bigg\|^2_{L_2(R^{n+1})}, \quad (5.6.16)$$

$$\langle\xi\rangle_- = \xi_n - i\sqrt{\sigma^2+\gamma^2+|\xi'|^2}.$$

We now use an assertion analogous to Lemma 5.6.1 and Corollary 5.6.1 (see [ChP]). We obtain, for an integer s,

$$|u_{q0}, G, \gamma|^2_{t_q+s,\tau}$$

$$\leq \frac{c_3}{\gamma} \sum_{j=1}^{N} \sum_{k:s_j+t_k\geq 1} \sum_{1\leq p\leq s_j+t_k} \int |\widehat{u}_{kp}(\sigma+i\gamma,\xi')|^2(\sigma^2+\gamma^2+|\xi'|^2)^{t_k+s-p+1+\tau}\, d\sigma d\xi'$$

$$\leq \frac{c_4}{\gamma} \sum_{k:t_k+æ\geq 1} \sum_{p=1}^{t_k+æ} |u_{k,p}, \partial G, \gamma|^2_{t_k+s+\tau-p+1} \quad (5.6.17)$$

$$(\forall q: t_q+s\geq 0).$$

This and (5.6.11) imply the estimate

$$|u_{q0}, G, \gamma|^2_{t_q+s,\tau} \leq \frac{c}{\gamma} \sum_{h=1}^{m} |\varphi_h, \partial G, \gamma|_{s-\sigma_h,\tau} \quad (5.6.18)$$

$$(\forall q: t_q+s\geq 0,\ s \text{ an integer}).$$

By interpolation we obtain estimate (5.6.18) also for non-integers s with $(t_q+s\geq 0)$.
Suppose that

$$s \geq æ = \max\{0, \sigma_1+1, \ldots, \sigma_m+1\}. \quad (5.6.19)$$

Then the element $u_0 = \mathcal{H}^{T+s,\tau}(G,\gamma)$ is a solution in G of the system $lu_0 = 0$ (this follows immediately from (5.5.60) and the smoothness of the restriction u_0 of u_{0+} to G). Then there hold the equalities obtained by replacing u_{jk} by $(D_n^{k-1}u_{j0})|_{x_n=0}$ in (5.5.60). The elements $u_{jk} = D_n^{k-1}u_{j0}|_{x_n=0}$ satisfy the system (5.5.70), (5.5.74), (5.5.75) with

$$\widehat{f}^{j\mu} = 0 \quad (j=1,\ldots,N,\ \mu=0,\ldots,r-m-1),$$

$$\widehat{\varphi}_h = 0 \quad (h=1,\ldots,m),$$

$$\widehat{f}_{kp} = 0 \quad (æ - s_k \geq 1, 1 \leq p \leq æ - s_k),$$

and the determinant of this system is nonzero. Therefore,

$$D_n^{k-1} u_{j0}|_{x_n=0} = u_{jk} \quad (j : t_j + æ \geq 1, \;\; k = 1, \ldots, t_j + æ). \tag{5.6.20}$$

In the general case we approximate the right hand sides by smooth elements, pass to the limit, and see that the equality $D_n^{k-1} u_{q0}|_{x_n=0} = u_{qk}$ holds if $t_q + s - k + 1/2 > 0$.

Let us now prove that estimate (5.6.18) is also true when $t_q + s < 0$. Indeed, if $t_q + s < 0$, $\tau \in \mathbf{R}$, and $\gamma \geq \gamma_0 > 0$, then

$$\begin{aligned}
&|u_{q0}, G, \gamma|^2_{t_q+s,\tau} \\
&= |u_{q0+}, \mathbf{R}^{n+1}, \gamma|^2_{t_q+s,\tau} \\
&= \int |\tilde{u}_{q0+}(\sigma + i\gamma, \xi', \xi_n)|^2 (\sigma^2 + \gamma^2 + |\xi'|^2)^{t_q+s} (\sigma^2 + \gamma^2 + |\xi'|^2)^\tau \, d\sigma d\xi \\
&\leq \int |\tilde{u}_{q0+}(\sigma + i\gamma, \xi', \xi_n)|^2 (\sigma^2 + \gamma^2 + |\xi'|^2)^{t_q+s+\tau} \, d\sigma d\xi \\
&= |u_{q0+}, \mathbf{R}^{n+1}, q|^2_{0,t_q+s+\tau}.
\end{aligned} \tag{5.6.21}$$

Therefore, estimate (5.6.18) with $t_q + s$ replaced by 0 and τ by $t_q + s + \tau$ gives us the following estimate

$$|u_{q0+}, R^{n+1}, q|^2_{0,t_q+s+\tau} \leq \frac{c}{\gamma} \sum_{h=1}^{m} |\varphi_h, \partial G, \gamma|_{s-\sigma_h+\tau}. \tag{5.6.22}$$

From (5.6.21) and (5.6.22) we obtain estimate (5.6.18) also for the case $t_q + s < 0$.

Thus, we have the next result.

Lemma 5.6.4. *Suppose that*

$$f = 0, \quad \varphi_h \in \mathcal{H}^{s-\sigma_h+\tau}(\partial G, \gamma)$$

$$\Big(h = 1, \ldots, m, \quad s, \tau \in \mathbf{R}, \quad \gamma \geq \gamma_0 > 0\Big),$$

and problem (5.1.10) is hyperbolic. Then there exist elements

$$u_0 = (u_{10}, \ldots, u_{N0}) \in \mathcal{H}^{T+s,\tau}(G, \gamma) = \prod_{1 \leq j \leq N} \mathcal{H}^{t_j+s,\tau}(G, \gamma), \tag{5.6.23}$$

$$u_{j\nu} \in \mathcal{H}^{t_j+s-\nu+1+\tau}(\partial G, \gamma) \quad (j : t_j + æ \geq 1, \ \nu = 1, \dots, t_j + æ)$$

such that equalities (5.5.60), (5.5.61), (5.5.68), (5.5.70) hold; elements (5.6.23) are uniquely determined, and estimates (5.6.11) and (5.6.18) are valid; if $t_j + s - k > 1/2$, *then*

$$u_{jk} = D_n^{k-1} u_{j0}|_{x_n=0}.$$

Moreover, if

$$supp\, \varphi \subset \partial G_+ = \{(t,x) \in \partial G : t \geq 0\},$$

then also

$$supp\, u_{j\nu} \subset \partial G_+ \quad (j : t_j + æ \geq 1, \quad \nu = 1, \dots, t_j + æ),$$
$$supp\, u_0 \subset \overline{G}_+ = \{(t,x) \subset \partial G_+ : t \geq 0\}.$$

5.6.5. In this subsection we investigate hyperbolic problem (5.1.10) for the case of a single equation ($N = 1$):

$$l(D_t, D_x)u = f \quad (\text{in } G), \quad b_j u|_{\partial G} = \varphi_j \quad (j = 1, \dots, m), \qquad (5.6.24)$$

where $\operatorname{ord} l = t_1$, $\operatorname{ord} b_j = t_j + \sigma_j$, and $\sigma_1, \dots, \sigma_m$ are given integers.

Lemma 5.6.5. *Let* $æ = \max\{0, \sigma_1 + 1, \dots, \sigma_m + 1\}$,

$$f = f_0, \dots, f_æ) \in \widetilde{\mathcal{H}}^{s,\tau+1,(æ)}(G, \gamma), \quad \varphi_j \in \mathcal{H}^{s-\sigma_j+\tau}(\partial G, \gamma) \quad (j = 1, \dots, m),$$
$$s, \tau \in \mathbf{R}, \quad \gamma \geq \gamma_0 > 0,$$

and let problem (5.6.24) be hyperbolic. Then there exists one and only one element

$$u = (u_0, \dots, u_{t_1+æ}) \in \mathcal{H}^{t_1+s,\tau,(t_1+æ)}(G, \gamma), \qquad (5.6.25)$$

which solves problem (6.24). The estimate

$$\gamma |u_0, G, \gamma|^2_{t_1+s,\tau} + \sum_{j=1}^{t_1+æ} |u_j, \partial G, \gamma|^2_{t_1+s-j+1+\tau}$$
$$\leq c\Big(\frac{1}{\gamma} |f_0, G, \gamma|^2_{s,\tau+1} + \sum_{j=1}^{æ} |f_j, \partial G, \gamma|^2_{s-j+1+\tau} + \sum_{h=1}^{m} |\varphi_h, \partial G, \gamma|^2_{s-\sigma_h+\tau}\Big) \quad (5.6.26)$$

holds with a constant $c > 0$ *independent of* f, φ, *and* γ. *If* $supp(f, \varphi) \subset \overline{G}_+$, *then also* $supp\, u \subset \overline{G}_+$.

Proof. We note first that if $f = 0$ then the solvability of problem (5.6.24) in space (5.6.25) and estimate (5.6.26) follow immediately from Lemma 5.6.4. Suppose that $f \neq 0$. By Lemma 5.6.3, there exists one and only one element $(u_0, \ldots, u_{t_1+æ})$ satisfying (5.6.24); furthermore,

$$u_j \in \mathcal{H}^{t+s-j+1+\tau}(\partial G, \gamma) \quad (j = 1, \ldots, t_1 + æ),$$

and

$$\sum_{j=1}^{t_1+æ} |u_j, \partial G, \gamma|^2_{t_1+s+\tau-j+1}$$

$$\leq c\Big(\frac{1}{\gamma}|f_0, G, \gamma|^2_{s,\tau+1} + \sum_{j=1}^{æ} |f_j, \partial G, \gamma|^2_{s-j+1+\tau} + \sum_{h=1}^{m} |\varphi_h, \partial G, \gamma|^2_{s-\sigma_h+\tau}\Big). \tag{5.6.27}$$

We prove now that $u_0 \in \mathcal{H}^{t_1+s,\tau}(G,\gamma)$ and that the estimate (5.6.26) holds. We consider the Dirichlet problem

$$l(D_t, D_x)u = f_0, \quad D_n^{j-1}u|_{x_n=0} = u_j \quad (j = 1, \ldots, m). \tag{5.6.28}$$

It is easy to see that this problem is hyperbolic, and hence it has one and only one solution $(u_0, u_1, \ldots, u_{t_1})$. Moreover, if $(u_0, u_1, \ldots u_{t_1+æ})$ is a solution of problem (5.6.24), then $(u_0, \ldots, u_{t_1})$ is a solution of problem (5.6.28). Therefore, we find the 'truncated vector' $(u_0, \ldots, u_{t_1})$ by solving problem (5.6.28). It is clear that the solution of problem (5.6.28) is the sum of the solution of (5.6.28) with $f_0 = 0$ and the solution of (5.6.28) with $u_1 = \ldots = u_m = 0$. Since the solvability of the first problem and the corresponding estimates were established in Lemma 5.6.4, it suffices to investigate the second problem

$$l(D_t, D_x)u = f_0, \qquad D_n^{j-1}u|_{x_n=0} = 0 \quad (j = 1, \ldots, m). \tag{5.6.29}$$

We also consider the problem formally adjoint to (5.6.29):

$$l^+(D_t, D_x)v = g_0, \qquad D_n^{j-1}v|_{x_n=0} = 0 \quad (j = 1, \ldots, t_1 - m). \tag{5.6.30}$$

It can be verified immediately that problem (5.6.30) is also hyperbolic and that if $u, v \in \mathcal{H}^{t_1,\tau}(G,\gamma)$ are solutions of respective problems (5.6.29) and (5.6.30), then

$$(lu, v) = (u, l^+v). \tag{5.6.31}$$

For each element

$$f_0 \in \mathcal{H}^{s,\tau+1}(G,\gamma) \quad (s \geq 0, \tau \in \mathbf{R})$$

problem (5.6.29) has a unique solution

$$u \in \mathcal{H}^{t_1+s,\tau}(G,\gamma),$$

and the operator

$$A_s: \quad f_0 \mapsto u_0 \quad (\mathcal{H}^{s,\tau+1}(G,\gamma) \to \mathcal{H}^{t_1+s,\tau}(G,\gamma)) \tag{5.6.32}$$

is continuous, with the norm $\|A_s\| \leq c/\gamma$ ([Sak], [ChP]).

Similarly, for each

$$g_0 \in \mathcal{H}^{s,\tau+1}(G,\gamma) \quad (s \geq 0, \tau \in \mathbf{R})$$

problem (5.6.30) has a unique solution

$$v_0 \in \mathcal{H}^{t_1+s,\tau}(G,\gamma),$$

and the operator

$$A_s^+: \quad g_0 \mapsto v_0 \quad (\mathcal{H}^{s,\tau+1}(G,\gamma) \to \mathcal{H}^{t_1+s,\tau}(G,\gamma)) \tag{5.6.33}$$

is continuous, with the norm $\|A_s^+\| \leq c\gamma$.

It follows from (5.6.31) that

$$(A_0 f_0, g_0) = (f_0, A_0^+ g_0) \quad (f_0, g_0 \in \mathcal{H}^{0,\tau+1}(G,\gamma)). \tag{5.6.34}$$

The operator $\widetilde{A}_s$ adjoint to the operator A_s^+ exists, acts continuously in the pair of spaces

$$\mathcal{H}^{-t_1-s,-\tau}(G,\gamma) \to \mathcal{H}^{-s,-\tau-1}(G,\gamma),$$

and is such that

$$(A_s^+ g_0, f_0) = (g_0, \widetilde{A}_s f_0) \quad (g_0 \in \mathcal{H}^{s,\tau+1}(G,\gamma), \quad f_0 \in \mathcal{H}^{-t_1-s,-\tau}(G,\gamma)) \tag{5.6.35}$$

$$|\widetilde{A}_s f_0, G, \gamma|_{-s,-\tau-1} \leq \frac{c}{\gamma} |f_0, G, \gamma|_{-t_1-s,-\tau}.$$

It follows from (5.6.34) and (5.6.35) that $\widetilde{A}_s$ $(s \geq 0)$ is an extension by continuity of the operator A_0. Consequently,

$$A_0: \quad \mathcal{H}^{0,\tau+1}(G,\gamma) \to \mathcal{H}^{t_1,\tau}(G,\gamma), \quad \|A_0\| \leq \frac{c}{\gamma},$$

$$\widetilde{A}_0: \quad \mathcal{H}^{-t_1,-\tau}(G,\gamma) \to \mathcal{H}^{0,-\tau-1}(G,\gamma), \quad \|\widetilde{A}_0\| \leq \frac{c}{\gamma}.$$

By the interpoloation theorem, the restriction $\widetilde{A}_0$ acts continuously in the pair of spaces

$$\mathcal{H}^{-t_1+\Theta t_1,-\tau+\Theta(2\tau+1)}(G,\gamma) \to \mathcal{H}^{\Theta t_1,-\tau-1+\Theta(2\tau+1)}(G,\gamma), \quad 0 \leq \Theta \leq 1, \tag{5.6.36}$$

and the norm of the corresponding operator does not exceed $\frac{c}{\gamma}$. It follows from (5.6.36) that the restriction of the operator $\widetilde{A}_0$ acts continuously in the pair of spaces

$$\mathcal{H}^{s,\tau+1}(G,\gamma) \to \mathcal{H}^{s+t,\tau}(G,\gamma) \quad (-t_1 \leq s \leq 0, \tau \in \mathbf{R}) \tag{5.6.37}$$

and that the norm of the corresponding operator is $\leq \frac{c}{\gamma}$.

Thus, foe any $s, \tau \in \mathbf{R}$ and $\gamma \geq \gamma_0 > 0$ the solution of problem (5.6.29) satisfies the inequality

$$|u_0, G, \gamma|_{s+t_1,\tau} \leq \frac{c}{\gamma} |f_0, G, \gamma|_{s,\tau+1}. \tag{5.6.38}$$

For $s \geq 0$ this inequality was established in [Sak] and [ChP], for $s \in [-t_1, 0]$ it follows from (5.6.37), and for $s < -t_1$ it follows from (5.6.35). This and Lemma 5.6.4 give us all the assertions of the lemma. □

5.6.6. We now study problem (5.1.10) in the case where the system is diagonal:

$$l(D_t, D_x) = (l_{jj}(D_t, D_x) : j = 1, \ldots, N\}, \quad \operatorname{ord} l_{jj} = t_j + s_j.$$

In this case the problem has the form

$$\begin{aligned} l_{jj}(D_t, D_x)u_j &= f_j \quad (j = 1, \ldots, N), \\ b_h u|_{x_n=0} &= \varphi_h \quad (h = 1, \ldots, m). \end{aligned} \tag{5.6.39}$$

The following asssertion is true.

Lemma 5.6.6. *Suppose that problem (5.6.39) is hyperbolic, $s, \tau, \gamma \in \mathbf{R}$, $\gamma \geq \gamma_0 > 0$, and*

$$f \in \widetilde{\mathcal{H}}^{s-S,\tau+1,(\text{æ}-S)}(G,\gamma), \quad \varphi_h \in \mathcal{H}^{s-\sigma_h+\tau}(\partial G,\gamma), \quad (h = 1, \ldots, m).$$

Then problem (5.6.39) has one and only one solution

$$u = (u_1, \ldots, u_N) \in \widetilde{\mathcal{H}}^{T+s,\tau,(T+\text{æ})}(G,\gamma),$$

$$u_j = (u_{j0}, u_{j1}, \ldots, u_{j,t_j+\text{æ}}) \in \widetilde{\mathcal{H}}^{t_j+s,\tau(t_j+\text{æ})}(G,\gamma).$$

There exists a constant $c > 0$ independent of f, φ, and γ ($\gamma \geq \gamma_0 > 0$), such that

$$\sum_{j=1}^{N}\Big(\gamma|u_{j0}, G, \gamma|^2_{t_j+s,\tau} + \sum_{k=1}^{t_j+æ} |u_{jk}, \partial G, \gamma|^2_{t_j+s-k+1+\tau}\Big)$$

$$\leq c\Big(\frac{1}{\gamma}\sum_{j=1}^{N} |f_{j0}, G, \gamma|^2_{s-s_j,\tau+1} + \sum_{j:æ-s_j\geq 1}\ \sum_{p=1}^{æ-s_j} |f_{jp}, \partial G, \gamma|_{s-s_j-p+1+\tau}$$

$$+ \sum_{h=1}^{m} |\varphi_h, \partial G, \gamma|^2_{s-\sigma_h+\tau}.\Big) \qquad (5.6.40)$$

If $supp(f, \varphi) \subset \overline{G}_+$ then $supp\, u \subset \overline{G}_+$.

Proof. The solvability of problem (5.6.39) and estimate (5.6.11) follow from Lemma 5.6.3. Then problem (5.6.39) is equivalent to the Dirichlet problems

$$l_{jj}(D_t, D_x)u_j = f_j; \quad D_n^{k-1}u_j|_{x_n=0} = u_{jk} \quad (k = 1, \ldots, n_j^-) \qquad (5.6.41)$$

$$(j = 1, \ldots, N)$$

(n_j^- is the number of roots of the polynomial $l_{jj}^-(\zeta) = l_{jj}^-(\sigma + i\gamma, \xi', \zeta)$ (see 5.5.2)). By Lemma 5.6.5, we deduce from (5.6.41) the required estimates (5.6.40). □

5.6.7.

The case where the system l is not diagonal can be reduced by diagonalization to the case already treated. We give the appropriate arguments.

First of all note that all the results given remain true (with the same proofs) if instead of (5.4.1) (or (5.5.63)) we assume that

$$æ \geq \max\{0, \sigma_1 + 1, \ldots, \sigma_m + 1\}. \qquad (5.6.42)$$

Further, linear systems (5.5.70) and (5.5.74) do not change, and in system (5.5.75) new equations are added which enable us to find and estimate the additional components $\widehat{u}_{j\beta}$ from the additional components $\widehat{f}_{kp}$.

The (unique) solvability of hyperbolic problem (5.1.10) follows from Lemma 5.6.3. It thus remains to show that $u \in \widetilde{\mathcal{H}}^{T+s,\tau,(T+æ)}(G, \gamma)$ and that estimates (5.6.40) hold. We first assume that

$$t_j + æ \geq r \quad (j = 1, \ldots, N). \qquad (5.6.43)$$

We write system (5.1.10) in the form

$$\sum_{j=1}^{N} l_{kj}(D_t, D_x)u_j = f_k \quad (k = 1, \ldots, N). \tag{5.6.44}$$

Let $L_{kj}(D_t, D_x)$ be the cofactor of the elements $l_{kj}(D_t, D_x)$ of the matrix $l(D_t, D_x)$. Owing to (5.6.43), we can apply the operator L_{kj} to both sides of (5.6.44). Then summing over k, we obtain

$$L(D_t, D_x)u_j = \sum_{k=1}^{N} L_{kj}(D_t, D_x)f_k = \Phi_j \quad (j = 1, \ldots, N). \tag{5.6.45}$$

Moreover, if

$$f_k \in \widetilde{\mathcal{H}}^{s-s_k,\tau+1,(æ-s_k)}(G,\gamma) \quad (k = 1, \ldots, N),$$

then

$$\Phi_j \in \widetilde{\mathcal{H}}^{s-r+t_j,\tau+1,(æ+t_j-r)}(G,\gamma) \quad (j = 1, \ldots, N),$$

and the operator $f \mapsto \Phi$ is linear and continuous. In this case, if

$$u = (u_1, \ldots, u_N), \quad u_j = (u_{j0}, \ldots, u_{j,t_j+æ}),$$

is a solution of problem (5.1.10), then also

$$L(D_t, D_x)u_j = \Phi_j, \qquad D_n^{p-1}u_j = u_{jp} \quad (p = 1, \ldots, m, \ j = 1, \ldots, N). \tag{5.6.46}$$

Lemma 5.6.5 now gives us that $u_j \in \widetilde{\mathcal{H}}^{t_j+s,\tau,(t_j+æ)}(G,\gamma)$ and that estimates (5.6.40) hold.

Thus, we have the following result.

Lemma 5.6.7. *Suppose that problem (5.1.10) is hyperbolic, $s, \tau, \gamma \in \mathbf{R}$, $\gamma \geq \gamma_0 > 0$, and f and φ are the same as in Lemma 5.6.6, and æ satisfies relations (5.6.42) and (5.6.43). Then problem (5.1.10) has one and only one solution $u \in \widetilde{\mathcal{H}}^{T+s,\tau,(T+æ)}(G,\gamma)$. There exists a constant $c > 0$ independent of f, φ, and γ ($\gamma \geq \gamma_0 > 0$) such that estimate (5.6.40) holds.*

Finally, we note that the assertion of Lemma 5.6.7 remains true even (5.6.43) fails to hold and æ is defined by (5.4.1).

Indeed, let

$$f = (f_1, \ldots, f_N) \in \widetilde{\mathcal{H}}^{s-S,\tau+1,(æ-S)}(G,\gamma).$$

We extend f_j by the new components to form an element

$$\widetilde{f}_j \in \widetilde{\mathcal{H}}^{s-s_j,\tau+1,(æ+æ_1-s_j)}(G,\gamma) \quad (j = 1, \ldots, N),$$

where $æ_1 > 0$ is chosen so that $t_j + æ + æ_1 \geq r$ $(j = 1, \dots, N)$. Then by Lemma 5.6.7, there exists one and only one element

$$\widetilde{u} \in \widetilde{\mathcal{H}}^{T+s,\tau,(T+æ+æ_1)}(G,\gamma)$$

that solves problem (5.1.10) with f replaced by $\widetilde{f}$. Then the 'truncated' vector $u \in \widetilde{\mathcal{H}}^{T+s,\tau,(T+æ)}(G,\gamma)$ is a solution of problem (5.1.10). It depends continuously on $\widetilde{f}$. Moreover, to different 'lengthenings' $\widetilde{f}$ of the vector f there corresponds one and the same solution $u \in \widetilde{\mathcal{H}}^{T+s,\tau,(T+æ)}(G,\gamma)$ (Lemma 5.6.3), and hence we obtain inequality (5.6.40) with $\widetilde{f}$ instead of f if we pass to the infimum over the different 'lengthenings' $\widetilde{f}$ of f.

The proof of Theorem 5.4.1 is thus complete.

GREEN'S FORMULA AND DENSITY OF SOLUTIONS FOR GENERAL PARABOLIC BOUNDARY VALUE PROBLEMS IN FUNCTIONAL SPACES ON MANIFOLDS

The aim of this chapter is to obtain the Green's formula for a general parabolic boundary value problem whose boundary conditions are not necessarily normal boundary conditions, to study problems formally adjoint to parabolic ones with respect to the Green's formula, and to investigate the density of the set of solutions of general parabolic boundary value problems and the derivatives of these solutions in functional spaces of vector-functions on a manifold inside of a domain. The main results of this chapter are published in [RR1] and [RR2].

The chapter consists of two sections. The main results are stated in the first section, and the proofs are given in the second section.

6.1. Statement of Main Results

6.1.1.

Let $G \in \mathbf{R}^n$ be a bounded domain with infinitely smooth boundary $\partial G \in C^\infty$, let $\Omega = G \times (0,T)$, $0 < T < \infty$, and let $\Omega' = \partial G \times (0,T)$. Denote by $(\cdot,\cdot)_G$, $(\cdot,\cdot)_\Omega$, $\langle\cdot,\cdot\rangle_{\Omega'}$, and $\langle\cdot,\cdot\rangle_{\partial G}$ the scalar products (or their extensions) in $L_2(G)$, $L_2(\Omega)$, $L_2(\Omega')$, and $L_2(\partial G)$, respectively.

In Ω we consider the general parabolic problem ([AgV], [Eid], [Sol], [EZh])

$$Lu \equiv L(x,t,D_x,\partial_t)u(x,t) = f(x,t), \quad (x,t) \in \Omega, \quad \operatorname{ord} L = 2m, \tag{6.1.1}$$

$$B_j u \equiv B_j(x,t,D_x,\partial_t)u(x,t)\big|_{x=x'} = \varphi_j(x',t), \tag{6.1.2}$$
$$j = 1,\dots,m, \; x' \in \partial G, \quad \operatorname{ord} B_j = m_j,$$

$$\partial_t^{k-1} u\big|_{t=0} = \varphi_{0k}, \quad k = 1,\dots,æ. \tag{6.1.3}$$

Here $D_x = (D_1 \cdots D_n)$, $D_j = i\partial/\partial x_j$, $\partial_t = \partial/\partial t$. The order of an operator is defined as the highest order of its terms, and the order of the term

$D_1^{\alpha_1}\cdots D_n^{\alpha_n}\partial_t^\beta = D_x^\alpha\partial_t^\beta$ is equal to $\alpha_1+\cdots+\alpha_n+2b\beta = |\alpha|+2b\beta$, where b is a divisor of m. The number $2b$ is called the parabolic weight of the problem; $æ = m/b$, and m_j $(j = 1,\ldots,m)$ are arbitrary nonnegative integers. For simplicity sake, the coefficients of all differential expressions are assumed to be infinitely smooth in $\overline{\Omega}$ and $\overline{\Omega'}$, respectively. Let r_j be the order of B_j with respect to the derivatives $D_\nu = i\partial/\partial\nu$, where ν is a normal to Ω', and let

$$r = \max\{2m, r_1+1,\ldots,r_m+1\}. \qquad (6.1.4)$$

Let us now formulate the main results of this chapter.

Theorem 6.1.1. *Let problem (6.1.1), (6.1.2), (6.1.3) be parabolic. Then there exist (in general, pseudo-differential) expressions* $C_j(x,t,D_x,\partial_t)$, $B_j'(x,t,D_x,\partial_t)$, *and* $C_k'(x,t,D_x,\partial_t)$ *of orders* l_j, m_j' $(j = 1,\ldots,m)$, *and* l_k' $(k = 1,\ldots,r-m)$, *respectively, such that the Green's formula*

$$(Lu,v)_\Omega + \sum_{1\le j\le m}\left\langle B_ju, C_j'v\right\rangle_{\Omega'} + \sum_{j:1\le j\le r-2m}\left\langle \left(D_\nu^{j-1}L\right)u, C_{m+j}v\right\rangle_{\Omega'}$$
$$+\sum_{0\le k\le æ-1}\left(\partial_t^k u(x,0), T_kv(x,0)\right)_G$$
$$= (u,L^+v)_\Omega + \sum_{1\le j\le m}\left\langle C_ju, B_j'v\right\rangle_{\Omega'} + \sum_{0\le k\le æ-1}\left((T_k'u)(x,T), \partial_t^kv(x,T)\right) \qquad (6.1.5)$$

holds. Here $u,v\in C^\infty(\overline{\Omega})$, *and* T_k *and* T_k' *are differential expressions of orders*

$$2m-(k+1)2b,\quad m_j+l_j' = l_j+m_j' = 2m-1\quad (j=1,\ldots,m),$$
$$l_{m+j}' = -j\quad (1\le j\le r-2m) \qquad (6.1.6)$$

(if $r = 2m$*, then the third therm of the left hand side of (6.1.5) equals to zero).*

Theorem 6.1.2. *The problem formally adjoint to problem (6.1.1)–(6.1.3) with respect to Green's formula (6.1.5), namely,*

$$L^+v = g\quad (in\,\Omega), \qquad (6.1.7)$$

$$B_j'v\big|_{\Omega'} = \psi_j\quad (j=1,\ldots,m); \qquad (6.1.8)$$

$$\partial_t^{k-1}v\big|_{t=T} = \psi_{0k}\quad (k=1,\ldots,æ) \qquad (6.1.9)$$

(with inverted time) is parabolic if and only if problem (6.1.1)–(6.1.3) is parabolic.

6.1.2. To obtain the results mentioned, we introduce some functional spaces.

For any $s \geq 0$ we denote by $H^s(G)$ the Sobolev–Slobodetskii space, and denote by $H^{-s}(G)$ the space dual to $H^s(G)$ with respect to the extension $(\cdot,\cdot)_G$ of the scalar product in $L_2(G)$. We also denote by $\|\cdot, G\|_s$ the norm of $H^s(G)$, $s \in \mathbf{R}$.

For $s \geq 0$ we denote by $\mathcal{H}^s(\Omega) = \mathcal{H}^{s,s/(2b)}_{x,t}(\Omega)$ the anisotropic space of functions $u(x,t)$ which are s times continuously differentiable with respect to x and $s/(2b)$ times continuously differentiable with respect to t. Denote by $C_0^\infty(\overline{\Omega})$ $[C^{\infty,0}(\overline{\Omega})]$ the set of functions from $C^\infty(\overline{\Omega})$ vanishing near $t = 0$ $(t = T)$, and denote by $\mathcal{H}_0^s(\Omega)$ $[\mathcal{H}^{s,0}(\Omega)]$ the closure of $C_0^\infty(\overline{\Omega})$ $[C^{\infty,0}(\overline{\Omega})]$ in $\mathcal{H}^s(\Omega)$.

Denote by $\mathcal{H}^{-s}(\Omega)$, $s > 0$, both the space dual to $\mathcal{H}^{s,0}(\Omega)$ (when studying problem (6.1.1)–(6.1.3)) and the space dual to $\mathcal{H}_0^s(\Omega)$ (when studying problem (6.1.7)–(6.1.9)). In both cases, the duality is considered with respect to the extension $(\cdot,\cdot)_\Omega$ of the scalar product in $L_2(\Omega)$. By $\|\cdot,\Omega\|_s$ we denote the norm in $\mathcal{H}^s(\Omega)$, $s \in \mathbf{R}$.

Similarly, $\mathcal{H}^s(\Omega')$, $s \geq 0$, is the anisotropic Sobolev-Slobodetskii space on Ω', and $\mathcal{H}_0^s(\Omega')$ $[\mathcal{H}^{s,0}(\Omega')]$ is the closure in $\mathcal{H}^s(\Omega')$ of the set of functions from $C^\infty(\overline{\Omega'})$ vanishing near the plane $t = 0$ $(t = T)$. We denote by $\mathcal{H}^{-s}(\Omega')$, $s > 0$, both the space dual to $\mathcal{H}^{s,0}(\Omega')$ and the space dual to $\mathcal{H}_0^s(\Omega')$. In both cases, the duality is considered with respect to the extension $\langle\cdot,\cdot\rangle_{\Omega'}$ of the scalar product in $L_2(\Omega')$. By $\langle\langle\cdot,\Omega'\rangle\rangle_s$ we denote the norm in $\mathcal{H}^s(\Omega')$, $s \in \mathbf{R}$.

For any s such that

$$s \in \mathbf{R} \setminus \left\{\frac{1}{2}, \frac{3}{2}, \ldots, \frac{2b-1}{2}, b, \ldots, (2æ-1)b\right\}$$

we denote by $\widetilde{\mathcal{H}}^{s,(r)}(\Omega)$ the completion of $C^\infty(\overline{\Omega})$ in the norm

$$|||u,\Omega|||_{s,(r)} := \Bigg(\|u,\Omega\|_s^2 + \sum_{1\leq j\leq r} \langle\langle D_\nu^{j-1}u, \Omega'\rangle\rangle_{s-j+1/2}^2 + \sum_{1\leq k\leq æ} \|\partial_t^{k-1}u(x,0), G\|^2_{s-(2k-1)b}\Bigg)^{1/2}; \quad (6.1.10)$$

here $D_\nu = i\partial/\partial\nu$, and ν is a normal to Ω'. The space $\widetilde{\mathcal{H}}^{s,(r)}(\Omega)$ and norm $|||u,\Omega|||_{s,(r)}$ for $s \in \{\frac{1}{2}, \ldots, (2æ-1)b\}$ are defined by interpolation. Finally, for $r = 0$ we set $\widetilde{\mathcal{H}}^{s,(0)}(\Omega) := \mathcal{H}^s(\Omega)$.

The spaces $\widetilde{\mathcal{H}}^{s,(r)}(\Omega)$ were inroduced for the istropic case in [R10] and were studied in detail in [R11], [R1], [RR1] (see also [Ber]). For the anisotropic case, these spaces were introduced in [EZh], [Zh2].

For each $s \in \mathbf{R}$ let us consider the closure S of the mapping

$$u \mapsto (u|_{\overline{\Omega}}, u|_{\Omega'}, \ldots, D_\nu^{r-1}u|_{\Omega'}, u|_{t=0}, \ldots, \partial_t^{æ-1}u|_{t=0}), \quad u \in C^\infty(\overline{\Omega}),$$

acting in the pair of spaces

$$\widetilde{\mathcal{H}}^{s,(r)}(\Omega) \to K^s,$$

$$K^s := \mathcal{H}^s(\Omega) \times \prod_{1 \le j \le r} \mathcal{H}^{s-j+1/2}(\Omega') \times \prod_{1 \le k \le æ} H^{s-(2k-1)b}(G).$$

This mapping is an isometry between the space $\widetilde{\mathcal{H}}^{s,(r)}(\Omega)$ and the subspace K_0^s of the space K^s (the subspace K_0^s coincides with the space K^s for $s < 1/2$). Therefore, we can identify an element $u \in \widetilde{\mathcal{H}}^{s,(r)}(\Omega)$ with the element $Su \in K_0^s$. This means that we can write

$$u = (u_0, u_1, \ldots, u_r, u_{10}, \ldots, u_{æ0}) \in K^s.$$

Below the components of the vector Su are called the components of the element $u \in \widetilde{\mathcal{H}}^{s,(r)}(\Omega)$. It is clear that if $s > r - 1/2$, then

$$u_j \in D_\nu^{j-1}u_0|_{\Omega'} \quad (j = 1, \ldots, r), \qquad u_{j0} = \partial_t^{j-1}u_0|_{t=0} \quad (j = 1, \ldots, æ).$$

For all other j, the elements u_j and u_{j0} do not depend on u_0. The closure of the mapping

$$u \mapsto (Mu|_{\overline{\Omega}}, M_1u|_{\Omega'}, u|_{t=0}, \ldots, \partial_t^{æ-1}u|_{t=0}), \quad u \in C^\infty(\overline{\Omega}),$$

acts continuously from the whole space $\widetilde{\mathcal{H}}^{s,(r)}(\Omega)$ into the direct product

$$\mathcal{H}^{s-q}(\Omega) \times \mathcal{H}^{s-q_1-1/2}(\Omega') \times \prod_{1 \le k \le æ} H^{s-(2k-1)b}(G)$$

for each differential expression $M(x, t, D_x, \partial_t)$ of order $q \le r$ in Ω and for each boundary relation $M_1(x, t, D_x, \partial_t)$ of order $q_1 \le r - 1$. Therefore, the closure $A = A_s$ of the mapping

$$u \mapsto (Lu, B_1u, \ldots, B_mu, u|_{t=0}, \ldots, \partial_t^{æ-1}u|_{t=0}), \quad u \in C^\infty(\overline{\Omega}),$$

acts continuously from the whole space $\widetilde{\mathcal{H}}^{s,(r)}(\Omega)$ into the direct product

$$\mathcal{F}^s := \mathcal{H}^{s-2m,(r-2m)}(\Omega) \times \prod_{1 \le j \le r} \mathcal{H}^{s-m_j-1/2}(\Omega') \times \prod_{1 \le k \le æ} H^{s-(2k-1)b}(G).$$

For each $s \in \mathbf{R}$, the operator $A = A_s$ realizes an isomorphism between the space $\widetilde{\mathcal{H}}^{s,(r)}(\Omega)$ and the subspace $\widetilde{\mathcal{F}}$ of $\mathcal{F}^s$ consisting of the elements which satisfy logical matching conditions ([EZh], [Zh2]). In particular, if $s < 1/2$, then there are no matching conditions, and we have that $\tilde{\mathcal{F}} = \mathcal{F}^s$. It follows from Theorem 6.1.2 that the theorem on isomorphisms holds also for the operator $A^+ = A_s^+$ generated by problem (6.1.7)–(6.1.9).

Now, by passing to the limit, it is easy to verify that Green's formula (6.1.5) holds for any elements $u \in \tilde{\mathcal{H}}^{s,(r)}(\Omega)$ and $v \in \tilde{\mathcal{H}}^{2m-s,(2m)}(\Omega)$. Indeed, let $u_n \in C^\infty(\overline{\Omega})$, $\lim_{n\to\infty} u_n = u$ in $\tilde{\mathcal{H}}^{s,(r)}(\Omega)$, and let $v_k \in C^\infty(\overline{\Omega})$, $\lim_{k\to\infty} v_k = v$ in $\tilde{\mathcal{H}}^{2m-s,(2m)}(\Omega)$. Now we write Green's formula (6.1.5) for $u = u_n$ and $v = v_k$, pass to the limit as $n \to \infty$, then as $k \to \infty$, and thus odtain the desired formula.

Green's formula (6.1.5) implies directly that if

$$u = (u_0, \ldots, u_r, u_{10}, \ldots, u_{\text{æ}0}) \in \tilde{\mathcal{H}}^{s,(r)}(\Omega),$$

then

$$Au = F = (f_0, \ldots, f_{r-2m}, \varphi_1, \ldots, \varphi_m, \varphi_{10}, \ldots, \varphi_{\text{æ}0}) \in \mathcal{F}^s$$

if and only if

$$(u, L^+v)_\Omega + \sum_{1\le j\le m} \left\langle C_j u, B_j' v\right\rangle_{\Omega'} + \sum_{0\le k\le \text{æ}-1} ((T_k' u)(x,T), \partial_t^k v(x,T))_G$$

$$= (f_0, v)_\Omega + \sum_{j:1\le j\le r-2m} \left\langle f_j, C_{m+j}' v\right\rangle_{\Omega'} + \sum_{1\le j\le m} \left\langle \varphi_j, C_j' v\right\rangle_{\Omega'}$$

$$+ \sum_{0\le k\le \text{æ}-1} (u_{k0}(x,0), (T_k v)(x,0))_G \qquad (\forall v \in C^\infty(\overline{G})).$$

6.1.3. We now consider problem (6.1.1), (6.1.2), (6.1.3) with $m_j < 2m$ ($j = 1, \ldots, m$), and $r = 2m$. Let $\gamma_1 \subset G$ be a smooth $(n-1)$-dimensional manifold without boundary, let γ be an open subset of the manifold γ_1, and let $\Sigma = \gamma \times (0,T)$. For any smooth solution u of problem (6.1.1), (6.1.2), (6.1.3) we set

$$\nu_k u = \left(u\big|_\Sigma, \ldots, D_\nu^{k-1} u\big|_\Sigma\right), \tag{6.1.11}$$

where $D_\nu = i\partial/\partial\nu$, and ν is a normal to Σ. Furthermore, let G_0 be a chosen subdomain of G with arbitrary small diameter, and let γ_0 be an arbitrary small open part of ∂G. Let us modify the function $f(x,t)$ on the cylinder $\Omega_0 = G_0 \times (0,T)$ and the functions $\varphi_1, \ldots, \varphi_m$ on $\Gamma_0 = \gamma_0 \times (0,T)$, in an arbitrary way. Then, let us try to approximate an arbitrary vector-functions $(\psi_1, \ldots, \psi_k)$ defined on Σ by the obtained vector-functions $\nu_k u$. Similar approximation problems were considered since 1960 for elliptic and

the simplest parabolic boundary value problems (see [SchW], [Ham], [HaW], [Rsh5], [Rsh6], [R14]) and the bibliography therein).
Let us set

$$\begin{aligned} M(\Omega_0) &= \Big\{ u \in C^\infty(\overline{\Omega}) : \operatorname{supp} Lv \subset \Omega_0,\ B_j u\big|_{\Omega'} = 0\ \ (j = 1, \ldots, m); \\ &\qquad\qquad \partial_t^{j-1} u\big|_{t=0} = 0\ \ (j = 1, \ldots, æ) \Big\}, \\ \nu_k M(\Omega_0) &= \Big\{ \nu_k u : u \in M(\Omega_0) \Big\}, \\ M(\Gamma_0) &= \Big\{ u \in C^\infty(\overline{\Omega}) : Lu = 0,\ \operatorname{supp} B_j u\big|_{\Omega'} \subset \Gamma_0,\ \ (j = 1, \ldots, m); \\ &\qquad\qquad \partial_t^{j-1} u\big|_{t=0} = 0\ \ (j = 1, \ldots, æ) \Big\}, \\ \nu_k M(\Gamma_0) &= \Big\{ \nu_k u : u \in M(\Gamma_0) \Big\}. \end{aligned}$$

Theorem 6.1.3. *If the set $G \setminus \overline{\gamma} \subset \mathbf{R}^n$ is connected, then $\nu_{2m} M(\Omega_0)$ is dense in the direct product*

$$\prod_{1 \le j \le 2m} \mathcal{H}_0^{s_j}(\Sigma)$$

for any integer $s_j \ge 0$ $(j = 1, \ldots, 2m)$.

Theorem 6.1.4. *Let $\gamma = \gamma_1$ bound a subdomain G_1 of the domain G (i.e., $G \setminus \Sigma$ is not connected). Then the set $\nu_m M(\Omega_0)$ is dense in the direct product*

$$\prod_{1 \le j \le m} \mathcal{H}_0^{m-j+1/2}(\Sigma).$$

Theorem 6.1.5. *If the set $G \setminus \overline{\gamma}$ is connected, then $\nu_{2m} M(\Gamma_0)$ is dense in the direct product*

$$\prod_{1 \le j \le 2m} \mathcal{H}^{s_j}(\Sigma)$$

for any integer $s_j \ge 0$ $(j = 1, \ldots, 2m)$.

Theorem 6.1.6. *Let $\gamma = \gamma_1$ bound a subdomain G_1 of the domain G. Then the set $\nu_m M(\Gamma_0)$ is dense in the direct product*

$$\prod_{1 \le j \le m} \mathcal{H}_0^{m-j+1/2}(\Sigma).$$

Remark 6.1.1. For simplicity sake, we present the L_2-theory. The L_p-theory ($1 < p < \infty$) also holds (cf. [R11], [R1], [R2]).

Remark 6.1.2. All the results remain true if the system L in (6.1.1) is parabolic in the Petrovskii sense (cf. [RSh6]).

6.2. Proofs. Elliptic Boundary Value Problems with a Parameter

As in [AgV], we first study elliptic boundary problems with a parameter in order to prove Theorems 6.1.1 and 6.1.2. Here we substantially apply results of Chapter 1, [R11], [R1], [R2], [R15], [RS2].

6.2.1. We now introduce some functional spaces.
Let $s, q \in \mathbf{R}$. Denote by $H^s(\mathbf{R}^n, q)$ the space of distributions f such that

$$\|f, \mathbf{R}^n, q\|_s = \left(\int\limits_{\mathbf{R}^n} (1 + |\xi|^2 + q^2)^s \, |(Ff)(\xi)|^2 \, d\xi \right)^{1/2} < \infty,$$

where $(Ff)(\xi)$ is the Fourier transform of the element f. If $s \geq 0$, then we denote by $H^s(G, q)$ the space of restrictions to G of functions from $H^s(\mathbf{R}^n, q)$ with norm $\|u, G, q\|_s = \inf \|v, \mathbf{R}^n, q\|_s$, where the infimum is taken over all functions $v \in H^s(\mathbf{R}^n, q)$ equal to v on G. If $s < 0$, then we denote by $H^s(G, q)$ the space dual to $H^{-s}(G, q)$ with respect to the extension $(\cdot, \cdot)_G$ of the inner product in $L_2(G)$. The norm in $H^s(G, q)$ is determinated as

$$\|u, G, q\|_s = \sup_{v \in H^{-s}(G,q)} \frac{|(u, v)_G|}{\|v, G, q\|_{-s}} \qquad (s < 0).$$

By means of partitions of unity and the local straightening of boundary, the norm $\|u, \mathbf{R}^{n-1}, q\|_s$ determines the norm $\langle\langle u, \partial G, q \rangle\rangle_s$ on ∂G and the space $H^s(\partial G, q)$ with this norm.

6.2.2. Let the differential expression

$$L = L(x, D_x, q) = \sum_{k+|\mu| \leq 2m} a_{k\mu}(x) \left(q e^{i\theta}\right)^k D^\mu$$

be given on G, and let the system of boundary conditions

$$B_j = B_j(x, D_x, q) = \sum_{k+|\mu| \leq m_j} b_{jk\mu}(x) \left(q e^{i\theta}\right)^k D^\mu \qquad (j = 1, \ldots, m)$$

be given on ∂G. Here q is a real parameter, $\theta_0 \le \theta \le \theta_1$ (the case $\theta_0 = \theta = \theta_1$ is also admissible), and $a_{k\mu}$ and $b_{jk\mu}$ are infinitely smooth complex-valued functions in $\overline{G}$ and ∂G, respectively. The problem

$$Lu = f \quad (\text{ in } G), \qquad B_j u\big|_{\partial G} = \varphi_j \quad (j = 1, \ldots, m) \tag{6.2.1}$$

is called an elliptic problem with a parameter if Conditions I and II of [AgV] are satisfied, i.e., if the relation

$$L(x, D_x, D_t) = \sum_{k+|\mu|\le 2m} a_{k\mu}(x) \left(e^{i\theta} D_t\right)^k D_x^\mu \qquad \left(D_t = i\frac{\partial}{\partial t}\right)$$

is regular elliptic on the cylinder $\overline{G} \times \mathbf{R}^n$ for each $\theta \in [\theta_0, \theta_1]$, and the boundary conditions

$$B_j(x, D_x, D_t) = \sum_{k+|\mu|\le m_j} b_{jk\mu}(x) \left(e^{i\theta} D_t\right)^k D_x^\mu \qquad (j = 1, \ldots, m)$$

cover this cylinder on $\partial G \times \mathbf{R}$.

6.2.3. In a neighborhood $U(\partial G)$ in $\overline{G}$ of the boundary ∂G, let us introduce special local coordinates. If

$$x = (x_1, \ldots, x_{n-1}, x_n) = (x', x_n) \in U(\partial G),$$

then $(x', 0)$ are local coordinates on ∂G, and x_n is the distance between the point x and ∂G. Using these coordinates, we can write L and B_j on $U(\partial G)$ in the following form:

$$L(x, D_x, q) = \sum_{0\le k\le 2m} \tau_{2m-k}(x, D', e^{i\theta} q) D_n^k, \tag{6.2.2}$$

$$B_j(x, D_x, q) = \sum_{1\le k\le m_j+1} b_{jk}(x, D', e^{i\theta} q) D_n^{k-1}, \tag{6.2.3}$$

where

$$D_x = (D_1, \ldots, D_{n-1}, D_n) = (D', D_n),$$

and $\tau_j(B_{jk})$ are tangential expressions of orders at most j and $m_j - k$, respectively, with respect to (D', q). Since (6.2.1) is an elliptic problem with a parameter, we obtain that the problem

$$L(x, D_x, D_t)u \equiv \sum_{0\le k\le 2m} \tau_{2m-k}\left(x, D', e^{i\theta} D_t\right) D_n^k u = f \quad (\text{ in } G \times \mathbf{R}), \tag{6.2.4}$$

$$B_j(x, D_x, D_t)u \equiv \sum_{1\le k\le m_j+1} b_{jk}\left(x, D', e^{i\theta} D_t\right) D_n^{k-1} u\big|_{x_n=0} = \varphi_j \qquad (6.2.5)$$

$$(j = 1, \ldots, m)$$

is elliptic in the cylinder $G \times \mathbf{R}$ for all $\theta \in [\theta_0, \theta_1]$.
Since problem (6.2.4), (6.2.5) is elliptic, it follows (see [R2]) that there exist (in general, pseudo-differential) operators

$$C_j(x, D_x, D_t) = \sum_{1\le k\le l_j+1} c_{jk}\left(x, D', e^{i\theta} D_t\right) D_n^{k-1} \qquad (j = 1, \ldots, m)$$

of order $l_j \le 2m - 1$ such that for each $(\xi', q) \ne 0$ the polynomials

$$B_j(\eta) = \sum_{1\le k\le m_j+1} b_{jk}(x, \xi', e^{i\theta} q)\eta^{k-1}, \qquad (6.2.6)$$

$$C_j(\eta) = \sum_{1\le k\le l_j+1} c_{jk}(x, \xi', e^{i\theta} q)\eta^{k-1} \quad (j = 1, \ldots, m), \qquad (6.2.7)$$

$$(D_\nu^{j-1} L)(\eta) = \sum_{0\le k\le 2m} \tau_{2m-k}(x, \xi', e^{i\theta} q)(\eta^{k+j+1} + \text{ lower terms }) \qquad (6.2.8)$$

$$(1 \le j \le r - 2m)$$

are linearly independent. We write

$$\widehat{B}_j(x, D_x, D_t) = \begin{cases} B_j & \text{for } j = 1, \ldots, m, \\ C_{j-m} & \text{for } j = m+1, \ldots, 2m, \\ (D_\nu^{j-2m-1} L) & \text{for } 2m+1 \le j \le r, \end{cases} \qquad (6.2.9)$$

where $\widehat{B}u = \left(\widehat{B}_1 u\big|_{\partial G}, \ldots, \widehat{B}_r u\big|_{\partial G}\right)$, and r is determined by (6.1.4). Here we have

$$\widehat{B}_j(x, D_x, D_t) = \sum_{1\le k\le \widehat{m}_j} \widetilde{\Gamma}_{jk}(x, D', D_t) D_n^{k-1} \qquad (j = 1, \ldots, r),$$

where $\widetilde{\Gamma}_{jk}$ are tangential operators of orders at most $\widehat{m}_j - k + 1$, and $\widetilde{\Gamma}_{jk}(x) \ne 0$ for $k = m_j + 1$.

Let $\Gamma(x, D', D_t) = (\Gamma_{jk}(x, D', D_t))_{j,k=1,\ldots,r}$ $(x \in \partial G)$ be a square matrix such that Γ_{jk} is equal to $\widehat{\Gamma}_{jk}$ for $k \le \widehat{m}_j + 1$ and is zero for $k > \widehat{m}_j + 1$. On $\partial G \times \mathbf{R}$ let us consider the system of equations

$$\sum_{1\le k\le r} \Gamma_{jk}(x, D', D_t) U_k = \Phi_j \qquad (x \in \partial G, \quad j = 1, \ldots, r), \qquad (6.2.10)$$

or, in the matrix form, $\Gamma(x, D', D_t)U = \Phi$. Here $U(x)$ is the required functional column of dimension r and $\Phi(x)$ is the given one. Since the polynomials in (6.2.6), (6.2.7), (6.2.8) are linearly independent, we obtain that system (6.2.10) is elliptic on $\partial G \times \mathbf{R}$ in the sense of Douglis–Nirenberg, of order

$$(t_1, \dots, t_r, s_1, \dots, s_r), \qquad t_j = r - j, \quad s_j = \widehat{m}_j + 1 - r.$$

Then the system $\Gamma(x, D', D_t)U = \Phi$, or, in more details, the system

$$\sum_{1 \le k \le r} \Gamma_{jk}(x, D', e^{i\theta}q)U_k = \Phi_j \qquad (x \in \partial G, \quad j = 1, \dots, r, \ \theta \in [\theta_0, \theta_1]) \tag{6.2.11}$$

is a Douglis–Nirenberg elliptic system with a parameter on ∂G. Therefore, there exists a number $q_0 \ge 0$ such that for $|q| \ge q_0$ the closure $\Gamma(x, D', q) = \Gamma_s(x, D', q)$ of the mapping

$$U \to \Gamma(x, D', q)U \qquad (U \in (C^\infty(\partial G))^r)$$

realizes an isomorphism

$$\Gamma(x, D', q) : \prod_{1 \le j \le r} H^{r+s-j}(\partial G, q) \to \prod_{1 \le k \le r} H^{r+s-\widehat{m}_k-1}(\partial G, q).$$

Moreover, there exists a constant $c > 0$ independent of U, θ and q ($\theta \in [\theta_0, \theta_1]$, $|q| \ge q_0 > 0$), such that

$$c^{-1}\|U\|_{\prod_{1 \le j \le r} H^{r+s-j}(\partial G, q)} \le \|\Gamma U\|_{\prod_{1 \le k \le r} H^{r+s-\widehat{m}_k-1}(\partial G, q)} \le c\|U\|_{\prod_{1 \le j \le r} H^{r+s-j}(\partial G, q)}.$$

Thus, in contrast to [R2], we obtain that problem (6.2.11) is uniquely solvable in $\prod_{1 \le j \le r} H^{r+s-j}(\partial G, q)$ for any $\Phi \in \prod_{1 \le k \le r} H^{r+s-\widehat{m}_k-1}(\partial G, q)$ and $s \in \mathbf{R}$.

Let $\Lambda = \Gamma^{-1}$. Let us represent Λ in the matrix form, namely, if

$$\Phi = (0, \dots, \Phi_k, \dots, 0), \qquad \Lambda\Phi = U = (U_1, \dots, U_r),$$

then we set $\Lambda_{jk}\Phi_k = U_j$. The operator Λ_{jk} acts continuously in the pair of spaces

$$H^{r+s-\widehat{m}_k-1}(\partial G, q) \to H^{r+s-j}(\partial G, q) \qquad (s \in \mathbf{R}).$$

This means that the order of the operator Λ_{jk} is at most $j - 1 - \widehat{m}_k$. If $\Lambda\Phi = U$, then

$$U_j = \sum_{1 \le k \le r} \Lambda_{jk}\Phi_k \qquad (j = 1, \dots, r). \tag{6.2.12}$$

Now let us derive the Green's formula. In a neighborhood of the boundary ∂G in $\overline{G}$, let us represent L as in (6.2.2). Then, by integrating by parts, we obtain

$$(Lu, v)_G - (u, L^+v)_G = \sum_{1\le j\le 2m} \langle D_\nu^{j-1}u, T_{2m-j+1}v\rangle \quad (u, v \in H^{2m}(G)),$$

where

$$T_{2m-j+1}v = \sum_{j\le k\le 2m} D_n^{k-j}\, \tau_{2m-k+1}^+ v$$

is an operator of order $2m - j$. Then we have the following formula

$$(Lu, v)_G - (u, L^+v)_G = \langle u, Tv\rangle_{(L_2(\partial G))^{2m}} \qquad (u, v \in H^{2m}(G)), \quad (6.2.13)$$

with

$$U = \left(u\big|_{\partial G}, \dots, D_\nu^{2m-1}u\big|_{\partial G}\right), \qquad Tv = (T_{2m}v, \dots, T_1 v).$$

Formulas (6.2.12) and (6.2.13) imply the validity of the Green's formula

$$\big(L(x, D_x, q)u, v\big)_G - \big(u, L^+(x, D_x, q)v\big)_G = \sum_{1\le k\le r} \langle \widehat{B}_k u, \widehat{B}'_k v\rangle_{\partial G} \quad (6.2.14)$$

$$(u, v \in H^r(G)),$$

where

$$\widehat{B}'_k = \sum_{1\le j\le t} \Lambda_{jk}^+ T_{2m-j+1}$$

is a boundary operator of order $\widehat{m}'_k = r - \widehat{m}_k - 1$. Furter, we use formulas (6.2.9) in order to rewrite Green's formula (6.2.14) in the form

$$\big(L(x, D_x, q)\, u, v\big)_G + \sum_{1\le j\le m} \Big\langle B_j(x, D_x, q)u, C'_j(x, D_x, q)v\Big\rangle_{\partial G}$$

$$+ \sum_{1\le j\le r-2m} \Big\langle \Big(D_\nu^{j-1} L(x, D_x, q)\Big)u, C'_{m+j}(x, D_x, q)v\Big\rangle_{\partial G}$$

$$= \Big(u, L^+(x, D_x, q)v\Big)_G + \sum_{1\le j\le m} \Big\langle C_j(x, D_x, q)u, B'_j(x, D_x, q)v\Big\rangle_{\partial G} \quad (6.2.15)$$

$$\Big(u \in H^r(G), \quad v \in H^{2m}(G)\Big).$$

Thus, we have proved the following statement.

Theorem 6.2.1. *Assume that (6.2.1) is an elliptic problem with a parameter. Then there exist (pseudo-differential) operators C_j, B'_j, and C'_k of* *orders l_j, m'_j, and l'_k $(j-1,\dots,m,\ k=1,\dots,r-m)$, respectively, with respect to (D_x, q), such that Green's formula (6.2.15) holds. Moreover, relations (6.1.6) are valid.*

Theorem 6.1.1 readily follows from Theorem 6.2.1. Indeed, integrating by parts we obtain

$$\Big(L(x,t,D_x,\partial_t)u,v\Big)_\Omega - \Big(u, L^+(x,t,D_x,\partial_t)v\Big)_\Omega$$

$$+\sum_{0<k<æ-1}\Big(\partial_t^k u(x,0), T_k v(x,0)\Big)_G - \sum_{0<k<æ-1}\Big((T'_k u)(x,T), \partial_t^k v(x,T)\Big)_G$$

$$=\sum_j \int_0^T \int_{\partial G} D_\nu^{j-1}u(x,t)\overline{T_{2m-j+1}v(x,t)}\,dx\,dt = \int_0^T \langle U(x,t), Tv(x,t)\rangle_{\partial G}\,dt$$

$$=\int_0^T \sum_{1\le k\le r}\Big\langle \widehat{B}_k(x,t,D_x,\partial_t)u, \widehat{B}'_k(x,t,D_x,\partial_t)v\Big\rangle_{\partial G}\,dt$$

(cf. (6.2.13) and (6.2.14)), and the Green's formula follows from (6.2.15) if we replace q by ∂_t and integrate over $t \in (0,T)$.

Theorem 6.2.2. *The problem*

$$L^+(x,D_x,q)v = g \qquad (\text{ in } G), \tag{6.2.16}$$

$$B'_j(x,D_x,q)v|_{\partial G} = \psi_j \qquad (j=1,\dots,m), \tag{6.2.17}$$

formally adjoint to problem (6.2.1) with respect to Green's formula (6.2.15), is an elliptic problem with a parameter if and only if problem (6.2.1) is a parameter-elliptic problem.

Proof. Indeed, (6.2.16), (6.2.17) is an elliptic problem with a parameter if and only if the problem

$$L^+(x,D_x,q)v(x,t) = g(x,t), \quad B'_j(x,D_x,\partial_t)v|_{\partial G} = \psi_j \quad (j=1,\dots,m) \tag{6.2.18}$$

is elliptic in the cylinder $g \times \mathbf{R}$. Problem (6.2.18) is elliptic if and only if problem (6.2.4), (6.2.5) is elliptic; this fact was proved in [Sche] (see also [Ber]) for normal boundary conditions and in Chapter 1, [RSh3], [RSh4] in the general case. Thus, the theorem is proved. □

Theorem 6.1.2 follows immediately from Theorem 6.2.1. Note that Theorems 6.1.1 and 6.1.2 are well-known for the case where boundary conditions are normal ([Iva]). Recall that the aim of the present chapter is to prove them for the other cases.

6.2.4. Proof of Theorem 6.1.3

It suffices to prove that the set $\nu_{2m} M(\Omega_0)$ is dense in $\prod_{1 \le j \le 2m} \mathcal{H}^{2m+s-j+1/2}(\Sigma)$ for any integral $s \ge 0$. To this end, we must verify that if $\mu_j \in \mathcal{H}^{-(2m+s-j+1/2)}(\Sigma)$ and

$$\sum_{1 \le j \le 2m} \left\langle \mu_j, D_\nu^{j-1} u \right\rangle_\Sigma = 0 \qquad (\forall u \in M(\Omega_0)), \tag{6.2.19}$$

then $\mu_1 = \cdots = \mu_{2m} = 0$. We rewrite relation (6.2.19) in the form

$$\left(\sum_{1 \le j \le 2m} D_\nu^{j-1}(\mu_j \times \delta_\Sigma), u \right)_\Omega = 0 \qquad (\forall u \in M(\Omega_0)), \tag{6.2.20}$$

where δ_Σ is the Dirac measure concentrated on Σ. Otherwise, we have that

$$\begin{aligned} \left| \left(D_\nu^{j-1}(\mu_j \times \delta_\Sigma), u \right)_\Omega \right| &= |\langle \mu, D_\nu u \rangle_\Sigma| \\ &\le \langle\langle \mu_j, \Sigma \rangle\rangle_{-(2m+s-j+1/2)} \langle\langle D_\nu^{j-1} u, \Sigma \rangle\rangle_{2m+s-j+1/2} \\ &\le c \langle\langle \mu_j, \Sigma \rangle\rangle_{-(2m+s-j+1/2)} \| u, \Omega \|_{2m+s}; \end{aligned}$$

therefore,

$$w := \sum_{1 \le j \le 2m} D_\nu^{j-1}(\mu_j \times \delta_\Sigma) \in \mathcal{H}^{-2m-s}(\Omega).$$

Let us consider problem (6.1.7), (6.1.8), (6.1.9) with $g = w$, $\psi_1 = \cdots = \psi_m = 0$, and $\psi_{0k} = 0$ ($k = 1, \ldots, æ$), namely,

$$L^+ v = w \quad (\text{in } \Omega), \tag{6.2.21}$$

$$B_j' v|_{\Omega'} = 0 \quad (j = 1, \ldots, m); \tag{6.2.22}$$

$$\partial_t^{k-1} v|_{t=T} = 0 \quad (k = 1, \ldots, æ). \tag{6.2.23}$$

It follows from the theorem on complete collection of isomorphisms for problem (6.1.7), (6.1.8), (6.1.9) (see Subsection 6.1.3) that this problem has a solution $v = (v_0, v_1, \ldots, v_{2m}, v_{10}, \ldots, v_{æ0}) \in \widetilde{\mathcal{H}}_{-s,(2m)}(\Omega)$. By the theorem on local increasing of the smoothness ([EZh], [Zh2]), we have that $v \in C^\infty(\Omega \setminus \Sigma)$ and

$$L^+ v = 0 \ \text{ in } \ \Omega \setminus E. \tag{6.2.24}$$

Let us write Green's formula (6.1.5) with $u \in M(\Omega_0)$ and $v \in C^\infty(\overline{\Omega})$. We obtain

$$(Lu, v)_{\Omega_0} = (u, L^+v)_\Omega + \sum_{1\le j\le m} \langle C_j u, B'_j v\rangle + \sum_{0\le k\le æ-1} \left(T'_k u(x,T), \partial_t^k v(x,T)\right)_G.$$

By passing to the limit (see Subsection 6.1.3) we obtain that this formula holds for $u \in M(\Omega_0)$ and for the solution $v \in \widetilde{\mathcal{H}}_{-s,(2m)}(\Omega)$ of problem (6.2.21), (6.2.22), (6.2.23). This, together with (6.2.20), implies that

$$(Lu, v)_{\Omega_0} = (u, L^+v)_\Omega = (u, w)_\Omega = 0 \qquad (\forall u \in M(\Omega_0)).$$

Thus, in particular, by taking u such that $Lu = v$ in Ω_0 (this can always be done fore a parabolic equation), we see that $v = 0$ in Ω_0. Since the Cauchy problem for the equation $L^+v = 0$ is uniquely solvable ([Mal]), by (6.2.24) the relation $v = 0$ in Ω_0 implies $v_0(x,t) = 0$ in $\Omega \setminus \Sigma$ (i.e., $\operatorname{supp} v_0 \subset \Sigma$), $v_j = 0$ $(j = 1,\dots,2m)$, and $v_{j0} = 0$ $(j = 1,\dots,æ)$. Thus,

$$v_0 \in \mathcal{H}^{-s}(\Omega), \quad \operatorname{supp} v_0 \subset \Sigma, \quad L^+v_0 = w.$$

Now we can readily verify that

$$v_0 = \sum_{0\le j\le k} D_\nu^i(\tau_j \times \delta_\Sigma) \qquad (\tau_j \in \mathcal{H}^{-s-j-1/2}(\Sigma), \quad j = 1,\dots,\overline{k}),$$

where $\overline{k} = [s - 1/2]$ (i.e., $\overline{k}$ is the closest integer to $s-1$). Then by (6.2.21) we have

$$L^+\left(\sum_{0\le j\le k} D_\nu^j(\tau_j \times \delta_\Sigma)\right) = \sum_{1\le j\le 2m} D_\nu^{j-1}(\mu_j \times \delta_\Sigma). \tag{6.2.25}$$

Since the Dirac measures and their derivatives are linearly independent and L^+ is parabolic, we subsequently find that $\tau_k = \dots = \tau_0 = 0$. Now it follows from (6.2.25) that $\mu_j = 0$ $(j = 1,\dots,2m)$. This completes the proof of the theorem. □

6.2.5. Proofs of Theorems 6.1.4, 6.1.5, and 6.1.6

Proof of Theorem 6.1.4. It suffices to prove that if $\mu_j \in \mathcal{H}^{-(m-j+1/2)}(\Sigma)$ $(j = 1,\dots,m)$ and

$$\sum_{1\le j\le m} \langle \mu_j, D_\nu^{j-1}u\rangle_\Sigma = 0 \qquad (\forall u \in M(\Omega_0)), \tag{6.2.26}$$

then $\mu_1 = \dots = \mu_m = 0$. Let us rewrite equality (6.2.26) as

$$\left(\sum_{1\le j\le m} D_\nu^{j-1}(\mu_j \times \delta_\Sigma), u\right) = 0 \qquad (\forall u \in M(\Omega_0)),$$

and, as in the proof of Theorem 6.1.3, let us verify that

$$w := \sum_{1\le j\le m} D_\nu^{j-1}(\mu_j \times \delta_\Sigma) \in \mathcal{H}^{-m}(\Omega).$$

Then problem (6.2.21), (6.2.22), (6.2.23) has a solution $v \in \widetilde{\mathcal{H}}^{m,(2m)}(\Omega)$. Arguing as in the proof of Theorem 6.1.3, we can prove that $v = 0$ in $\overline{\Omega} \setminus \overline{\Omega}_1$, where $\Omega_1 = G_1 \times (0,T)$. Then v is a solution of the parabolic problem

$$L^+v = 0 \quad (\text{in } \Omega_1), \qquad D_\nu^{j-1}v\big|_\Sigma = 0 \quad (j = 1,\dots,m);$$

$$\partial_t^{k-1}v\big|_{t=0} = 0 \quad (k = 1,\dots,æ).$$

in Ω_1, and hence $v \equiv 0$. Now it follows from (6.2.21) that $w = 0$ and $\mu_1 = \dots = \mu_m = 0$. The theorem is thus proved. □

Proof of Theorem 6.1.5. Assume that $G_2 \subset \mathbf{R}^n$ be a domain with an arbitrary small diameter. Let this domain border on the domain G along γ_0, $\gamma_0 \subset \partial G_2$, and $G_2 \cap G = \emptyset$. Let also $\Omega_2 = G_2 \times (0,T)$. We assume that the coefficients of the corresponding formulas are extended to Ω_2 and the obtained problem of the form (6.1.1) is parabolic on $\Omega \cap \Omega_2$. Let us define $M(\Omega_2)$ like $M(\Omega_0)$. By Theorem 6.1.3, the set $\nu_{2m}M(\Omega_2)$ is dense in the direct product $\prod_{1\le j\le 2m} \mathcal{H}_0^{s_j}(\Sigma)$ for any $s_j > 0$ $(j = 1,\dots,2m)$. Since $\nu_{2m}M(\Gamma_0) \supset \nu_{2m}M(\Omega_2)$, the statement of Theorem 6.1.5 follows immediately. □

Theorem 6.1.6 follows from Theorem 6.1.4 just like Theorem 6.1.5 follows from Theorem 6.1.3.

REFERENCES

Agmon, S., Douglis, A., and Nirenberg, L.

[ADN] *Estimates near the boundary for solutions of elliptic partial differential equations satysfying general boundary conditions, I, II*, Comm.Pure Appl.Math, **12** (1959), p.623–727; **17** (1964), p.35–92.

Agranovich, M. S.

[Agr1] *Boundary value problems for systems with a parameter*, Mat. sb. **84** (1971), No. 1, 27–65;
English transl. in Math USSR Sb. **13** (1971).

[Agr2] *Some Asymptotic Formulas for Elliptic Pseudo-Differential Operators*, Functional Analysis and Its Applications, **21**, No. 1 (1987), 53–56.

[Agr3] Elliptic Operators on Closed Manifolds, Encyclopedia Math. Sci. **63**, Springer-Verlag, 1994.

Agranovich, M. S., and Vishik, M.I.

[AgV] *Elliptic problems with a parameter and parabolic problems of the general form*, Uspekhi Mat. Nauk, **19**, No. 3, 53–161 (1964).

Aslanyan, A.G., Vassiliev, D.G., and Lidskii V.B.

[AVL] *Frequencies of free oscillations of thin shell interacting with fluid*, Functional Anal. Appl., **15**, No.3, 1–9 (1981).

Berezanskii Yu.M.

[Ber] Expansions in eigenfunctions of selfadjoint operators, Naukova Dumka, Kiev, 1965; English transl., Amer. Math. Soc., Providence, RI, 1968.

Berezanskii Yu.M., Krein, S. G., and Roitberg, Ya. A.

[BKR] *Theorem on homeomorphisms and local increase in smoothness of solutions of elliptic equations up to the boundary*, Dokl. Akad. Nauk SSSR **148** (1963), No. 4, 745–748.

Boimatov, K. Kh.

[Boi] *Spectral Asymptotics of Linear and Nonlinear Pencils of Pseudo-Differential Operators That are Elliptic in the Douglis–Nirenberg Sense*, Soviet Math. Dokl. **45**

(1992), No. 1, 99–104.

Chazarain, J., and Piriou A.

[ChP] *Caractérisation des problèmes mixted hyperboliques bien posés* Ann. Inst. Fourier. (Grenoble) **22** (1972), No. 4, 193–237.

Denk, R., Mennicken, R., and Volevich, L.

[DMV] *The Newton Polygon and Elliptic Problems with Parameter*, Math. Nachr. **192** (1998), 125–157.

Doropienko, E.

[Dor] *Boundary value and mixed problems for general hyperbolic systems with multiple characteristics in complete scale of spaces of Sobolev type*, Dokl. Akad. Nauk Ukraine **6** (1995), 14–17.

Eidelman, S. D.

[Eid] *Parabolic equations*, Sovrem. Problemy Mat. Fundafental'nye Mapravlenija, **63** (1990), 201–313.

Eidelman, S. D., and Zhitarashu, N. V.

[EZh] Parabolic Boundary Value Problems, Operator Theory: Advanses and Appl., Vol. 101, Birkhäuser Verlag, 1998.

Eskin, G. I.

[Esk] Boundary-value problems for elliptic pseudodifferential equations, Nauka, Moscow, 1973;
English transl., Amer. Math. Soc., Provodence, RI, 1981.

Garlet P.G.

[Gar] Plates and junctions in ellastic multistructures, An asymptotic analysis, Masson: Paris, Milan, Barselona, Mexica, 1990.

Grubb, G.

[Gr1] *Remainder Estimates for Eigenvalues and Kernels of Pseudo-Differential Elliptic Systems*, Math. Scand. **43** (1978), 275–307.

[Gr2] *Partial diffrential problems in L_p spaces*, Comm. Part. Diff. Eq-s., **15(3)** (1990), 289–340.

Hamann, U.

[Ham] *Approximation durch Normalableitungen von Lösungen elliptischer Randwertprobleme in beliebigen Sobolev-Räumen*, Math. Nachr., **128** (1986), 199–214.

Hamann, U., and Wildenhain, G.

[HaW] Partial Diff. Equations, Banach Center Publ., **19** (1987), 113–119.

Ivasyshen, S. D.

[Iva] *On normal parabolic boundary-value problem*, Vestnik Kiev. Univ., Matematika i Telemekhanika, **25** (1983), 77–82.

Kostarchuk, Yu. V., and Roitberg, Ya. A.

[KosR] *Theorems on isomorphisms for elliptic boundary-value problems without the assumption that boundary conditions are normal*, Ukr. Mat. Zh. **25** (1973), No. 2, 277–283.

Kozhevnikov, A.

[Kozh1] *Spectral Problems for Pseudo-Differential Systems Elliptic in the Douglis–Nirenberg Sense, and Their Applications*, Mat. USSR Sb. **21** (1973), 63–90.

[Kozh2] *Asymptotics of the Spectrum of Douglis–Nirenberg Elliptic Operators on a Compact Manifold*, Math. Nachr. **182** (1996), 261–293.

Kozhevnikov, A., Yakubov, S.

[KoY] *On Operators Generated by Elliptic Boundary Problems with a Spectral Parameter in Boundary Conditions*, Integral Equations and Operator Theory **23** (1995), 205–231.

Kreiss, H.-O.

[Kreis] Initial boundary value problems for hyperbolic systems, Comm. Pure Appl. Math. **23** (1970), 277–298.

Lapa, T. V, and Movsha, E. N.

[LaMo] *The Cauchy problem for general hyperbolic equations and systems with multiple characteristics in complete scale of spaces of Sobolev type*, Dokl. Akad. Nauk Ukraine, to appear.

Lions, J.-L., and Magenes, E.

[LiM] Problèmes aux limites non homogènes et applications, Vol. 1, Dunod, Paris, 1968; Russian transl., Moscow, Mir, 1971.

Los, V. N., and Roitberg, Ya. A.

[LR1] *Sobolev's problem*, Dokl. Akad. Nauk Ukraine, **8** (1998), 31–38.

[LR2] *Sobolev's problem*, Ukr.Mat. Zh., to appear.

[LR3] *Sobolev's problem*, Spectral and Evolutionary problems: Proc. of the Eight Crimean Autumn Mathematical School-Symposium. Vol. 8, 1998. 197–203

L'vin, S. Ya.

[L'v] *Green's formula and solvability of Douglis–Nirenberg elliptic problems with the boundary conditions of arbitrary order*, Dep. VINITI, No. 3318-78, 1–30. (Russian)

Malamud, M. M.

[Mal] *On one analog of Nelson's theorem*, Funkz. Analiz, **19:2** (1985), 82–83.

Nazarov S., and Pileckas K.

[NaP] *On noncompact free boundary problems for the plane stationary Navier–Stokes equations*, J. Reine Angew. Math., **438**, 103–141 (1993).

Roitberg, I. Ya.

[RI1] *Boundary value problems for Douglis–Nirenberg (T, S)-systems in complete scales of Banach spaces*, IWOTA-95, University of Regensburg, Book of Abstracts, Regensburg, p.60 (1995).

[RI2] *Boundary value problems with additional unknown functions on the boundary for general elliptic systems in complete scales of Banach spaces,* International Conference 'Nonlinear Differential Equations', Kiev, August 21-27, 1995, Book of Abstracts, Kiev, p.139 (1995).

[RI3] *Elliptic Boundary Problems for General Systems of Equations in Complete Scales of Banach Spaces.* Doklady Akad Nauk Russia, **354**, No. 1, 25–29 (1997). Engl. trans.: Doklady Mathematics, **55**, No. 3, 335–339 (1997).

[RI4] *Elliptic boundary value problems for general elliptic systems in complete scales of Banach spaces.* Nonlinear Boundary Value Problems, Donetsk, Issue 7 , 159–164 (1997).

[RI5] *Elliptic boundary value problems for general elliptic systems in complete scales of Banach spaces*, Operator Theory: Advances and Applications, Vol. 102, 231–241(1998).

[RI6] *Green's formula for general systems of equations*, 2nd European Congress of Mathematics in Budapest, Satellite Conference 'Aspects of Spectral Theory: Operator methods in boundary value problems, Schrodinger operators', Vienna, July 15 to 18, 1996, Book of Abstracts, 1996, p. 48.

[RI7] *Green's formula for general systems of equations*, Abstracts Sommerfeld'96 Workshop, Techniche Hochschule Darmstadt, 1996, p. 35.

Roitberg, I. Ya., and Roitberg, Ya. A.

[RR1] *Green's formula for general parabolic boundary-value problems*, Dokl. Akad. Nauk Russia, **333** (1994), No. 1, 24–28.

[RR2] *Green's formula and density of solutions of general parabolic boundary-value problems in functional spaces on manifolds*, Differntsial'nyje Uravnenija **31** (1995), No. 8, 1437–1444;
English transl. in Differential Equations, **31:8** (1995), 1382–1390.

[RR3] *The Green Formula for General Elliptic Boundary-Value Problems for Systems of the Douglis–Nirenberg Structure*, Dokl. Akad. Nauk Russia, **359:6** (1998), 739–743;

English transl. in Russian Acad. Sci. Math. Dokl. **57:2**(1998), 290–294.

[RR4] *Theorems on complete collection of isomorphisms for general elliptic systems*, Dokl. Akad. Nauk Russia, 1998, to appear .

Roitberg, Ya. A.

[R1] Elliptic Boundary Value Problems in the Spaces of Distributions, Kluwer Academic Publishers, Dordrecht/Boston/London, 1996.

[R2] *Theorems on homeomorphisms and Green's formula for general elliptic boundary value problems whose boundary conditions are not nolmal*, Mat. Sb. **83** (125) (1970), No. 2 (10), 181–213.

[R3] *Theorem on complete collection of isomorphisms for general elliptic system*, in: Abstracts of the Eight Voronezh Winter Mathematical School, Voronezh (1974), 92–93.

[R4] *On the existence of the limiting values of generalized solutions of elliptic equations on the boundary of the domain*, Sib. Mat. Zh. **20** (1979), No. 2, 386–396.

[R5] *Theorem on homeomorphisms realized in L_p by elliptic operators and local increasing of smoothness of generalized solutions*, Ukr. Mat. Zh. **17** (1965), No. 5, 122–129.

[R6] *Theorem on complete collection of isomorphisms for Douglis – Nirenberg elliptic systems*, Ukrain. Mat. Zh., **27**, (1975), No. 4, 554–548;
English transl. in Ukrainian Math. J. **27** (1975).

[R7] *The Cauchy problem for hyperbolic equations in the full scale of spaces of Sobolev type*, Dokl. Akad. Nauk SSSR **290** (1986), No. 2, 296–300;
English transl. in Soviet Math. Dokl. **34** (1987).

[R8] *Solvability of the Cauchy problem in the full scale of spaces of Sobolev type*, Spectral Theory of Differential-Operator Equations (Yu. M. Berezanskii, ed.), Inst. Mat. Akad. Nauk Ukrain. SSR, Kiev, 1986, 33–52;
English transl., Selecta Math. Sovietica **10** (1991), 278–295.

[R9] *Boundary-value and mixed problems for homogeneous hyperbolic equations in the full scale of spaces of Sobolev type*, Ukrain. Mat. Zh. **41** (1989), No. 6, 686–690;
English transl. in Ukrainian Math. J. **41** (1989).

[R10] *Elliptic problems with nonhomogeneous boundary conditions*, Dokl. Akad. Nauk SSSR **157** (1964), No. 4, 798–801;
English transl. in Soviet Math. Dokl. **5** (1964).

[R11] On the values on the boundary of a domain for generalized solutions of elliptic equations, Mat. Sb. **86** (1971), No. 2 (10), 248–267;
English transl. in Math. USSR Sb. **15** (1971).

[R12] *Theorems on homeomorphisms established by elliptic operators*, Dokl. Akad. Nauk SSSR **180** (1968), No. 3, 542–545.

[R13] *Theorem on homeomorphisms for general elliptic boundary-value problems with boundary conditions that are not normal*, Dokl. Akad. Nauk SSSR **191** (1970), No. 6, 1228–1231.

[R14] *On the density of solutions of parabolic boundary-value problems in the functional spaces on the manifolds*, Dokl. Akad. Nauk USSR, **3** (1993), 22–26.

[R15] *Elliptic problems with a parameter in classes of generalized functions*, Sib. Mat. Zh. **22:2** (1981), 214–218.

Roitberg, Ya. A., and Serdjuk, V. A.

[RS1] Elliptic problems with a parameter in L_2 spaces of generalized functions for general systems of equations, Preprint No. 82.30, Inst. Mat. Akad. Nauk Ukrain. SSR, Kiev, 1982, 3–60. (Russian) **MR 84b:** 35041.

[RS2] *Theory of solvability in generalized functions of problems with a parameter for general systems of equations*, Trudy Mosk. Mat. Obshch. **53** (1990), 229–258; English transl. in Trans. Moscow Math. Soc., 1991.

Roitberg, Ya. A., and Sheftel, Z. G.

[RSh1] *General boundary-value problems for elliptic equations with discontinuous coefficients*, Dokl. Akad. Nauk SSSR, **148** (1963), No. 5, 1034–1037.

[RSh2] *Green's formula and theorem on homeomorphisms for elliptic systems*, Uspekhi Mat. Nauk, **22** (1967), No. 5, 181–182.

[RSh3] *Theorem on homeomorphisms for elliptic systems and its applications*, Mat. Sb. **78** (1969), No. 3, 446–472.

[RSh4] *Homeomorphism theorems for elliptic systems with the boundary conditions which are not normal*, Matem. Issledov., Kishinjov, **7** (1972), No. 2, 143–157.

[RSh5] *On density of solutions of elliptic problems with localized right-hand sides in the functional spaces on manifold*, Dokl. Akad. Nauk SSSR, **305:6** (1989), 1317–1320.

[RSh6] *On density of solutions of boundary value problems for Petrovskii elliptic systems*, Dokl. Akad. Nauk Ukrain. SSR, Ser. A, **9** (1990), 17–20.

Roitberg, Ya. A., and Sklyarets, A. V.

[RSk1] *Sobolev's problem in complete scales of Banach spaces*, Ukr. Math. J., **48** (1996), No. 11, 1555–1563.

[RSk2] *Sobolev's problem in complete scales of Banach spaces*, Dokl. Ukrain. Akad. Nauk, **48** (1996), No. 1.

Sakamoto, R.

[Sak] Mixed problems for hyperbolic equations, I, II, J. Math. Kyoto Univ. **10** (1970), No. 2, 349–373, No. 3, 403–417.

Schechter, M.

[Sche] *Coerciveness of linear partial differential operators and functions satisfying zero Dirichlet-type boundary data*, Comm. Pure Appl. Math, **11** (1958), No. 2, 153–

174.

Schulze, B.-W., and Wildenhain, G.

[SchW] Methoden der Potential-teorie für elliptische Differentialgleichungen beliebiger Ordnung, Berlin, 1977.

Seeley, R.

[See] *Complex Powers of an Elliptic Operator*, Amer. Math. Soc. Proc. Symp. Pure Math., **10** (1967), 288–307.

Sobolev, S. L.

[Sob] Some Applications of Functional Analysis in Mathematical Physics, Izd. Leningrad Univ., Leningrad, 1950.

Solonnikov, V.A.

[Sol] On general boundary value problems elliptic according to Douglis - Nirenberg, I, II; I: Izv.Akad.Nauk SSSR, Ser.Matematika, **29**, No.3, 665–706 (1964); II: Trudy Mat. Inst. Akad. Nauk SSSR, **92**, 233–297 (1966).

Sternin, B. Yu.

[St1] *Elliptic and padabolic problems on manofolds whose boundary consist of components of various dimensions*, Trudy Mosk. Mat. Obsc., **15** (1966), 346–382.

[St2] *Relative elliptic theory and problem of S. L. Sobolev*, Dokl. Akad. Nauk SSSR, **230** (1976), No. 2, 287–290.

Volevich, L.R.

[Vol] *Solvability of boundary-value problems for general elliptic systems*, Mat. Sb. **68** (1965), No. 3, 373–416;
English transl., Amer. Math. Soc. Transl. (2) **67** (1968), 182–225.

Volevich, L. R., and Gindikin, S. G.

[VG] *The method of energy estimates in the mixed problem*, Uspekhi Mat. Nauk **35** (1980), No. 5, 53–120;
English transl. in Russian Math. Surveys **35** (1980), No. 5.

Volevich, L. R., and Ivrii, V. Ya.

[VIv] *Hyperbolic equations,* Selected works of I. G. Petrovskii: Systems of Partial Differential Equations. Algebraic Geometry, 1986, 395–418. (Russian)

Volevich, L. R., and Shirikjan, A.

[VSh1] Remarks on Calderón Projections for Elliptic Equation with Parameter, MV. Keldysh Institute of Appl. Math., Russian Acad. of Sc. Preprint No. 48 (1997).

[VSh2] Stable and Unstable Manifolds for Non-linear Elliptic Equations with Parameter, MV. Keldysh Institute of Appl. Math., Russian Acad. of Sc. Preprint No. 49

(1997).

Wloka, J. T., Rowley, B., and Lawruk, B.

[WRL] Boundary value problems for elliptic systems, Cambridge University Press, 1995.

Zhitarashu, N. V.

[Zh2] Dokl. Akad. Nauk SSSR **260** (1981), No. 5, 1054–1058.

[Zh2] *Theorem on complete collection of isomorphisms in the L_2-theory of generalized solutions of boundary-value problems for one parabolic equation in the Petrovskii sense*, Mat. Sb. **128** (1985), 451–473;
English transl., in Math USSR Sb. **56** (1987).

SUBJECT INDEX

Other *Mathematics and Its Applications* titles of interest:

J. Chaillou: *Hyperbolic Differential Polynomials and their Singular Perturbations.* 1979, 184 pp. ISBN 90-277-1032-5

S. Fučik: *Solvability of Nonlinear Equations and Boundary Value Problems.* 1981, 404 pp. ISBN 90-277-1077-5

V.I. Istrăţescu: *Fixed Point Theory. An Introduction.* 1981, 488 pp. *out of print,* ISBN 90-277-1224-7

F. Langouche, D. Roekaerts and E. Tirapegui: *Functional Integration and Semiclassical Expansions.* 1982, 328 pp. ISBN 90-277-1472-X

N.E. Hurt: *Geometric Quantization in Action. Applications of Harmonic Analysis in Quantum Statistical Mechanics and Quantum Field Theory.* 1982, 352 pp. ISBN 90-277-1426-6

F.H. Vasilescu: *Analytic Functional Calculus and Spectral Decompositions.* 1983, 392 pp. ISBN 90-277-1376-6

W. Kecs: *The Convolution Product and Some Applications.* 1983, 352 pp. ISBN 90-277-1409-6

C.P. Bruter, A. Aragnol and A. Lichnerowicz (eds.): *Bifurcation Theory, Mechanics and Physics. Mathematical Developments and Applications.* 1983, 400 pp. *out of print,* ISBN 90-277-1631-5

J. Aczél (ed.) : *Functional Equations: History, Applications and Theory.* 1984, 256 pp. *out of print,* ISBN 90-277-1706-0

P.E.T. Jørgensen and R.T. Moore: *Operator Commutation Relations.* 1984, 512 pp. ISBN 90-277-1710-9

D.S. Mitrinović and J.D. Keckić: *The Cauchy Method of Residues. Theory and Applications.* 1984, 376 pp. ISBN 90-277-1623-4

R.A. Askey, T.H. Koornwinder and W. Schempp (eds.): *Special Functions: Group Theoretical Aspects and Applications. 1984, 352 pp.* ISBN 90-277-1822-9

R. Bellman and G. Adomian: *Partial Differential Equations. New Methods for their Treatment and Solution.* 1984, 308 pp. ISBN 90-277-1681-1

S. Rolewicz: *Metric Linear Spaces.* 1985, 472 pp. ISBN 90-277-1480-0

Y. Cherruault: *Mathematical Modelling in Biomedicine.* 1986, 276 pp. ISBN 90-277-2149-1

R.E. Bellman and R.S. Roth: *Methods in Approximation. Techniques for Mathematical Modelling.* 1986, 240 pp. ISBN 90-277-2188-2

R. Bellman and R. Vasudevan: *Wave Propagation. An Invariant Imbedding Approach.* 1986, 384 pp. ISBN 90-277-1766-4

Other *Mathematics and Its Applications* titles of interest:

A.G. Ramm: *Scattering by Obstacles.* 1986, 440 pp. ISBN 90-277-2103-3

C.W. Kilmister (ed.): *Disequilibrium and Self-Organisation.* 1986, 320 pp. ISBN 90-277-2300-1

A.M. Krall: *Applied Analysis.* 1986, 576 pp. ISBN 90-277-2328-1 (hb), ISBN 90-277-2342-7 (pb)

J.A. Dubinskij: *Sobolev Spaces of Infinite Order and Differential Equations.* 1986, 164 pp. ISBN 90-277-2147-5

H. Triebel: *Analysis and Mathematical Physics.* 1987, 484 pp. ISBN 90-277-2077-0

B.A. Kupershmidt: *Elements of Superintegrable Systems. Basic Techniques and Results.* 1987, 206 pp. ISBN 90-277-2434-2

M. Greguš: *Third Order Linear Differential Equations.* 1987, 288 pp. ISBN 90-277-2193-9

M.S. Birman and M.Z. Solomjak: *Spectral Theory of Self-Adjoint Operators in Hilbert Space.* 1987, 320 pp. ISBN 90-277-2179-3

V.I. Istrăţescu: *Inner Product Structures. Theory and Applications.* 1987, 912 pp. ISBN 90-277-2182-3

R. Vich: *Z Transform Theory and Applications.* 1987, 260 pp. ISBN 90-277-1917-9

N.V. Krylov: *Nonlinear Elliptic and Parabolic Equations of the Second -Order.* 1987, 480 pp. ISBN 90-277-2289-7

W.I. Fushchich and A.G. Nikitin: *Symmetries of Maxwell's Equations.* 1987, 228 pp. ISBN 90-277-2320-6

P.S. Bullen, D.S. Mitrinović and P.M. Vasić (eds.): *Means and their Inequalities.* 1987, 480 pp. ISBN 90-277-2629-9

V.A. Marchenko: *Nonlinear Equations and Operator Algebras.* 1987, 176 pp. ISBN 90-277-2654-X

Yu.L. Rodin: *The Riemann Boundary Problem on Riemann Surfaces.* 1988, 216 pp. ISBN 90-277-2653-1

A. Cuyt (ed.): *Nonlinear Numerical Methods and Rational Approximation.* 1988, 480 pp. ISBN 90-277-2669-8

D. Przeworska-Rolewicz: *Algebraic Analysis.* 1988, 640 pp. ISBN 90-277-2443-1

V.S. Vladimirov, YU.N. Drozzinov and B.I. Zavialov: *Tauberian Theorems for Generalized Functions.* 1988, 312 pp. ISBN 90-277-2383-4

G. Moroşanu: *Nonlinear Evolution Equations and Applications.* 1988, 352 pp. ISBN 90-277-2486-5

A.F. Filippov: *Differential Equations with Discontinuous Righthand Sides.* 1988, 320 pp. ISBN 90-277-2699-X

Other *Mathematics and Its Applications* titles of interest:

A.T. Fomenko: *Integrability and Nonintegrability in Geometry and Mechanics.* 1988, 360 pp. ISBN 90-277-2818-6

G. Adomian: *Nonlinear Stochastic Systems Theory and Applications to Physics.* 1988, 244 pp. ISBN 90-277-2525-X

A. Tesár and Ludovt Fillo: *Transfer Matrix Method.* 1988, 260 pp. ISBN 90-277-2590-X

A. Kaneko: *Introduction to the Theory of Hyperfunctions.* 1989, 472 pp. ISBN 90-277-2837-2

D.S. Mitrinović, J.E. Pećarič and V. Volenec: *Recent Advances in Geometric Inequalities.* 1989, 734 pp. ISBN 90-277-2565-9

A.W. Leung: *Systems of Nonlinear PDEs: Applications to Biology and Engineering.* 1989, 424 pp. ISBN 0-7923-0138-2

N.E. Hurt: *Phase Retrieval and Zero Crossings: Mathematical Methods in Image Reconstruction.* 1989, 320 pp. ISBN 0-7923-0210-9

V.I. Fabrikant: *Applications of Potential Theory in Mechanics. A Selection of New Results.* 1989, 484 pp. ISBN 0-7923-0173-0

R. Feistel and W. Ebeling: *Evolution of Complex Systems. Selforganization, Entropy and Development.* 1989, 248 pp. ISBN 90-277-2666-3

S.M. Ermakov, V.V. Nekrutkin and A.S. Sipin: *Random Processes for Classical Equations of Mathematical Physics.* 1989, 304 pp. ISBN 0-7923-0036-X

B.A. Plamenevskii: *Algebras of Pseudodifferential Operators.* 1989, 304 pp. ISBN 0-7923-0231-1

N. Bakhvalov and G. Panasenko: *Homogenisation: Averaging Processes in Periodic Media. Mathematical Problems in the Mechanics of Composite Materials.* 1989, 404 pp. ISBN 0-7923-0049-1

A.Ya. Helemskii: *The Homology of Banach and Topological Algebras.* 1989, 356 pp. ISBN 0-7923-0217-6

M. Toda: *Nonlinear Waves and Solitons.* 1989, 386 pp. ISBN 0-7923-0442-X

M.I. Rabinovich and D.I. Trubetskov: *Oscillations and Waves in Linear and Nonlinear Systems.* 1989, 600 pp. ISBN 0-7923-0445-4

A. Crumeyrolle: *Orthogonal and Symplectic Clifford Algebras. Spinor Structures.* 1990, 364 pp. ISBN 0-7923-0541-8

V. Goldshtein and Yu. Reshetnyak: *Quasiconformal Mappings and Sobolev Spaces.* 1990, 392 pp. ISBN 0-7923-0543-4

I.H. Dimovski: *Convolutional Calculus.* 1990, 208 pp. ISBN 0-7923-0623-6

Other *Mathematics and Its Applications* titles of interest:

Y.M. Svirezhev and V.P. Pasekov: *Fundamentals of Mathematical Evolutionary Genetics.* 1990, 384 pp. ISBN 90-277-2772-4

S. Levendorskii: *Asymptotic Distribution of Eigenvalues of Differential Operators.* 1991, 297 pp. ISBN 0-7923-0539-6

V.G. Makhankov: *Soliton Phenomenology.* 1990, 461 pp. ISBN 90-277-2830-5

I. Cioranescu: *Geometry of Banach Spaces, Duality Mappings and Nonlinear Problems.* 1990, 274 pp. ISBN 0-7923-0910-3

B.I. Sendov: *Hausdorff Approximation.* 1990, 384 pp. ISBN 0-7923-0901-4

A.B. Venkov: *Spectral Theory of Automorphic Functions and Its Applications.* 1991, 280 pp. ISBN 0-7923-0487-X

V.I. Arnold: *Singularities of Caustics and Wave Fronts.* 1990, 274 pp. ISBN 0-7923-1038-1

A.A. Pankov: *Bounded and Almost Periodic Solutions of Nonlinear Operator Differential Equations.* 1990, 232 pp. ISBN 0-7923-0585-X

A.S. Davydov: *Solitons in Molecular Systems. Second Edition.* 1991, 428 pp. ISBN 0-7923-1029-2

B.M. Levitan and I.S. Sargsjan: *Sturm-Liouville and Dirac Operators.* 1991, 362 pp. ISBN 0-7923-0992-8

V.I. Gorbachuk and M.L. Gorbachuk: *Boundary Value Problems for Operator Differential Equations.* 1991, 376 pp. ISBN 0-7923-0381-4

Y.S. Samoilenko: *Spectral Theory of Families of Self-Adjoint Operators.* 1991, 309 pp. ISBN 0-7923-0703-8

B.I. Golubov A.V. Efimov and V.A. Scvortsov: *Walsh Series and Transforms.* 1991, 382 pp. ISBN 0-7923-1100-0

V. Laksmikantham, V.M. Matrosov and S. Sivasundaram: *Vector Lyapunov Functions and Stability Analysis of Nonlinear Systems.* 1991, 250 pp. ISBN 0-7923-1152-3

F.A. Berezin and M.A. Shubin: *The Schrdinger Equation.* 1991, 556 pp. ISBN 0-7923-1218-X

D.S. Mitrinović, J.E. Pečarić and A.M. Fink: *Inequalities Involving Functions and their Integrals and Derivatives.* 1991, 588 pp. ISBN 0-7923-1330-5

Julii A. Dubinskii: *Analytic Pseudo-Differential Operators and their Applications.* 1991, 252 pp. ISBN 0-7923-1296-1

V.I. Fabrikant: *Mixed Boundary Value Problems in Potential Theory and their Applications.* 1991, 452 pp. ISBN 0-7923-1157-4

A.M. Samoilenko: *Elements of the Mathematical Theory of Multi-Frequency Oscillations.* 1991, 314 pp. ISBN 0-7923-1438-7

Other *Mathematics and Its Applications* titles of interest:

Yu.L. Dalecky and S.V. Fomin: *Measures and Differential Equations in Infinite-Dimensional Space.* 1991, 338 pp. ISBN 0-7923-1517-0

W. Mlak: *Hilbert Space and Operator Theory.* 1991, 296 pp. ISBN 0-7923-1042-X

N.Ja. Vilenkin and A.U. Klimyk: *Representation of Lie Groups and Special Functions. Volume 1: Simplest Lie Groups, Special Functions, and Integral Transforms.* 1991, 608 pp. ISBN 0-7923-1466-2

N.Ja. Vilenkin and A.U. Klimyk: *Representation of Lie Groups and Special Functions. Volume 2: Class I Representations, Special Functions, and Integral Transforms.* 1992, 630 pp. ISBN 0-7923-1492-1

N.Ja. Vilenkin and A.U. Klimyk: *Representation of Lie Groups and Special Functions. Volume 3: Classical and Quantum Groups and Special Functions.* 1992, 650 pp ISBN 0-7923-1493-X
(Set ISBN for Vols. 1, 2 and 3: 0-7923-1494-8)

K. Gopalsamy: *Stability and Oscillations in Delay Differential Equations of Population Dynamics.* 1992, 502 pp. ISBN 0-7923-1594-4

N.M. Korobov: *Exponential Sums and their Applications.* 1992, 210 pp. ISBN 0-7923-1647-9

Chuang-Gan Hu and Chung-Chun Yang: *Vector-Valued Functions and their Applications.* 1991, 172 pp. ISBN 0-7923-1605-3

Z. Szmydt and B. Ziemian: *The Mellin Transformation and Fuchsian Type Partial Differential Equations.* 1992, 224 pp. ISBN 0-7923-1683-5

L.I. Ronkin: *Functions of Completely Regular Growth.* 1992, 394 pp. ISBN 0-7923-1677-0

R. Delanghe, F. Sommen and V. Soucek: *Clifford Algebra and Spinor-valued Functions. A Function Theory of the Dirac Operator.* 1992, 486 pp. ISBN 0-7923-0229-X

A. Tempelman: *Ergodic Theorems for Group Actions.* 1992, 400 pp. ISBN 0-7923-1717-3

D. Bainov and P. Simenov: *Integral Inequalities and Applications.* 1992, 426 pp. ISBN 0-7923-1714-9

I. Imai: *Applied Hyperfunction Theory.* 1992, 460 pp. ISBN 0-7923-1507-3

Yu.I. Neimark and P.S. Landa: *Stochastic and Chaotic Oscillations.* 1992, 502 pp. ISBN 0-7923-1530-8

H.M. Srivastava and R.G. Buschman: *Theory and Applications of Convolution Integral Equations.* 1992, 240 pp. ISBN 0-7923-1891-9

A. van der Burgh and J. Simonis (eds.): *Topics in Engineering Mathematics.* 1992, 266 pp. ISBN 0-7923-2005-3

Other *Mathematics and Its Applications* titles of interest:

F. Neuman: *Global Properties of Linear Ordinary Differential Equations.* 1992, 320 pp. ISBN 0-7923-1269-4

A. Dvurečenskij: *Gleason's Theorem and its Applications.* 1992, 334 pp. ISBN 0-7923-1990-7

D.S. Mitrinović, J.E. Pečarić and A.M. Fink: *Classical and New Inequalities in Analysis.* 1992, 740 pp. ISBN 0-7923-2064-6

H.M. Hapaev: *Averaging in Stability Theory.* 1992, 280 pp. ISBN 0-7923-1581-2

S. Gindinkin and L.R. Volevich: *The Method of Newton's Polyhedron in the Theory of PDE's.* 1992, 276 pp. ISBN 0-7923-2037-9

Yu.A. Mitropolsky, A.M. Samoilenko and D.I. Martinyuk: *Systems of Evolution Equations with Periodic and Quasiperiodic Coefficients.* 1992, 280 pp. ISBN 0-7923-2054-9

I.T. Kiguradze and T.A. Chanturia: *Asymptotic Properties of Solutions of Nonautonomous Ordinary Differential Equations.* 1992, 332 pp. ISBN 0-7923-2059-X

V.L. Kocic and G. Ladas: *Global Behavior of Nonlinear Difference Equations of Higher Order with Applications.* 1993, 228 pp. ISBN 0-7923-2286-X

S. Levendorskii: *Degenerate Elliptic Equations.* 1993, 445 pp. ISBN 0-7923-2305-X

D. Mitrinović and J.D. Kečkić: *The Cauchy Method of Residues, Volume 2.* Theory and Applications. 1993, 202 pp. ISBN 0-7923-2311-8

R.P. Agarwal and P.J.Y Wong: *Error Inequalities in Polynomial Interpolation and Their Applications.* 1993, 376 pp. ISBN 0-7923-2337-8

A.G. Butkovskiy and L.M. Pustyl'nikov (eds.): *Characteristics of Distributed-Parameter Systems.* 1993, 386 pp. ISBN 0-7923-2499-4

B. Sternin and V. Shatalov: *Differential Equations on Complex Manifolds.* 1994, 504 pp. ISBN 0-7923-2710-1

S.B. Yakubovich and Y.F. Luchko: *The Hypergeometric Approach to Integral Transforms and Convolutions.* 1994, 324 pp. ISBN 0-7923-2856-6

C. Gu, X. Ding and C.-C. Yang: *Partial Differential Equations in China.* 1994, 181 pp. ISBN 0-7923-2857-4

V.G. Kravchenko and G.S. Litvinchuk: *Introduction to the Theory of Singular Integral Operators with Shift.* 1994, 288 pp. ISBN 0-7923-2864-7

A. Cuyt (ed.): *Nonlinear Numerical Methods and Rational Approximation II.* 1994, 446 pp. ISBN 0-7923-2967-8

G. Gaeta: *Nonlinear Symmetries and Nonlinear Equations.* 1994, 258 pp. ISBN 0-7923-3048-X

Other *Mathematics and Its Applications* titles of interest:

V.A. Vassiliev: *Ramified Integrals, Singularities and Lacunas*. 1995, 289 pp. ISBN 0-7923-3193-1

N.Ja. Vilenkin and A.U. Klimyk: *Representation of Lie Groups and Special Functions. Recent Advances*. 1995, 497 pp. ISBN 0-7923-3210-5

Yu. A. Mitropolsky and A.K. Lopatin: *Nonlinear Mechanics, Groups and Symmetry*. 1995, 388 pp. ISBN 0-7923-3339-X

R.P. Agarwal and P.Y.H. Pang: *Opial Inequalities with Applications in Differential and Difference Equations*. 1995, 393 pp. ISBN 0-7923-3365-9

A.G. Kusraev and S.S. Kutateladze: *Subdifferentials: Theory and Applications*. 1995, 408 pp. ISBN 0-7923-3389-6

M. Cheng, D.-G. Deng, S. Gong and C.-C. Yang (eds.): *Harmonic Analysis in China*. 1995, 318 pp. ISBN 0-7923-3566-X

M.S. Livšic, N. Kravitsky, A.S. Markus and V. Vinnikov: *Theory of Commuting Nonselfadjoint Operators*. 1995, 314 pp. ISBN 0-7923-3588-0

A.I. Stepanets: *Classification and Approximation of Periodic Functions*. 1995, 360 pp. ISBN 0-7923-3603-8

C.-G. Ambrozie and F.-H. Vasilescu: *Banach Space Complexes*. 1995, 205 pp. ISBN 0-7923-3630-5

E. Pap: *Null-Additive Set Functions*. 1995, 312 pp. ISBN 0-7923-3658-5

C.J. Colbourn and E.S. Mahmoodian (eds.): *Combinatorics Advances*. 1995, 338 pp. ISBN 0-7923-3574-0

V.G. Danilov, V.P. Maslov and K.A. Volosov: *Mathematical Modelling of Heat and Mass Transfer Processes*. 1995, 330 pp. ISBN 0-7923-3789-1

A. Laurinčikas: *Limit Theorems for the Riemann Zeta-Function*. 1996, 312 pp. ISBN 0-7923-3824-3

A. Kuzhel: *Characteristic Functions and Models of Nonself-Adjoint Operators*. 1996, 283 pp. ISBN 0-7923-3879-0

G.A. Leonov, I.M. Burkin and A.I. Shepeljavyi: *Frequency Methods in Oscillation Theory*. 1996, 415 pp. ISBN 0-7923-3896-0

B. Li, S. Wang, S. Yan and C.-C. Yang (eds.): *Functional Analysis in China*. 1996, 390 pp. ISBN 0-7923-3880-4

P.S. Landa: *Nonlinear Oscillations and Waves in Dynamical Systems*. 1996, 554 pp. ISBN 0-7923-3931-2

A.J. Jerri: *Linear Difference Equations with Discrete Transform Methods*. 1996, 462 pp. ISBN 0-7923-3940-1

Other *Mathematics and Its Applications* titles of interest:

I. Novikov and E. Semenov: *Haar Series and Linear Operators*. 1997, 234 pp.
ISBN 0-7923-4006-X

L. Zhizhiashvili: *Trigonometric Fourier Series and Their Conjugates*. 1996, 312 pp.
ISBN 0-7923-4088-4

R.G. Buschman: *Integral Transformation, Operational Calculus, and Generalized Functions*. 1996, 246 pp. ISBN 0-7923-4183-X

V. Lakshmikantham, S. Sivasundaram and B. Kaymakcalan: *Dynamic Systems on Measure Chains*. 1996, 296 pp. ISBN 0-7923-4116-3

D. Guo, V. Lakshmikantham and X. Liu: *Nonlinear Integral Equations in Abstract Spaces*. 1996, 350 pp. ISBN 0-7923-4144-9

Y. Roitberg: *Elliptic Boundary Value Problems in the Spaces of Distributions*. 1996, 427 pp.
ISBN 0-7923-4303-4

Y. Komatu: *Distortion Theorems in Relation to Linear Integral Operators*. 1996, 313 pp.
ISBN 0-7923-4304-2

A.G. Chentsov: *Asymptotic Attainability*. 1997, 336 pp. ISBN 0-7923-4302-6

S.T. Zavalishchin and A.N. Sesekin: *Dynamic Impulse Systems*. Theory and Applications. 1997, 268 pp. ISBN 0-7923-4394-8

U. Elias: *Oscillation Theory of Two-Term Differential Equations*. 1997, 226 pp.
ISBN 0-7923-4447-2

D. O'Regan: *Existence Theory for Nonlinear Ordinary Differential Equations*. 1997, 204 pp.
ISBN 0-7923-4511-8

Yu. Mitropolskii, G. Khoma and M. Gromyak: *Asymptotic Methods for Investigating Quasiwave Equations of Hyperbolic Type*. 1997, 418 pp. ISBN 0-7923-4529-0

R.P. Agarwal and P.J.Y. Wong: *Advanced Topics in Difference Equations*. 1997, 518 pp.
ISBN 0-7923-4521-5

N.N. Tarkhanov: *The Analysis of Solutions of Elliptic Equations*. 1997, 406 pp.
ISBN 0-7923-4531-2

B. Riečan and T. Neubrunn: *Integral, Measure, and Ordering*. 1997, 376 pp.
ISBN 0-7923-4566-5

N.L. Gol'dman: *Inverse Stefan Problems*. 1997, 258 pp. ISBN 0-7923-4588-6

S. Singh, B. Watson and P. Srivastava: *Fixed Point Theory and Best Approximation: The KKM-map Principle*. 1997, 230 pp. ISBN 0-7923-4758-7

A. Pankov: G-*Convergence and Homogenization of Nonlinear Partial Differential Operators*. 1997, 263 pp. ISBN 0-7923-4720-X

Other *Mathematics and Its Applications* titles of interest:

S. Hu and N.S. Papageorgiou: *Handbook of Multivalued Analysis.* Volume I: Theory. 1997, 980 pp. ISBN 0-7923-4682-3 (Set of 2 volumes: 0-7923-4683-1)

L.A. Sakhnovich: *Interpolation Theory and Its Applications.* 1997, 216 pp. ISBN 0-7923-4830-0

G.V. Milovanović: *Recent Progress in Inequalities.* 1998, 531 pp. ISBN 0-7923-4845-1

V.V. Filippov: *Basic Topological Structures of Ordinary Differential Equations.* 1998, 530 pp. ISBN 0-7293-4951-2

S. Gong: *Convex and Starlike Mappings in Several Complex Variables.* 1998, 208 pp. ISBN 0-7923-4964-4

A.B. Kharazishvili: *Applications of Point Set Theory in Real Analysis.* 1998, 244 pp. ISBN 0-7923-4979-2

R.P. Agarwal: *Focal Boundary Value Problems for Differential and Difference Equations.* 1998, 300 pp. ISBN 0-7923-4978-4

D. Przeworska-Rolewicz: *Logarithms and Antilogarithms.* An Algebraic Analysis Approach 1998, 358 pp. ISBN 0-7923-4974-1

Yu. M. Berezansky and A.A. Kalyuzhnyi: *Harmonic Analysis in Hypercomplex Systems.* 1998, 493 pp. ISBN 0-7923-5029-4

V. Lakshmikantham and A.S. Vatsala: *Generalized Quasilinearization for Nonlinear Problems.* 1998, 286 pp. ISBN 0-7923-5038-3

V. Barbu: *Partial Differential Equations and Boundary Value Problems.* 1998, 292 pp. ISBN 0-7923-5056-1

J. P. Boyd: *Weakly Nonlocal Solitary Waves and Beyond-All-Orders Asymptotics.* Generalized Solitons and Hyperasymptotic Perturbation Theory. 1998, 610 pp. ISBN 0-7923-5072-3

D. O'Regan and M. Meehan: *Existence Theory for Nonlinear Integral and Integrodifferential Equations.* 1998, 228 pp. ISBN 0-7923-5089-8

A.J. Jerri: *The Gibbs Phenomenon in Fourier Analysis, Splines and Wavelet Approximations.* 1998, 364 pp. ISBN 0-7923-5109-6

C. Constantinescu, W. Filter and K. Weber, in collaboration with A. Sontag: *Advanced Integration Theory.* 1998, 872 pp. ISBN 0-7923-5234-3

V. Bykov, A. Kytmanov and M. Lazman, with M. Passare (ed.): *Elimination Methods in Polynomial Computer Algebra.* 1998, 252 pp. ISBN 0-7923-5240-8

W.-H. Steeb: *Hilbert Spaces, Wavelets, Generalised Functions and Modern Quantum Mechanics.* 1998, 234 pp. ISBN 0-7923-5231-9

Other *Mathematics and Its Applications* titles of interest:

X. Xu: *Introduction to Vertex Operator Superalgebras and Their Modules.* 1998, 356 pp.
ISBN 0-7923-5242-4

E.E. Rosinger: *Parametric Lie Group Actions on Global Generalised Solutions of Nonlinear PDFs* including a Solution to Hilbert's Fifth Problem. 1998, 234 pp. ISBN 0-7923-5232-7

A. Khrennikov: *Superanalysis.* 1999, 370 pp. ISBN 0-7923-5607-1

V. Koshmanenko: *Singular Quadratic Forms in Perturbation Theory.* 1999, 316 pp.
ISBN 0-7923-5625-X

Z. Kamont: *Hyperbolic Functional Differential Inequalities and Applications.* 1999, 318 pp.
ISBN 0-7923-5791-4

A.G. Kusraev and S.S. Kutateladze: *Boolean Valued Analysis.* 1999, 334 pp.
ISBN 0-7923-5921-6

Y. Roitberg: *Boundary Value Problems in the Spaces of Distributions.* 1999, 288 pp.
ISBN 0-7923-6025-7